I0056376

Materials Handbook:
An Integrated Reference

Materials Handbook: An Integrated Reference

Edited by
Reece Hughes

WILLFORD PRESS

www.willfordpress.com

Published by Willford Press,
118-35 Queens Blvd., Suite 400,
Forest Hills, NY 11375, USA

Copyright © 2019 Willford Press

This book contains information obtained from authentic and highly regarded sources. Copyright for all individual chapters remain with the respective authors as indicated. All chapters are published with permission under the Creative Commons Attribution License or equivalent. A wide variety of references are listed. Permission and sources are indicated; for detailed attributions, please refer to the permissions page and list of contributors. Reasonable efforts have been made to publish reliable data and information, but the authors, editors and publisher cannot assume any responsibility for the validity of all materials or the consequences of their use.

Trademark Notice: Registered trademark of products or corporate names are used only for explanation and identification without intent to infringe.

ISBN: 978-1-68285-608-6

Cataloging-in-Publication Data

Materials handbook : an integrated reference / edited by Reece Hughes.
    p. cm.
Includes bibliographical references and index.
ISBN 978-1-68285-608-6
1. Materials science. 2. Materials. I. Hughes, Reece.
TA403 .M38 2019
620.11--dc23

For information on all Willford Press publications
visit our website at www.willfordpress.com

Contents

Preface

Materials are substances, either solid or in a condensed phase, that are intended for use for various applications. Materials can either be crystalline or non-crystalline and can be classified as metals, semiconductors, ceramics and polymers. An understanding of the structure of materials and their properties helps to design applications suitable for each class of materials. The structure of materials is investigated at different scales like atomic scales, microscales or nano scales. Modern developments in this field are occurring in the areas of nanomaterials, biomaterials, spintronics, metamaterials, etc. This book contains some path-breaking studies in the field of materials science. The topics included herein are of the utmost significance and bound to provide incredible insights to readers. Scientists and students actively engaged in this field will find this book full of crucial and unexplored concepts.

This book is a result of research of several months to collate the most relevant data in the field.

When I was approached with the idea of this book and the proposal to edit it, I was overwhelmed. It gave me an opportunity to reach out to all those who share a common interest with me in this field. I had 3 main parameters for editing this text:

1. Accuracy – The data and information provided in this book should be up-to-date and valuable to the readers.

2. Structure – The data must be presented in a structured format for easy understanding and better grasping of the readers.

3. Universal Approach – This book not only targets students but also experts and innovators in the field, thus my aim was to present topics which are of use to all.

Thus, it took me a couple of months to finish the editing of this book.

I would like to make a special mention of my publisher who considered me worthy of this opportunity and also supported me throughout the editing process. I would also like to thank the editing team at the back-end who extended their help whenever required.

Editor

Dysprosium Acetylacetonato Single-Molecule Magnet Encapsulated in Carbon Nanotubes

Ryo Nakanishi [1,*], **Mudasir Ahmad Yatoo** [1], **Keiichi Katoh** [1], **Brian K. Breedlove** [1] **and Masahiro Yamashita** [1,2,*]

[1]   Department of Chemistry, Graduate School of Science, Tohoku University, 6-3 Aza-Aoba, Aoba-ku, Sendai, Miyagi 980-8578, Japan; muda.amu@gmail.com (M.A.Y.); kkatoh@m.tohoku.ac.jp (K.K.); breedlove@m.tohoku.ac.jp (B.K.B.)

[2]   WPI Research Center, Advanced Institute for Materials Research, Tohoku University, 2-1-1 Katahira, Aoba-ku, Sendai 980-8577, Japan

*   Correspondence: r.nakanishi@m.tohoku.ac.jp (R.N.); yamasita.m@gmail.com (M.Y.)

Academic Editor: Wolfgang Linert

Abstract: Dy single-molecule magnets (SMMs), which have several potential uses in a variety of applications, such as quantum computing, were encapsulated in multi-walled carbon nanotubes (MWCNTs) by using a capillary method. Encapsulation was confirmed by using transmission electron microscopy (TEM). In alternating current magnetic measurements, the magnetic susceptibilities of the Dy acetylacetonato complexes showed clear frequency dependence even inside the MWCNTs, meaning that this hybrid can be used as magnetic materials in devices.

Keywords: single-molecule magnet; carbon nanotube

---

1. Introduction

Single-molecule magnets (SMMs) [1–4], which are composed of isolated molecules, usually with large spin angular momenta (S) in the ground state and strong uniaxial magnetic anisotropies (D), exhibit an extensive range of functional properties, like magnetic bistability [1], quantum tunneling of magnetization [5–8], and quantum coherence [9]. Thus, they can be considered as not only molecular equivalents of classical bulk ferromagnets but also as next-generation quantum magnets. Therefore, SMMs are being developed for application in memory storage and in the processing of quantum information [10,11]. Moreover, novel applications of SMMs, including their use in molecular spintronics [12] and quantum computing [13], are being explored.

To use SMMs, we must be able to exploit the functionality of individual SMM molecules and combine them with other functional materials. There have been a few reports on combining SMMs with materials. For example, SMMs have been combined with carbon nanotubes (CNTs) [14] and graphene [15]. From these examples, when lanthanoid SMMs interact with nanocarbon materials, their electronic properties are affected. Another example involves the encapsulation of SMMs into nanoscopic one-dimensional pores, such as the internal nano-space of CNTs [16] and metal-organic frameworks [17], in which SMMs become aligned and their magnetic properties are controlled. SMM-nanomaterial hybrids may have new structures and unique physical properties. If SMMs are encapsulated in one-dimensional pores, the stacking structure can be controlled, and the SMM properties should be enhanced. Furthermore, when SMMs are encapsulated in CNTs, they are protected from the surrounding environment, and thus, the hybrids are easier to use in real applications. However, little has been reported on lanthanoid SMMs encapsulated inside CNTs. In this work, we encapsulated Dy acetylacetonato SMMs [18] in multi-walled CNTs (MWCNTs) by using a capillary method [19,20]. Encapsulation was verified by using transmission electron microscopy (TEM). It was shown that Dy complexes maintained their SMM-like properties in the MWCNTs.

2. Results and Discussion

2.1. Synthesis

MWCNTs with an internal diameter of ~5 nm were purified by using centrifugation [21], and then the end-caps were opened by heating in air. The impurities in the internal nano-space were removed by heating in a vacuum. Next, $Dy(acac)_3(H_2O)_2$ was dissolved in 1,2-dichloroethane, and the solution was heated at 65 °C for 2 h to obtain a saturated solution. Cap-opened MWCNTs were added to the saturated solution and dispersed by using ultrasonication. Then the solution was left to stand for 3 d in order to encapsulate $Dy(acac)_3(H_2O)_2$ into the MWCNTs via a capillary phenomenon [19,20]. After filtering and washing the surfaces with 1,2-dichloroethane, $Dy(acac)_3(H_2O)_2$ encapsulated in MWCNTs ($Dy(acac)_3(H_2O)_2$@MWCNTs) were obtained.

2.2. Transmission Electron Microscopy, Elemental Analysis and Thermogravimetry

TEM was used to view the interior of the MWCNT hybrids; the structure images are illustrated in Figure 1a. In the TEM images, only $Dy(acac)_3(H_2O)_2$@MWCNTs as free-standing entities were observed, and there were no complexes on the external surfaces of the MWCNTs (Figure 1b). In enlarged images, a stark contrast between the $Dy(acac)_3(H_2O)_2$@MWCNT (Figure 1c) and the empty MWCNTs was observed, as shown in Supplementary Materials Figure S1, showing that $Dy(acac)_3(H_2O)_2$ was encapsulated. In order to confirm the encapsulation and characterize the material present inside the MWCNTs, energy-dispersive X-ray (EDX) spectroscopy was used to detect the Dy ions (Figure 1d). The results clearly indicate that $Dy(acac)_3(H_2O)_2$ is encapsulated in the MWCNTs. Thermogravimetric analysis (TGA) was performed on pristine MWCNTs and $Dy(acac)_3(H_2O)_2$@MWCNT (Figure 2). For the pristine MWCNTs, when $T > 600$ °C, all of the carbon was lost as CO_2. However, in the case of $Dy(acac)_3(H_2O)_2$@MWCNT, 22.3 wt % of a white compound remained even when $T > 1000$ °C. This material is thought to be Dy_2O_3. From the TGA data, the amount of $Dy(acac)_3(H_2O)_2$ was estimated to be 1.2 mmol in 1 g of $Dy(acac)_3(H_2O)_2$@MWCNT.

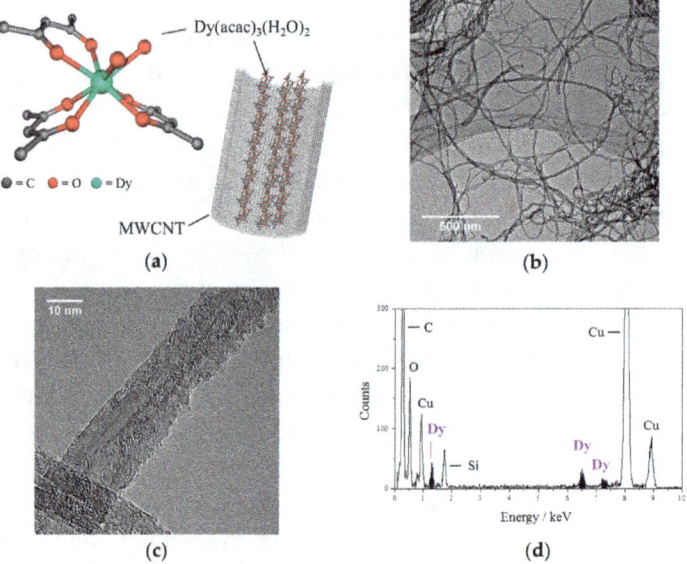

Figure 1. (**a**) Drawings of $Dy(acac)_3(H_2O)_2$ complex and the complexes encapsulated in multi-walled carbon nanotubes (MWCNT); (**b**) Low magnification and (**c**) high magnification transmission electron microscopy (TEM) images of $Dy(acac)_3(H_2O)_2$@MWCNTs; (**d**) energy dispersive X-ray spectroscopy (EDX) spectrum acquired for the sample in (**c**).

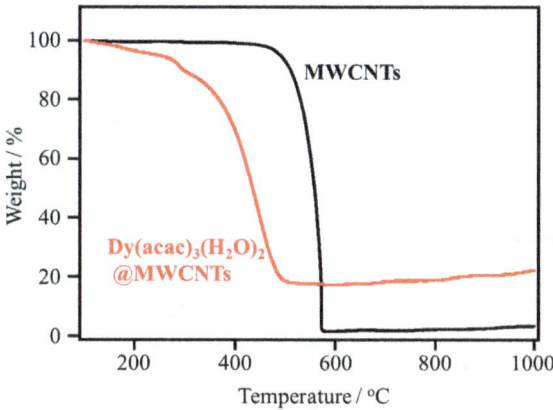

Figure 2. Thermogravimetric analyses of empty MWCNTs (black) and $Dy(acac)_3(H_2O)_2$@MWCNTs (red).

2.3. Magnetic Properties

To determine the effects of encapsulation of the SMMs in MWCNTs on the magnetic properties, both static and dynamic magnetic measurements on $Dy(acac)_3(H_2O)_2$@MWCNTs were performed, and the results were compared with those for free Dy complexes. Direct current (DC) measurements were used to obtain molar magnetic susceptibilities (χ_m), which depended on T and the magnetic field (H). $\chi_m T$-T plots for $Dy(acac)_3(H_2O)_2$@MWCNTs and pure $Dy(acac)_3(H_2O)_2$ are shown in Figure 3a. After correcting the diamagnetism of the MWCNTs (see Supplementary Materials Figure S2), we determined the χ_m values for $Dy(acac)_3(H_2O)_2$@MWCNTs by using the ratio obtained from TGA, and the resulting $\chi_m T$ value at 300 K agrees with that for an isolated Dy(III) ion (14.2 cm^3·K·mol^{-1}), which suggests that the estimated amount of $Dy(acac)_3(H_2O)_2$ is reliable. $\chi_m T$ values for $Dy(acac)_3(H_2O)_2$@MWCNTs decreased with a decrease in T, whereas those for pure $Dy(acac)_3(H_2O)_2$ did not. This difference was ascribed to depopulation of high energy m_J states due to configurational and orientational changes in the ligands upon encapsulation [22,23].

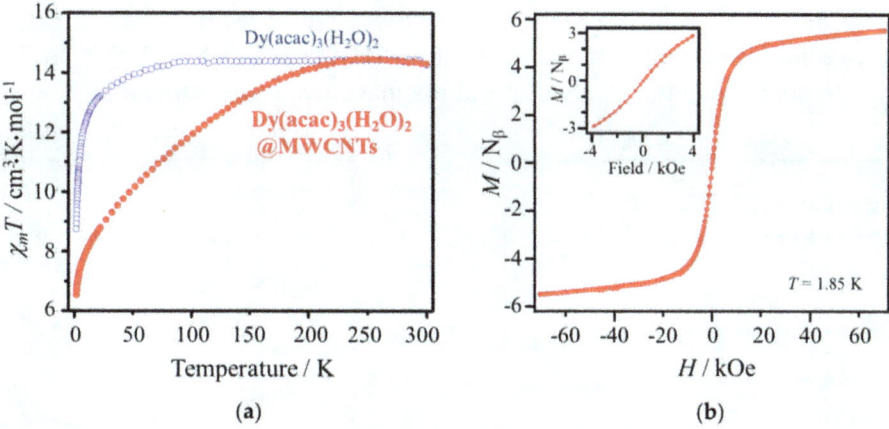

(a)   (b)

Figure 3. (a) $\chi_m T$ vs. T plots for $Dy(acac)_3(H_2O)_2$@MWCNTs (red filled circles) and pure $Dy(acac)_3(H_2O)_2$ (blue open circles); (b) M vs. H plots for $Dy(acac)_3(H_2O)_2$@MWCNTs at 1.85 K. The inset shows magnified curve in the range of −4–4 kOe.

In magnetization (M) vs. H plots, shown in Figure 3b, magnetic hysteresis was not observed. In the case of $Dy(acac)_3(H_2O)_2$ diluted with 20 equivalents of $Y(acac)_3(H_2O)_2$, slight hysteresis has been observed at 2 K because the distance between each $Dy(acac)_3(H_2O)_2$ is large and quantum tunneling of the magnetization (QTM) is suppressed [18]. Therefore, QTM is not suppressed for the $Dy(acac)_3(H_2O)_2$@MWCNTs. In addition, it is possible that the coordination environment of

Dy(acac)$_3$(H$_2$O)$_2$ changed upon encapsulation in the MWCNTs, which promotes the QTM process and shortens the relaxation time. Similar behavior for Mn$_{12}$-acetate SMMs encapsulated in MWCNTs has been reported [16]. In other words, no hysteresis was observed for the Dy hybrids. Thus, by controlling the coordination environment via encapsulation in CNTs, the relaxation time of the SMMs can be tuned.

Next, the dynamic magnetic properties were studied, and the results are shown in Figure 4. For Dy(acac)$_3$(H$_2$O)$_2$@MWCNTs, an out-of-phase (χ'') signal, which is indicative of slow relaxation of M, was observed. Furthermore, both the in-phase (χ') and χ'' signals were frequency dependent. This dependence is due to the Dy(acac)$_3$(H$_2$O)$_2$ complexes because the susceptibilities of the MWCNTs themselves are not frequency dependent (Supplementary Materials Figure S3). These results indicate that the observed slow relaxation is due to SMM behavior, that is, there is an energy barrier for relaxation of the magnetic moment even inside the MWCNTs. However, there was no peak top for the Dy(acac)$_3$(H$_2$O)$_2$@MWCNTs in the frequency range of 1–1000 Hz, whereas a clear peak top was observed for the pure complex (Supplementary Materials Figure S4). As seen in Figure 4b, peak top values of χ'' shifted towards higher frequencies. This indicates that the relaxation times for the hybrids are faster than those for the pure complex. In the χ'' versus T plots shown in Figure 5a, a peak top was still observed in the T region below 2 K, indicating that the magnetic moment was not frozen and that a different relaxation process, like QTM process, was dominant in the low-T region. We estimated the pre-exponential factor τ_0 and the activation energy ΔE from χ''/χ' versus T^{-1} (6–10 K) plots, shown in Figure 5b, in the ν range of 240–1103 Hz by using the Kramers-Kronig equation [23–27]:

$$\chi''/\chi' = \omega\tau \tag{1}$$

$$\chi''/\chi' = \omega\tau_0 + \exp\left(\Delta E/k_B T\right) \tag{2}$$

$$\ln(\chi''/\chi') = \ln(\omega\tau_0) + \Delta E/k_B T \tag{3}$$

where ω (=2$\pi\nu$) is the angular frequency. By fitting the data, the τ_0 and ΔE for Dy(acac)$_3$(H$_2$O)$_2$@MWCNTs were estimated to be in the range of 10^{-6}–10^{-7} s and 4–5 cm^{-1}, respectively (Supplementary Materials Table S1). For pure Dy(acac)$_3$(H$_2$O)$_2$, τ_0 and ΔE were determined to be 8.0×10^{-7} s and 45.9 cm^{-1}, respectively [18]. We think that ΔE for the hybrids is lower because of a conformational change in Dy(acac)$_3$(H$_2$O)$_2$ inside the MWCNTs. The values are consistent with the decrease in the $\chi_m T$ value and magnetic hysteresis behavior.

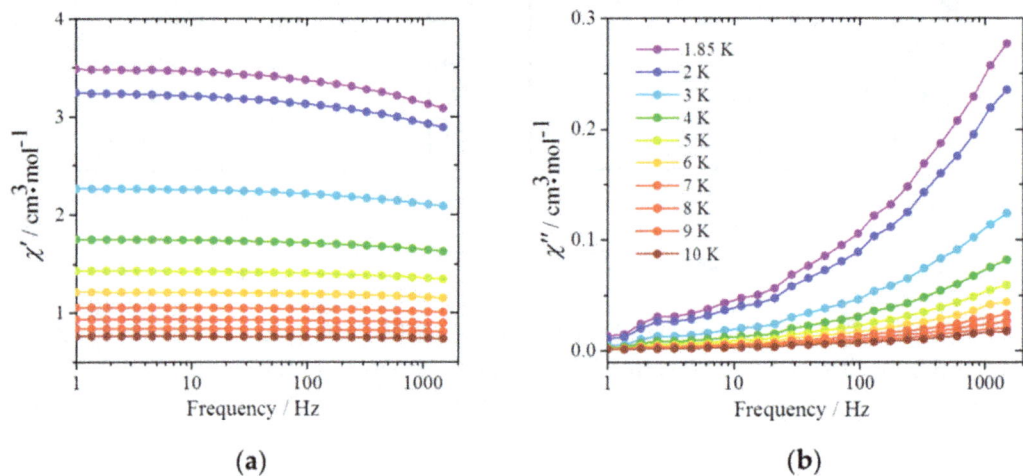

(a)   (b)

Figure 4. Frequency dependence of the (a) in-phase (χ') and (b) out-of-phase (χ'') AC magnetic susceptibilities of Dy(acac)$_3$(H$_2$O)$_2$@MWCNTs. The measurements were performed in an H_{DC} of 0 Oe and H_{AC} of 3 Oe in the T range of 1.85–10 K. The solid lines are guides for eyes.

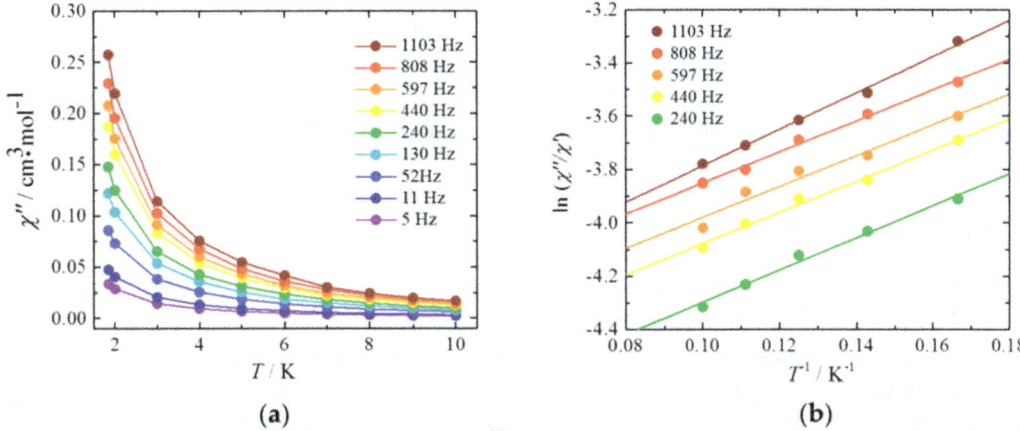

Figure 5. (a) χ'' vs. T plots for Dy(acac)$_3$(H$_2$O)$_2$@MWCNTs. The solid lines are guides for eyes; (b) χ''/χ' versus T^{-1} (6–10 K) plot in the ν range of 240–1103 Hz. The solid lines were fitted as described in Supplementary Materials Table S1.

3. Materials and Methods

3.1. General

Distilled water was obtained from a EYELA STILL ACE SA-2100E deionizer (Tokyo Rikakikai Co., Ltd., Tokyo, Japan). Dy(acac)$_3$(H$_2$O)$_2$ (STREM Chemicals, Inc., Newburyport, MA, USA), 1,2-dichloroethane and methanol (Wako Pure Chemical Industries, Ltd., Osaka, Japan) were used as received. MWCNTs synthesized by using the CoMoCAT™ catalytic chemical vapor deposition method with outer diameters of 10 ± 0.1 nm, inner diameters of 4.5 ± 0.5 nm, and lengths of 3–6 µm (Sigma-Aldrich Co. LLC., St. Louis, MO, USA) were purchased and used after removing the magnetic impurities by using a centrifugation method [21]. The MWCNTs (30 mg) were dispersed with 60 mL of 1 wt % sodium cholate in water by using ultrasonication with a tip-type sonicator (UP200S, Hielscher Ultrasonics GmbH, Teltow, Germany). The obtained black suspension was centrifuged at 18,500 rpm for ~1 h using a tabletop centrifuge (AS185, AS ONE Co., Osaka, Japan), and the upper 80% of the supernatant was collected. The well-dispersed MWCNTs were aggregated by adding methanol and filtered over a Kiriyama filter (Kiriyama glass Co., Tokyo, Japan) having a pore size of 1 µm. The aggregates were then washed with excess methanol and dried at 200 °C in a vacuum overnight, affording 15 mg of purified MWCNT buckypaper.

3.2. Synthesis

Purified MWCNTs were decapped by heating at 550 °C for 5 min in air and degassed by heating in a vacuum just before using. To a saturated solution of Dy(acac)$_3$(H$_2$O)$_2$ in 10 mL of 1,2-dichloroethane, which was heated at 65 °C for about 2 h to ensure that Dy(acac)$_3$(H$_2$O)$_2$ dissolved as much as possible, 10 mg of decapped MWCNTs were added. After 5 min of ultrasonication using a bath-type sonicator and letting stand for 3 d, MWCNTs were collected by filtration and washed with 1,2-dichloroethane to completely remove the Dy(acac)$_3$(H$_2$O)$_2$ from the surfaces of the MWCNTs.

3.3. TEM Observation

High-resolution transmission electron microscopy (TEM) and energy dispersive X-ray spectroscopy (EDX) were carried out using a JEM2100F (acceleration voltage; 200 kV, JEOL Ltd., Tokyo, Japan) with dry SD30GV detector (JEOL Ltd., Tokyo, Japan). The sample was dispersed in methanol and deposited on a carbon-coated Cu grid, which was dried by heating overnight at 100 °C in a 10^{-4} Pa vacuum before TEM was performed.

3.4. Thermogravimetric Analysis

Thermogravimetric analysis (TGA) was performed on a SHIMADZU DTG-60 (Shimadzu Corporation, Kyoto, Japan) using aluminum oxide powder as a standard material. Several milligrams of the sample were put in an aluminum cell, and the cell was heated to 1000 °C with a heating rate of 2 °C/min.

3.5. Magnetic Susceptibility Measurement

Magnetic susceptibility measurements were performed on a SQUID magnetometer (model MPMS-XL SQUID magnetometer, Quantum Design, Inc., San Diego, CA, USA). Samples were put into gelatin capsules, and eicosane was added to fix the samples during the measurement. DC measurements for $Dy(acac)_3(H_2O)_2$ were performed in an H_{DC} of 500 Oe, and those for the purified MWCNTs and $Dy(acac)_3(H_2O)_2$@MWCNTs were recorded in H_{DC} of 1000 Oe. T was changed from 300 K to 1.85 K with a sweep rate of 1 K/min. Field dependent DC measurements were performed at 1.85 K while changing the magnetic field as follows: 0 Oe → 70 kOe → −70 kOe → 70 kOe. AC measurements were recorded in an H_{AC} of 3 Oe in the frequency range of 1–1500 Hz and T range of 1.85–10 K. Diamagnetic contributions from the eicosane and $Dy(acac)_3(H_2O)_2$ were corrected by using Pascal's constants, and then the magnetic susceptibility for the purified MWCNTs was subtracted from that for $Dy(acac)_3(H_2O)_2$@MWCNTs. Magnetic moments χ_{CNT}, χ_{CNT}' and χ_{CNT}'' (Supplementary Materials Figures S2 and S3) were obtained by normalizing the obtained magnetic moments with the mass of CNT after applying the diamagnetic corrections.

4. Conclusions

In this work, we encapsulated $Dy(acac)_3(H_2O)_2$ SMMs in the internal nanospace of MWCNTs by using a capillary method. Encapsulation was confirmed by using TEM. From AC magnetic susceptibility measurements, both the in-phase and out-of-phase signals were clearly frequency dependent, indicating that $Dy(acac)_3(H_2O)_2$ complexes still exhibited SMM-like properties. To the best of our knowledge, this is the first example of a lanthanoid SMM encapsulated in CNTs. Although the encapsulation of $Dy(acac)_3(H_2O)_2$ into MWCNTs did not enhance the SMM properties, this work shows that it is possible to control the coordination environment and tune the magnetic properties of SMMs via encapsulation. In addition, we believe that the magnetic and electronic properties of lanthanoid SMM-CNT hybrids can be combined to bring about new applications in devices, like spintronic devices.

Supplementary Materials
Figure S1: TEM image and EDX spectrum of empty MWCNT, Figure S2: χ_{CNT} and $\chi_{CNT}T$ vs. T plots for MWCNT and $Dy(acac)_3(H_2O)_2$@MWCNTs without correction for the diamagnetism of the MWCNTs, Figure S3: Temperature-dependence of the in-phase (χ') and out-of-phase (χ'') AC magnetic susceptibilities of MWCNT and $Dy(acac)_3(H_2O)_2$@MWCNTs, Figure S4: Frequency-dependence of χ' and χ'' AC magnetic susceptibilities of $Dy(acac)_3(H_2O)_2$, Table S1: Selected values of ΔE and τ_0 for $Dy(acac)_3(H_2O)_2$@MWCNTs.

Acknowledgments: This work was supported by CREST, JST, a Grant-in-Aid for Scientific Research (S) (grant No. 20225003, Masahiro Yamashita), Grant-in-Aid for Scientific Research (C) (grant No. 15K05467, Keiichi Katoh), Grant-in-Aid for Young Scientists (B) (grant No. 24750119, Keiichi Katoh) from the Ministry of Education, Culture, Sports, Science, and Technology, Japan (MEXT). Ryo Nakanishi thanks Shorai Foundation for Science and Technology. We thank Takamichi Miyazaki (Technical Division, Department of Engineering, Tohoku University) for the support in the TEM and EDX analyses.

Author Contributions: Ryo Nakanishi, Keiichi Katoh and Masahiro Yamashita conceived and designed the experiments; Mudasir Ahmad Yatoo performed the experiments; Ryo Nakanishi, Mudasir Ahmad Yatoo and Keiichi Katoh analyzed the data; Ryo Nakanishi, Brian K. Breedlove and Masahiro Yamashita wrote the paper.

Conflicts of Interest: The authors declare no conflict of interest. The funding sponsors had no role in the design of the study; in the collection, analyses, and interpretation of data, in the writing of the manuscript, and in the decision to publish the results.

References

1. Sessoli, R.; Gatteschi, D.; Caneschi, A.; Novak, M.A. Magnetic bistability in a metal-ion cluster. *Nature* **1993**, *365*, 141–143. [CrossRef]

2. Ishikawa, N.; Sugita, M.; Ishikawa, T.; Koshihara, S.-Y.; Kaizu, Y. Lanthanide double-decker complexes functioning as magnets at the single-molecular level. *J. Am. Chem. Soc.* **2003**, *125*, 8694–8695. [CrossRef] [PubMed]

3. Woodruff, D.N.; Winpenny, R.E.P.; Layfield, R.A. Lanthanide single-molecule magnets. *Chem. Rev.* **2013**, *113*, 5110–5148. [CrossRef] [PubMed]

4. Horii, Y.; Katoh, K.; Yasuda, N.; Breedlove, B.K.; Yamashita, M. Effects of f–f interactions on the single-molecule magnet properties of terbium(III)–phthalocyaninato quintuple-decker complexes. *Inorg. Chem.* **2015**, *54*, 3297–3305. [CrossRef] [PubMed]

5. Thomas, L.; Lionti, F.; Ballou, R.; Gatteschi, D.; Sessoli, R.; Barbara, B. Macroscopic quantum tunnelling of magnetization in a single crystal of nanomagnets. *Nature* **1996**, *383*, 145–147. [CrossRef]

6. Friedman, J.R.; Sarachik, M.P.; Tejada, J.; Ziolo, R. Macroscopic measurement of resonant magnetization tunneling in high-spin molecules. *Phys. Rev. Lett.* **1996**, *76*, 3830–3833. [CrossRef] [PubMed]

7. Gatteschi, D.; Sessoli, R. Quantum tunneling of magnetization and related phenomena in molecular materials. *Angew. Chem. Int. Ed.* **2003**, *42*, 268–297. [CrossRef] [PubMed]

8. Mannini, M.; Pineider, F.; Danieli, C.; Totti, F.; Sorace, L.; Sainctavit, P.; Arrio, M.A.; Otero, E.; Joly, L.; Cezar, J.C.; et al. Quantum tunnelling of the magnetization in a monolayer of oriented single-molecule magnets. *Nature* **2010**, *468*, 417–421. [CrossRef] [PubMed]

9. Ardavan, A.; Rival, O.; Morton, J.J.L.; Blundell, S.J.; Tyryshkin, A.M.; Timco, G.A.; Winpenny, R.E.P. Will spin-relaxation times in molecular magnets permit quantum information processing? *Phys. Rev. Lett.* **2007**, *98*, 057201. [CrossRef] [PubMed]

10. Mannini, M.; Pineider, F.; Sainctavit, P.; Danieli, C.; Otero, E.; Sciancalepore, C.; Talarico, A.M.; Arrio, M.-A.; Cornia, A.; Gatteschi, D.; et al. Magnetic memory of a single-molecule quantum magnet wired to a gold surface. *Nat. Mater.* **2009**, *8*, 194–197. [CrossRef] [PubMed]

11. Komeda, T.; Isshiki, H.; Liu, J.; Zhang, Y.-F.; Lorente, N.; Katoh, K.; Breedlove, B.K.; Yamashita, M. Observation and electric current control of a local spin in a single-molecule magnet. *Nat. Commun.* **2011**, *2*, 217. [CrossRef] [PubMed]

12. Bogani, L.; Wernsdorfer, W. Molecular spintronics using single-molecule magnets. *Nat. Mater.* **2008**, *7*, 179–186. [CrossRef] [PubMed]

13. Leuenberger, M.N.; Loss, D. Quantum computing in molecular magnets. *Nature* **2001**, *410*, 789–793. [CrossRef] [PubMed]

14. Urdampilleta, M.; Klyatskaya, S.; Cleuziou, J.P.; Ruben, M.; Wernsdorfer, W. Supramolecular spin valves. *Nat. Mater.* **2011**, *10*, 502–506. [CrossRef] [PubMed]

15. Candini, A.; Klyatskaya, S.; Ruben, M.; Wernsdorfer, W.; Affronte, M. Graphene spintronic devices with molecular nanomagnets. *Nano Lett.* **2011**, *11*, 2634–2639. [CrossRef] [PubMed]

16. Del Carmen Giménez-López, M.; Moro, F.; La Torre, A.; Gómez-García, C.J.; Brown, P.D.; van Slageren, J.; Khlobystov, A.N. Encapsulation of single-molecule magnets in carbon nanotubes. *Nat. Commun.* **2011**, *2*, 407. [CrossRef] [PubMed]

17. Aulakh, D.; Pyser, J.B.; Zhang, X.; Yakovenko, A.A.; Dunbar, K.R.; Wriedt, M. Metal–organic frameworks as platforms for the controlled nanostructuring of single-molecule magnets. *J. Am. Chem. Soc.* **2015**, *137*, 9254–9257. [CrossRef] [PubMed]

18. Jiang, S.-D.; Wang, B.-W.; Su, G.; Wang, Z.-M.; Gao, S. A Mononuclear dysprosium complex featuring single-molecule-magnet behavior. *Angew. Chem. Int. Ed.* **2010**, *49*, 7448–7451. [CrossRef] [PubMed]

19. Ajayan, P.M.; Iijima, S. Capillarity-induced filling of carbon nanotubes. *Nature* **1993**, *361*, 333–334. [CrossRef]

20. Yudasaka, M.; Ajima, K.; Suenaga, K.; Ichihashi, T.; Hashimoto, A.; Iijima, S. Nano-extraction and nano-condensation for C_{60} incorporation into single-wall carbon nanotubes in liquid phases. *Chem. Phys. Lett.* **2003**, *380*, 42–46. [CrossRef]

21. Yu, A.; Bekyarova, E.; Itkis, M.E.; Fakhrutdinov, D.; Webster, R.; Haddon, R.C. Application of centrifugation to the large-scale purification of electric arc-produced single-walled carbon nanotubes. *J. Am. Chem. Soc.* **2006**, *128*, 9902–9908. [CrossRef] [PubMed]

22. Bi, Y.; Guo, Y.N.; Zhao, L.; Guo, Y.; Lin, S.Y.; Jiang, S.D.; Tang, J.; Wang, B.W.; Gao, S. Capping ligand perturbed slow magnetic relaxation in dysprosium single-ion magnets. *Chem. Eur. J.* **2011**, *17*, 12476–12481. [CrossRef] [PubMed]

23. Katoh, K.; Breedlove, B.K.; Yamashita, M. Symmetry of octa-coordination environment has a substantial influence on dinuclear Tb^{III} triple-decker single-molecule magnets. *Chem. Sci.* **2016**, *7*, 4329–4340. [CrossRef]

24. Luis, F.; Bartolomé, J.; Fernández, J.F.; Tejada, J.; Hernández, J.M.; Zhang, X.X.; Ziolo, R. Thermally activated and field-tuned tunneling in $Mn_{12}Ac$ studied by ac magnetic susceptibility. *Phys. Rev. B* **1997**, *55*, 11448–11456. [CrossRef]

25. Bartolomé, J.; Filoti, G.; Kuncser, V.; Schinteie, G.; Mereacre, V.; Anson, C.E.; Powell, A.K.; Prodius, D.; Turta, C. Magnetostructural correlations in the tetranuclear series of $\{Fe_3LnO_2\}$ butterfly core clusters: Magnetic and Mössbauer spectroscopic study. *Phys. Rev. B* **2009**, *80*, 014430. [CrossRef]

26. Ferrando-Soria, J.; Cangussu, D.; Eslava, M.; Journaux, Y.; Lescouëzec, R.; Julve, M.; Lloret, F.; Pasán, J.; Ruiz-Pérez, C.; Lhotel, E.; et al. Rational enantioselective design of chiral heterobimetallic single-chain magnets: Synthesis, crystal structures and magnetic properties of oxamato-bridged $M^{II}Cu^{II}$ chains (M=Mn, Co). *Chem. Eur. J.* **2011**, *17*, 12482–12494. [CrossRef] [PubMed]

27. Gass, I.A.; Moubaraki, B.; Langley, S.K.; Batten, S.R.; Murray, K.S. A π-π 3D network of tetranuclear $\mu2/\mu3$-carbonato Dy(III) bis-pyrazolylpyridine clusters showing single molecule magnetism features. *Chem. Commun.* **2012**, *48*, 2089–2091. [CrossRef] [PubMed]

2

Supercritical CO_2-Assisted Spray Drying of Strawberry-Like Gold-Coated Magnetite Nanocomposites in Chitosan Powders for Inhalation

Marta C. Silva [1,2], **Ana Sofia Silva** [1,3], **Javier Fernandez-Lodeiro** [2,4], **Teresa Casimiro** [1], **Carlos Lodeiro** [2,4,*] **and Ana Aguiar-Ricardo** [1,*]

[1] LAQV-REQUIMTE, Departamento de Química, Faculdade de Ciências e Tecnologia, Universidade NOVA de Lisboa, Campus de Caparica, Caparica 2829-516, Portugal; martasilva686@gmail.com (M.C.S.); asm.silva@campus.fct.unl.pt (A.S.S.); teresa.casimiro@fct.unl.pt (T.C.)

[2] BIOSCOPE Research Group, UCIBIO@REQUIMTE, Chemistry Department, Faculty of Science and Technology, University NOVA of Lisbon, Caparica Campus, Caparica 2829-516, Portugal; j.lodeiro@fct.unl.pt

[3] CICS-UBI, Health Sciences Research Center, Faculdade de Ciências da Saúde, Universidade da Beira Interior, Av. Infante D. Henrique, Covilhã 6200-506, Portugal

[4] PROTEOMASS Scientific Society, Rua dos Inventores, Madam Parque, Caparica Campus, Caparica 2829-516, Portugal

* Correspondence: cle@fct.unl.pt (C.L.); air@fct.unl.pt (A.A.-R.)

Academic Editor: Maryam Tabrizian

Abstract: Lung cancer is one of the leading causes of death worldwide. Therefore, it is of extreme importance to develop new systems that can deliver anticancer drugs into the site of action when initiating a treatment. Recently, the use of nanotechnology and particle engineering has enabled the development of new drug delivery platforms for pulmonary delivery. In this work, POXylated strawberry-like gold-coated magnetite nanocomposites and ibuprofen (IBP) were encapsulated into a chitosan matrix using Supercritical Assisted Spray Drying (SASD). The dry powder formulations showed adequate morphology and aerodynamic performances (fine particle fraction 48%–55% and aerodynamic diameter of 2.6–2.8 μm) for deep lung deposition through the pulmonary route. Moreover, the release kinetics of IBP was also investigated showing a faster release of the drug at pH 6.8, the pH of lung cancer. POXylated strawberry-like gold-coated magnetite nanocomposites proved to have suitable sizes for cellular internalization and their fluorescent capabilities enable their future use in in vitro cell based assays. As a proof-of-concept, the reported results show that these nano-in-micro formulations could be potential drug vehicles for pulmonary administration.

Keywords: lung diseases; dry powders; magnetic nanoparticles; nanocomposites; SASD; pulmonary delivery

1. Introduction

Lung cancer is one of the most common and leading causes of cancer death worldwide [1,2]. To address this problem, different systems that can easily carry active drugs into the site of action and thus initiate the respective cancer treatments have been widely investigated [3]. A suitable drug carrier, or drug delivery system (DDS), should protect the drug against degradation until it reaches the desired site of action [4]. Therefore, the treatment efficacy mostly depends on the system by which the therapeutic dose of the drug is delivered (carrier and administration route) [5,6].

As local administration to the lungs has become one of the best alternatives to improve lung cancer outcomes [6], pulmonary delivery has been widely investigated as a primary route of administration, as it enables direct targeting to the lungs for both local and systemic treatment [7]. Pulmonary delivery together with controlled release systems allows drug protection from rapid degradation or clearance; it enhances the accumulation in the cells by increasing its bioavailability, therefore lowering the amount of drug required, reducing harmful side effects and reducing costs [8].

Recently, nanotechnology has been combined with pulmonary delivery systems, and nanoparticles (NPs) can be used as efficient drug carriers that provide a controlled release of the drug and accurately recognize different cellular types [9–11]. In fact, over the last two decades, nanotechnology has provided extraordinary outcomes in both drug delivery and cell targeting [3,12]. Thus, the merging of nanotechnology, particle engineering, cancer biology, and pulmonary delivery can bring new improvements for the diagnosis and therapy of lung cancer [6,13]. Due to their small size (1–150 nm), NPs can easily penetrate the deep tissue and be taken up by cells more quickly while protecting the drugs at both extracellular and intracellular levels [14,15]. These achievements have been shown to improve drug half-lives and retention, which will substantially decrease unwanted side effects. Besides their therapeutic capabilities, nanoparticles can also be designed to exhibit diagnostic features, thus creating theragnostic nanosystems. Nanoparticles with theragnostic capabilities can offer many benefits for lung disease patients including real-time diagnosis, adequate therapy for individual patients (personalized medicine), and reduction of adverse drug effects [6,13].

In an attempt to produce theragnosis systems, different types of nanoparticles have been investigated: (i) metal (such as gold (Au), ferric (Fe), and silver (Ag)) [16]; (ii) polymeric, (such as polyethylene-glycol (PEG) [17], silica (Si) [18], and poly(lactic-co-glycolic acid) (PLGA)) [19]. In fact, metallic NPs have proven to be great drug carriers and contrast agents in cancer treatment [16]. Also, they are readily functionalized, which is a matter of extreme importance since NPs lacking functionalization are more likely to interact with plasma proteins or be uptaken by macrophages, and can be used as active targeting nanocarriers to target specific cellular types. By controlling nanoparticles' surface properties, a stealth character can be conferred to the nanosystems, preventing agglomeration and opsonization, therefore protecting them and thus increasing their residence time and ensuring the delivery of the drug to the desired site of action [20–22].

Magnetite NPs have been studied for several applications, in particular for medical diagnosis due to their magnetic properties by using MRI (magnetic resonance imaging) therapy and targeted drug delivery. These nanoparticles also show superparamagnetic behavior when working above their blocking temperature, which allows them a fast response to applied magnetic fields. However, in the majority of cases, naked particles tend to be very unstable over long periods of time and lean towards agglomeration [23,24]. Moreover, although they show low toxicity, they can also cause cell damage by forming reactive oxygen species. Some strategies comprising grafting or coating are used to prevent this toxicity and the possible aggregation [25].

Gold nanoparticles (AuNPs) have many advantages due to their inertness, non-toxic core, and excellent metal stability. Also, AuNPs are readily functionalized, biocompatible, have high light absorption, and are highly stable. Concerning these significant advantages, this work is focused on the synthesis of gold-coated magnetic nanocomposites as gold shells able to confer stability to the magnetic core, functionality, and biocompatibility properties. Also, the magnetic core will confer to the NPs magnetism that is capable of being used in magnetic resonance imaging and hyperthermia [23]. POXylation (grafting with oligo(2-ethyl-oxazolines), a similar strategy to PEGylation, currently under investigation to circumvent PEG hepatoxicity issues, has also been employed. Besides the ability to protect nanoparticles, POXylation also add fluorescent capabilities to the nanoparticles enabling an in vitro real time analysis of particles' course within cell environments [26–28]. However, nanoparticles are not in a suitable size for deep lung deposition, as they can be easily exhaled or muccociliary cleared out [29]. It has been established that the optimal aerodynamic diameter for particle deposition into the deep lung region is between 0.5 and 5 μm [30–32]. Therefore, to overcome this limitation, dry powder formulations containing both nanoparticles presented as nanocomposites and micro particles (nano-in-micro formulations) are the focus of study. These dry powder formulations make use of excipients capable of carrying the nanoparticles into the site of action.

Chitosan (CHT), as a biodegradable, non-toxic [33–35], mucoadhesive, and biocompatible polymer, has been chosen by many researchers as a microcarrier for therapeutic agent delivery through the pulmonary route [32,33,36,37]. Herein, CHT is used as a microcarrier for the POXylated strawberry-like gold-coated magnetite nanocomposites [20]. The encapsulation process is performed using a sustainable technology, SASD, that makes use of Supercritical Carbon Dioxide (scCO$_2$) which is mixed and solubilized in the liquid solution (thus minimizing the use of organic and not so environmentally friendly solvents) and further expansion of this mixture through a nozzle, a process patented by E. Reverchon (patent US 7,276,190 B2) [32,38,39]. This process will allow us to obtain particles with smaller sizes than conventional techniques such as spray-drying or spray freeze-drying, among others [40].

This work is focused on the development of new engineered POxylated strawberry-like partially gold-coated magnetite nanocomposites micronized into CHT [20,39,41], to produce dry powders suitable for deep lung deposition [32]. This mechanism is represented in Figure 1. Additionally, IBP was encapsulated to evaluate if this system can be used as a synergetic platform [27,32]. The final nano-in-micro drug delivery system is herein studied as a proof-of-concept and therefore, in order to understand the effectiveness of this system, a more detailed study is being taken by our group to prove its viability.

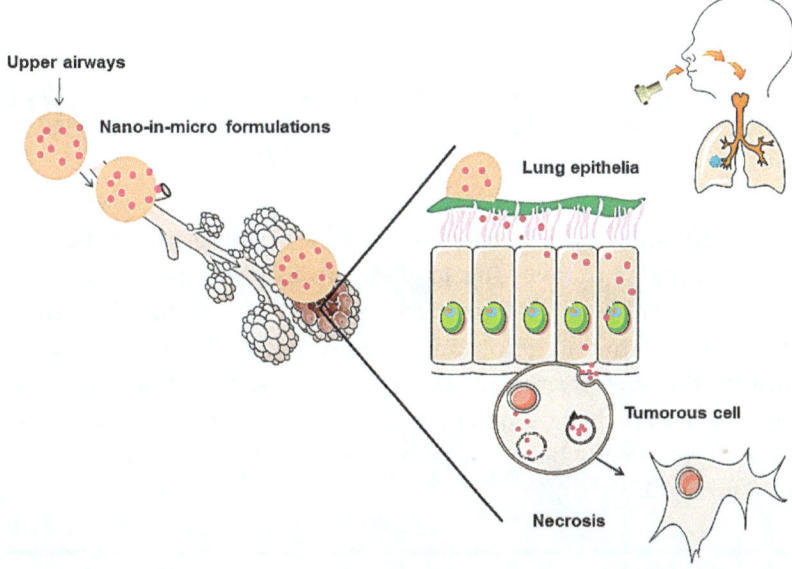

Figure 1. Mechanism of nanoparticle delivery through an aerosol of nano-in-micro formulations. Adapted from A. Silva et al. [20].

2. Results and Discussion

2.1. Nanoparticle Characterization

Size measurements of the strawberry magnetic@gold composites were obtained by TEM images, and Zeta Potential. Both data were reported previously [25].

Magnetic@gold nanocomposites were synthesized using a layer-by-layer deposition by adding three different layers of two polymers (one negative and one positive) onto the surface of the magnetic core, and then the partial gold shell was made in situ onto the magnetic nanoparticle. As depicted in Figure 2, the deposition of poly-diallyldimethylammonium chloride (PDADMAC) was clearly demonstrated through a complete reversal of surface charge, from negative to positive. As a layer of poly-sodium 4-styrenesulfonate (PSS) was deposited onto the particles, zeta potential was again reversed (to negative), corroborating this new layer deposition. As the process of PDADMAC deposition was repeated, the last particles exhibiting three polymeric layers showed a positive surface charge. The addition of a gold shell coated with the fluorescent oligo(2-ethyl-2-oxazoline) showed that the zeta potential of the particles had maintained its positive value of approximately 33.6 mV.

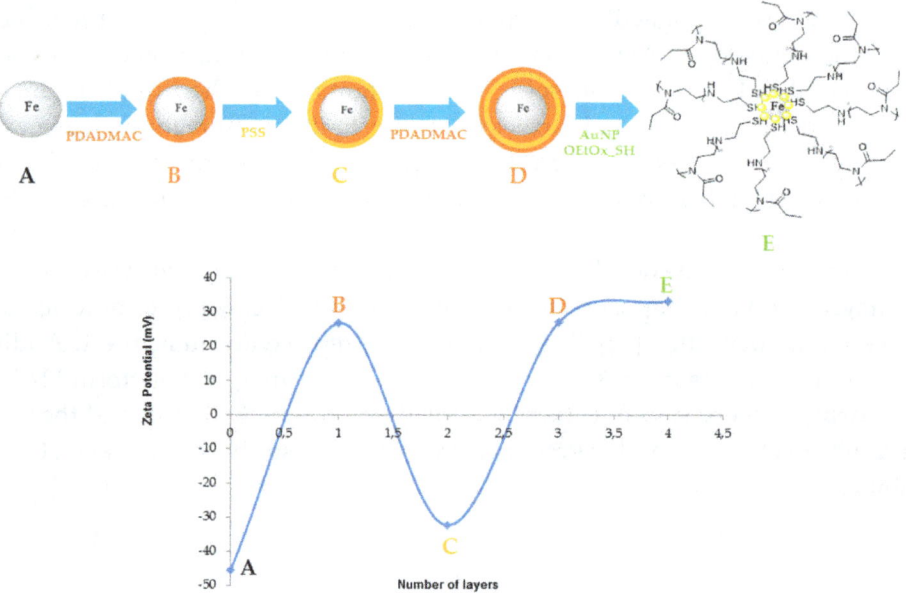

Figure 2. Zeta potential of the different layers, where the fourth layer corresponds to the final nanoparticles. PDADMAC—poly-diallyldimethylammonium chloride; PSS—poly-sodium 4-tyrenesulfonate.

The positive zeta potential exhibited at the surface of these new types of nanocomposites is expected to increase cellular uptake due to their electrostatic interaction with the negatively charged phospholipidic membrane. Their sizes are well suited for delivery into tumor cells. TEM images (Figure 3) show that all gold nanoparticles are anchored to the magnetic surfaces of magnetite nanoparticles, therefore forming stable nanocomposites. Nanocomposites with a diameter ranging within 50–200 nm were then obtained, which had a suitable size for cellular internalization. These nanocomposites have already been tested in a previous work applied to Proteomics, and have presented excellent stability [25].

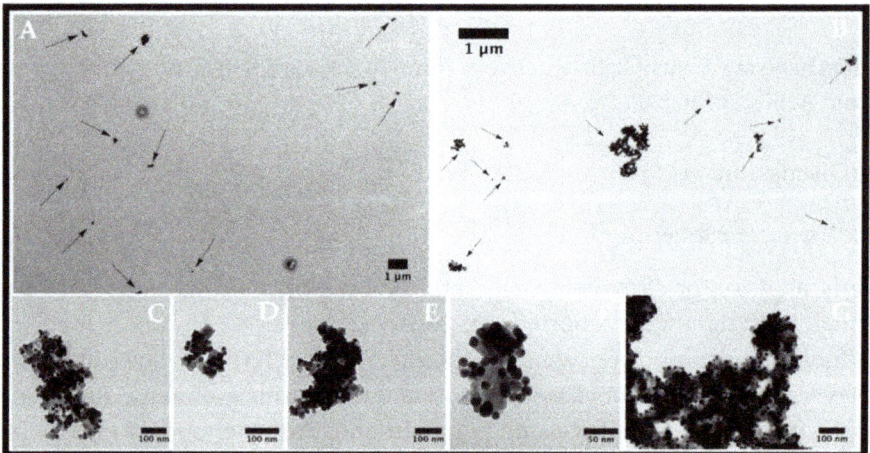

Figure 3. Transmission electron micrographs of strawberry Fe/Au nanomaterials functionalized with oligo(2-ethyl-2-oxazoline) in different magnifications. (**A**,**B**) magnetite nanoparticles; (**C–G**) Strawberry-like gold magnetite nanocomposites functionalized with oligo(2-ethyl-2-oxazoline).

Since magnetite gold coated nanoparticles do not present any intrinsic fluorescence, oligo(2-ethyl-2-oxazoline) end terminated with cysteamine (POx-SH) was conjugated to the nanoparticles in order to produce systems able to be tracked within cell environments using fluorescent microscopy assays. In fact, UV-Vis and fluorescent assays confirmed the grafting of POx-SH to the surface of the nanoparticles. Figure 4 shows both the emission and excitation spectra of POxylated nanoparticles. The representative bands of POx-SH are maintained after particle grafting ($\lambda_{exc} = 300$; $\lambda_{emm} = 384$ nm) [26,27,42].

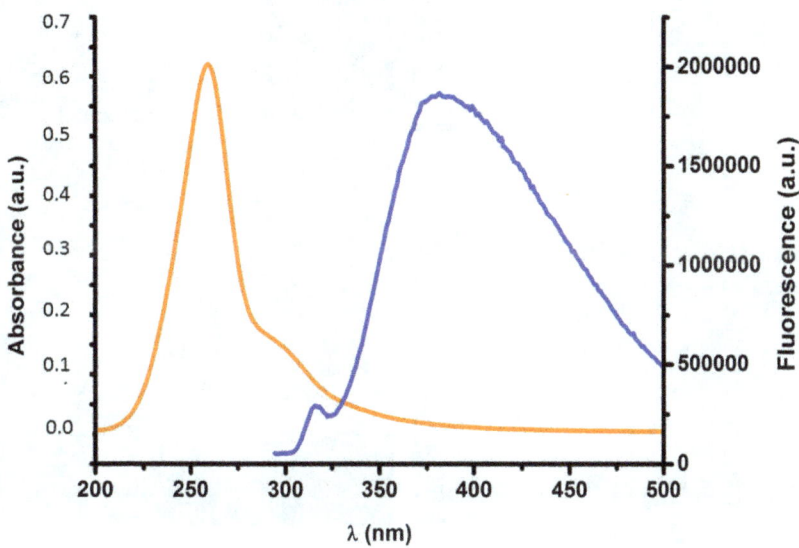

Figure 4. Fluorescence vs. Absorbance of the functionalized nanoparticles. The emission band can be denoted at 384 nm (blue line) and the absorption band of the polymer can be seen between 240–320 nm (orange line).

2.2. Characterization of Nano-in-Micro Formulations

2.2.1. Morphological and Physical-Chemical Properties

The morphological analysis of the particles was performed by scanning electron microscopy (SEM) and using Morphologi G3 equipment, where a population of 30,000 particles was analyzed. The micrographs displaying the shape of the particles are presented in Figure 5, and the results obtained regarding the volume mean diameter ($D_{v,50}$) and relative width of the distribution (Span) are summarized in Table 1. The water content of the particles was evaluated using Karl Fischer coulometric titration and is also included in Table 1. Both formulations had particles with spherical shapes (although some had smooth shapes and others had some indentations) with a volume mean diameter varying from 2.6 to 2.9 μm, suitable for deep lung deposition, and 11% water content, enough to promote a great and controlled swelling of the particles when in contact with the lung epithelia.

Table 1. Microparticle composition and properties.

Sample	$D_{v,50}$ (μm)	Span	Water Content (%)
CHT	2.6	0.9	11.1
CHT_Fe@Au_POX-SH	2.9	0.8	11.7

$D_{v,50}$—mean volume diameter; CHT–Chitosan; CHT_Fe@Au_POX-SH–Nano-in-micro formulations of the nanoparticles coated with oligo(2-ethyl-2-oxazoline) end terminated with cysteamine (POx-SH).

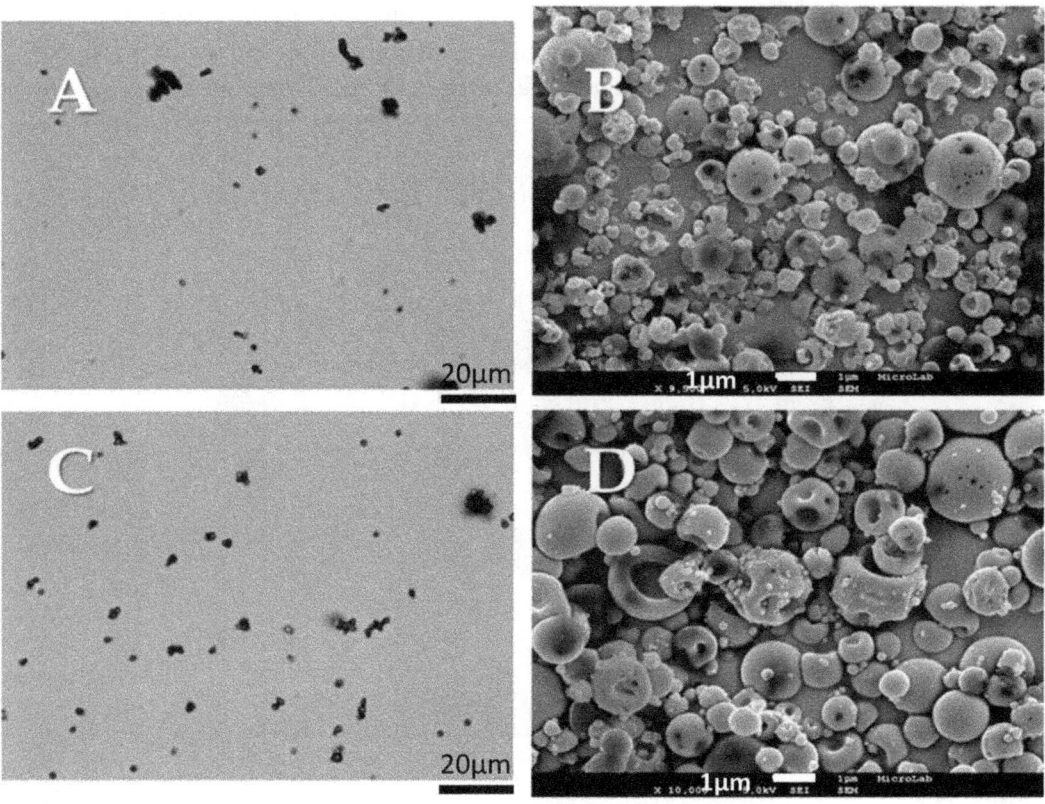

Figure 5. Morphological analysis of the produced particles: CHT: (**A**) Morphologi G3 image; (**B**) SEM image; CHT_Fe@Au_POX-SH: (**C**) Morphologi G3 image; (**D**) SEM image.

The presence of amorphous or crystalline states of the materials was investigated through X-ray diffraction (XRD), as demonstrated in Figure 6. The XRD diffractogram identifies the amorphous state of CHT_Fe@Au_POX-SH, similar to the XRD diffractogram reported for CHT alone [32]. No crystalline peaks were observed.

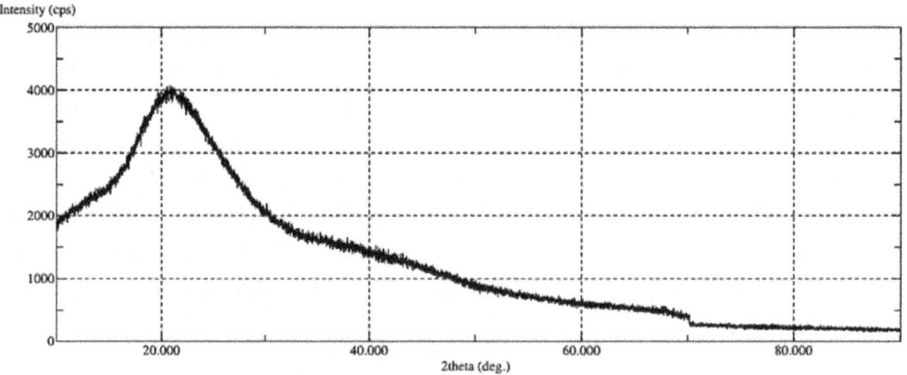

Figure 6. XRD results of the produced CHT_Fe@Au_POX-SH particles.

The Fourier Transform Infra-Red (FT-IR) spectra for the Fe@Au_POX-SH and CHT were performed to verify if any new bond was created through the CHT and nanoparticle interaction. The FT-IR spectrum shows no significant changes between both formulations, which can be justified by the large weight ratio of CHT over the nanoparticles.

2.2.2. Aerodynamic Performance

The in vitro aerodynamic performance of all powder formulations was investigated through Andersen Cascade impact measurements, a device that mimics every stage of the respiratory tract from upper airways to alveolus. Also, by determining specific parameters such as emitted fraction (EF), fraction of particles leaving the capsule; fine particle fraction (FPF), the percentage of particles with aerodynamic diameter below 5 μm; mass median aerodynamic diameter (MMAD); and geometric standard deviation (GSD), it is possible to predict particle deposition in the lungs.

It is well established that the microparticles should have aerodynamic diameters between 0.5 and 5 μm to successfully reach the deep lung. Particles with aerodynamic diameters larger than 5 μm will be trapped in the upper airways and particles with diameters lower than 0.5 μm are exhaled and fail to deposit. Also, particles with sizes below 5 μm may be easily taken up by macrophages [43]. As demonstrated in Table 2, an MMAD of approximately 1.5 μm was obtained and both microparticles, with and without the nanocomposites embedded, reached the final stages of the Anderson Cascade Impactor (ACI), representing the deep area of the lung. By comparing bare chitosan microparticles and the nano-in-micro formulations, it is possible to verify in all the assays that the EF (%) obtained was above 96%, indicating that almost all powder is released from the capsule. In Figure 7, it is possible to perceive that a higher ratio of the powder was collected in the induction port, which simulates the upper airways and the inhaler. This may be due to the formation of turbulent eddies in this zone, which may prevent the aggregated particles from reaching further stages, and to static electricity, respectively [32,44]. It was also found that the FPF (%), i.e., the respirable fraction that is most likely to deposit in the deep lung, was about 55%, exceeding the majority of marketed dry powder inhalers available [45].

Table 2. Microparticles' aerodynamic properties. Comparison between CHT alone and CHT_Fe@Au_POX-SH (CHT_NP), with and without the use of a magnet applied at the last stage of the ACI (Anderson Cascade Impactor).

Assay	MMAD (μm)	RF (%)	FPF (%)	GSD (%)	EF (%)	Yield, η (%)
CHT_NP	1.5 ± 0.1	70.2 ± 0.3	55 ± 2	2.5 ± 0.1	97.50 ± 0.04	38
CHT	1.2 ± 0.2	65 ± 2	48 ± 2	3.4 ± 0.3	96.3 ± 0.1	60

MMAD—Mass Median Aerodynamic Diameter; RF—Respirable Fraction; FPF—Fine Particle Fraction; GSD—Geometric Standard Deviation; EF—Emitted Fraction.

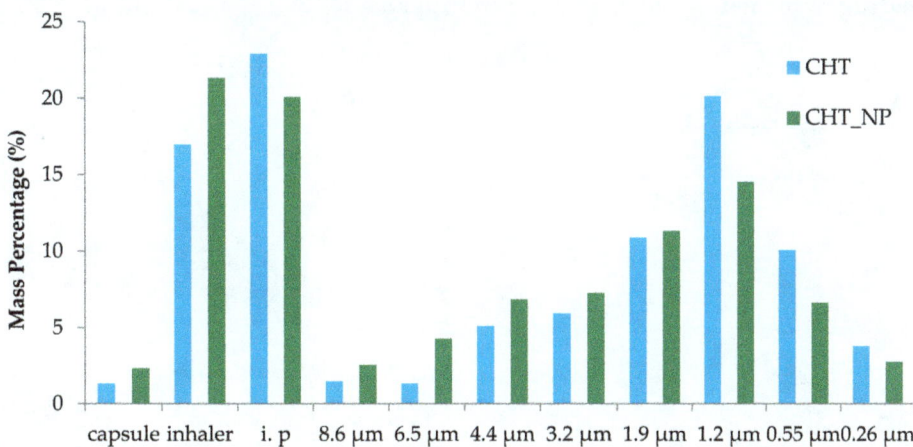

Figure 7. Percentage of mass powder entrapped in the different stages of the ACI. The ACI experiment was performed in a Coppley Scientific instrument at 60 L/min and four kPa. I.p stands for the induction port.

2.2.3. Entrapment Efficiency

The entrapment efficiency was calculated to estimate the number of nanoparticles and IBP that remained in the CHT matrix after particle engineering. For microparticles with IBP entrapped and for microparticles with nanoparticles embedded, the entrapment efficiency was 71.7% for IBP and 68.3% for the nanoparticles.

2.2.4. In Vitro Cumulative Studies

To access particles' ability to release the entrapped contents, an in vitro study was carried using two different pHs, mimicking the conditions of healthy alveolar epithelial cells (pH 7.4) and cancerous alveolar epithelial cells (pH 6.8). The IBP release mechanism from the nano-in-micro formulations was used to mimic all contents release.

The release profile was adjusted to the Korsmeyer-Peppas mathematical model and the n-value was obtained from the slope of the Korsmeyer-Peppas plot and represents different release mechanisms. For spherical geometries: $n = 0.43$ for Fickian diffusion; $0.43 < n < 0.85$ for anomalous non-Fickian diffusion; $n = 0.85$ for Case-II transport. In both cases herein described, we have obtained a not-Fickian diffusion, since $n = 0.59$ for pH 7.4 and $n = 0.61$ for pH 6.8 (Table 3). Having an anomalous transport, the release rate has the contribution of both diffusion and relaxation mechanisms-stresses during polymer swelling [32,46].

Table 3. Peppas' Adjustment for IBP (Ibuprofen) release.

	n	$k\ (h^{-1})$	% Mass Released
pH 7.4	0.6	23.1	68.9
pH 6.8	0.6	19.2	81.7

n—diffusional exponent; k—constant.

As expected, the amount of IBP released from the particles is higher at pH 6.8 than pH 7.4 (Figure 8 and Table 3). Since CHT pKa is around 6.5, at pH 6.8 the CHT amino groups are partially protonated, leading to particle swelling and thus, a faster release. Nevertheless, the sustained and controlled release profile observed at both pHs proves that CHT is a suitable carrier to be used in lung disease situations (such as cancer) through pulmonary administration. After reaching the lung epithelium and upon contact with the diseased cells, the engineered formulations are expected to promote a proper release of the drug or the nanosystems that are entrapped.

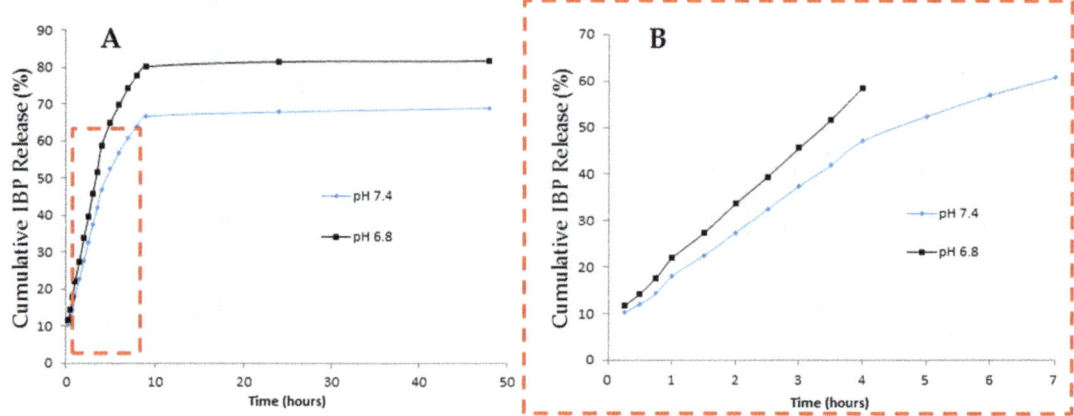

Figure 8. (A) In vitro cumulative liberation studies at various pHs; **(B)** Korsmeyer-Peppas adjustment for 60% of the release at different pHs.

3. Materials and Methods

3.1. Materials

All components were used as received without any further purification. Chitosan (viscosity 5–20 mPa·s) was purchased from Tokyo Chemical Industry (Tamil Nadu, India). The monomer 2-ethyl-2-oxazoline (EtOx, $\geq$99% purity), boron trifluoride diethyl etherate (BF$_3$·OEt$_2$), acetone (99.8% purity), acetonitrile (ACN, 99.9% purity), and acetic acid glacial (99.7% purity) were purchased from Sigma-Aldrich (Saint Louis, MO, USA). Ethanol (EtOH) (96% purity) was purchased from Panreac (Barcelona, Spain), Industrial carbon dioxide (CO$_2$, purity $\geq$ 99.93%) was obtained from Air Liquide (Paris, France), and SnakeSkin™ Dialysis Tubing 3500 MWCO (ThermoFisher, Waltham, MA, USA), 22 mm $\times$ 35 feet dry diameter, 34 mm dry flat width, 3.7 mL/cm, was purchased from Thermo Fisher Scientific (Waltham, MA, USA).

3.2. Synthesis of the Gold Coated Magnetite (Fe@Au)

The synthesis of Fe@Au nanocomposite was obtained using magnetite Fe@SO4^{2-} as the initial support material [25], by polyelectrolyte/gold layer by layer deposition processes developed by Caruso et al. [47]. AuNPs can be adsorbed to form a hybrid nanosystem. In a previous study, Araújo and co-workers proved that by using these layer-by-layer deposition techniques directly on the Fe@SO$_4$$^{2-}$, it is possible to obtain a robust Magnetic@gold hybrid with potential applications in biomarker discovery [25].

The synthesis of Fe@Au magnetic nanocomposites was conducted in several simple steps as was already reported by Araújo and co-workers [25], which was performed following a previous procedure for layer-by-layer polyelectrolytes described by Caruso et al. [47–49]. PDADMAC and PSS were alternately adsorbed onto the negative surface of the magnetic nanoparticles, Fe@SO$_4$$^{2-}$. A solution of 5 mL of the previous Fe@SO$_4$$^{2-}$ synthesized NPs was re-suspended in 100 mL of milli-Q water for a duration of 2 min assisted with ultrasonic energy. 100 mL of a water solution containing 1 mg/mL of PDADMAC was added. The solution was maintained in an ultrasonic bath for 1 min, and under magnetic stirring for 2 h. The NPs obtained were separated from the supernatant by magnetic separation and were washed five times with milli-Q water. This process was repeated for the adsorption of the PSS layer, and for the second monolayer of PDADMAC. The three-layer polyelectrolyte magnetite nanoparticles, washed previously, were re-suspended in 100 mL of milli-Q water and were mixed with 5 mL of NaCl 0.2 M solution. This solution was kept for 1 minute in an ultrasonic bath. After this time, 20 mL of previously synthesized gold nanoparticles was added, and the entire solution was maintained with magnetic stirring for 2 h. The final nanoparticles were separated from the supernatant by magnetic decantation, washed five times with milli-Q water, and were finally re-suspended in 100 mL of Milli-Q water. The formation and growth of the partial gold shell was performed based on a previous report [50] reducing aliquots of HAuCl$_4$ (100 μL, 5×10^{-4} M) with ascorbic acid (100 μL, 0.34×10^{-3} M) under ultrasonic stimulation to help nanoparticle separation during the growth process. This gold polymerization process was repeated 15 times. The Strawberry-like nanostructured Fe@Au NPs were separated by magnetic decantation and washed five times with milli-Q water. Subsequently, they were re-suspended in 20 mL of milli-Q water.

3.3. Synthesis of the Living Oligooxazoline

The polymerization was carried out following the procedure described by Macedo et al. [51]. Thus, 2-ethyl-2-oxazoline monomer and the initiator BF$_3$·OEt$_2$ were both added to a stainless-steel reactor, with a monomer/initiator ratio of [M]/[I] = 1/12 [26]. The reaction was carried out under stirring conditions, and the reactor cell was placed in a water bath at 60 °C. Carbon dioxide was introduced into the reactor up to 160 bar. After 24 h of reaction, a continuous washing process was undertaken using CO$_2$ for 1 h. Then the pressure was slowly released and viscous foam was obtained inside the reactor.

3.3.1. End-Capping of Living Oligo(2-ethyl-2-oxazoline)

The living Oligo(2-ethyl-2-oxazoline) was end-capped with cysteamine. This process is performed by adding a tenfold excess of cysteamine (POX-SH) (regarding the amount of initiator) solubilized in anhydrous DMF. The mixtures were kept at 70 °C using an oil bath under stirring for 24 h. The oily oligomer solubilized in dry DMF was purified through dialysis against pure milli-Q water. The resulting mixture was dried under vacuum and the resulting oily polymer presented a yellow brownish color. The polymer was soluble in water and shows a blue fluorescence at 384 nm [26].

3.3.2. Nanocomposite Functionalization with OPOX-SH

The Fe@Au NPs were suspended in 10 mL of milli-Q water and were placed in a flask reaction. Under stirring in a US bath, 5 mL of the POX-SH solution in MeOH (5 mg/mL) was added drop-wise. After finishing the addition, the NPs were keep on magnetic stirring overnight. After that time, the nanoparticles were washed in MeOH twice. The final material showed high solubility in MeOH. After repeated magnetic cycles of purification, we observed for the first supernatant a subtle red colour, mainly due to a small percentage of gold nuclei released during functionalization. However, as proven by TEM microscopy (Figure 3), the final nanomaterial exhibits the strawberry structure with the AuNPs anchored to the surface of the magnetic cores.

3.4. Nanoparticle Characterization

The size measurements were performed with the end-capped nanoparticles diluted in 1.5 mL of methanol in a Zetasizer Nano ZS instrument (Malvern Instruments, Malvern, UK) in the PROTEOMASS facilities. Zeta potential quantification was carried out in the same Zetasizer Nano ZS instrument using a zeta dip cell. Samples for TEM were prepared by pipetting a drop of the colloidal dispersion onto an ultrathin carbon coated copper grid and allowing the solvent to evaporate. TEM analysis was performed in Spain, at the University of Vigo, CACTI (Center for Researcher and Technical Assistance).

The UV-Vis spectra of the nanoparticles herein developed were acquired using a Perkin Elmer Lambda 25 UV/Vis Spectrometer with a slit width of 5 nm at a scan rate of 240 nm·min^{-1} at 25 °C, in a wavelength range from 350 to 750 nm, and were then analyzed with PerkinElmer UV WinLab™ software (PerkinElmer, Rotterdam, The Netherlands). Fluorescence assays were also performed on a PerkinElmer LS 45 Luminescence Spectrometer with a slit width of 5 nm at a scan rate of 240 nm·min^{-1} using a 10 mm path quartz cell and analyzed using FL WinLab™ software. The excitation wavelength was fixed at 300 nm.

3.5. Production and Characterization of the Nano-in-Micro Formulations

3.5.1. SASD Apparatus: Particle Production

The dry powder formulations were prepared by supercritical CO_2-assisted spray drying. The apparatus is described in detail by Cabral et al. in a previous publication [32]. The SASD is an atomization technology where a solution of the product to be micronized is mixed with $scCO_2$ and forced through a nozzle. Therefore, two solutions were prepared: one containing 1% (w/v) CHT in a mixture of 1% acidic water and ethanol (3:2) and the other one containing both 1% (w/v) CHT and the previously synthesized nanoparticles (10:1). In order to test the release profile, a model drug (IBP) was used, adding 100 mg of IBP to the solution of CHT and NPs. CO_2 is liquefied in a cryogenic bath (-20 °C) and pumped using a liquid pump (HPLC pump K-501, Knauer, Berlin, Germany), which is then heated in an oil bath (70 °C) and sent to the static mixer (model 37-03-075, Kenics-Chemineer, 4.8 mm ID × 191 mm L, 27 helical mixing elements). A high-pressure pump (HPLC 305, Gilson, Middleton, WI, USA) was used to pressurize the solution into the static mixer, allowing the solubilization of the CO_2 into the liquid solution due to its high surface packing, promoting the near equilibrium. In order to maintain CO_2 in its supercritical state (above 73.8 bar and 31 °C), the static mixer is enrolled by heating tapes controlled by a Shinko FCS-13A temperature controller (0.2 °C resolution). The pressure in the static mixer is measured using a Setra pressure transducer (0.1 psig stability). The mixture was then sprayed through a 150 μm diameter nozzle into an aluminum precipitator (±0.1 °C), where the primary droplets are formed, and which operates at near-atmospheric conditions. At the same time, a continuously heated air flow enters into the precipitator and dries the particles by evaporating the liquid solvent and promoting the expansion of CO_2 from the inside of these primary droplets, originating the secondary droplets. The compressed air flow is heated before entering the precipitator. The formed and dried particles exit the precipitator from the bottom side and enter into a high-efficiency (Bucchi) cyclone where they are separated from the gas stream and collected in a glass flask, placed under the cyclone [32,46].

3.5.2. Particle Size Analysis and Morphological Assessment

Morphology G3 Analysis

Particle size analysis (sieve and Feret's diameter) was determined using an optical particle analyzer system (Morphologi G3 Essentials, from Malvern Instruments Ltd., (Malvern, UK)). Both particle size and particle size distributions were measured considering more than 30,000 particles. This characterization was performed in terms of the volume mean diameter (D_v) and the relative width of the distribution (span). The span is calculated using three measures, $D_{v,10}$, $D_{v,50}$, and $D_{v,90}$ (particle volume diameter corresponding to 10%, 50%, and 90% of the population, respectively) by the following equation:

$$\mathbf{Span} = \frac{\mathbf{D_{v,90} - D_{v,10}}}{\mathbf{D_{v,50}}} \tag{1}$$

Scanning Electron Microscopy

Microparticles' morphology was also determined via SEM (Hitachi, S-2400 instrument; Tokyo, Japan) with an accelerating voltage set to 15 kV and magnifications of 500 and 10 k. All samples were mounted on aluminum stubs using carbon tape and were gold-coated prior to analysis.

Fourier Transformed Infrared Spectra

FT-IR spectra were carried out on a PerkinElmer spectrum (Rotterdam, The Netherlands) 1000 FT-IR coupled with Opus Spectroscopy Software with potassium bromide (KBr) tablets containing 20% sample.

X-ray Diffraction (XRD)

Samples of CHT_Fe@Au_POX-SH were subjected to XRD analysis (RIGAKU X-ray diffractometer (Ettlingen, Germany), model Miniflex II). Samples were placed in a sample holder and analyzed through CuKα radiation (30 kV/15 mA), with a 2θ angle ranging between 10° and 90° with a scan rate of 1°/min.

Karl Fischer Coulometric Titration

The moisture content of each sample was determined using Karl Fischer coulometric titration. 1.5 mg of each sample was placed into the titration vessel and titrated with Karl Fischer reagent, which reacts quantitatively and selectively with water. The instrument was composed of a 831 KF Coulometer and a 728 stirrer both from Metrohm and of a Pt/−20–70 °C electrode.

3.5.3. Aerodynamic Performance

Aerodynamic performance was characterized using an Andersen Cascade Impactor (ACI) (Coopley Scientific, Nottingham, UK). Hydroxypropylmethylcellulose capsules (Aerovaus, USA) were loaded with 20 mg of the dry powder formulation (chitosan with the magnetic nanocomposites).

The total percentage of the powder that is released from the capsule and inhaler, the emitted fraction (EF), was determined gravimetrically and can be expressed as shown in Equation (2):

$$EF(\%) = \frac{m_{full} - m_{empty}}{m_{powder}} \times 100 \tag{2}$$

where m_{full} and m_{empty} are the weights (mg) of the capsule before and after simulating the inhalation and m_{powder} is the initial weight (mg) of the powder in the capsule. The respirable fraction (RF) was calculated as the amount of powder reaching the respiratory tract. Fine particle fraction (FPF) was determined by the interpolation of the amount of particles (in percentage) with a diameter below 5 μm emitted from the capsules in each experiment. Mass Median Aerodynamic Diameter (MMAD) was calculated as the particle diameter corresponding to 50% of the cumulative distribution. Geometric standard deviation (GSD) was determined considering the d_{84} and d_{16} measures, which are the diameters corresponding to 84% and 16% of the cumulative distribution, respectively, and can be calculated by the following equation:

$$GSD = \sqrt{\frac{d_{84}}{d_{16}}} \tag{3}$$

3.5.4. Entrapment Efficiency

The drug encapsulation was determined by milling 20 mg of the produced powder and by then adding a known amount of PBS. The solution was left under stirring for 3 h and was then centrifuged at 15,000 rpm for 7 min. Then, the supernatant was used and the amount of IBP was determined by UV spectroscopy at 225 nm. In order to know the entrapment efficiency of the nanoparticles into the chitosan microparticles, a sample of 20 mg of powder was resuspended in aqua regia and the amount of gold and iron was measured by ICP analysis. The encapsulation (E%) was determined by:

$$E\% = \frac{m_r}{m_i} \times 100 \tag{4}$$

where m_r is the remaining mass and m_i is the initially uploaded mass.

3.5.5. In Vitro Cumulative Release Studies

To evaluate the percentage of IBP that was released from the microparticles, 20 mg of the CHT_Fe@Au_POX-SH_IBP powders were transferred into a snakeskin membrane with a cut off size of 1 μm and were incubated into 5 mL of different pH solutions in a shaking bath (36 rpm and 37 °C): 7.4 and 6.8, pH of an healthy cell and pH of a cancer cell, respectively. At different time intervals, 1 mL of each solution was taken and another mL of fresh solution was replaced. The absorbance from each sample was read using a UV-Vis spectrophotometer at 225 nm.

4. Conclusions

In the present work, CHT microparticles containing the functionalized magnetic NPs and IBP were successfully micronized. These powders exhibit appropriate characteristics, either in terms of their size, morphology, and aerodynamic properties, all suitable for inhalation. The most common excipient used in DPIs is lactose (since it is FDA approved) [52] with FPFs ranging from 15% to 45% [53]. PLGA has also been used as a carrier [19], demonstrating high FPF values (30%), however it does not have benefits regarding drug absorption in the lungs. Therefore, CHT with even higher FPF values (approximately 55%) has proven to be a potential carrier when working with pulmonary delivery [53,54]. Also, these values are promising when compared with marketed DPIs, such as the Turbohaler (FPF between 30% and 50%) [55,56].

In this work, CHT has also proven to be a potential candidate to be used as a carrier for therapeutic agents in a lung cancer situation, since it has a higher release percentage (80% release after 10 h) at an acidic pH condition (lung cancer pH).

It is envisaged that fluorescent and magnetic nanoparticles can be used in lung cancer imaging and hyperthermia. However, further studies concerning the use of an applied magnetic field should be performed in order to prove their efficacy.

ACI studies showed an increased fraction of particles reaching the deepest stage of the apparatus. Nevertheless, new experiments need to be performed in order to confirm the statistical significance of the data. To conclude, these proof-of-concept studies suggest that the developed system is a conceivable candidate for pulmonary administration of therapeutic agents for local cancer theragnosis.

Acknowledgments: This work was supported by the Associate Laboratory Research Unit for Green Chemistry—Clean Processes and Technologies—LAQV and the Unidade de Ciências Biomoleculares Aplicadas—UCIBIO which are financed by national funds from FCT/MEC (UID/QUI/50006/2013) and (UID/Multi/04378/2013) respectively, and co-financed by the ERDF under the PT2020 Partnership Agreement (POCI-01-0145-FEDER—007265). A.S.S. and J.F.-L. thank FCT/MEC (Portugal) for their doctoral and postdoctoral grant reference SFRH/BD/51584/2011 and SFRH/BPD/93982/2013 respectively. T.C. also acknowledges FCT-Lisbon for the IF/00915/2014 contract. J.F.-L. and C.L. are grateful to the Scientific Society PROTEOMASS (Portugal) for funding support (General Funding Grant).

Author Contributions: M.C.S., A.S.S., A.A.-R., C.L., and T.C. conceived and designed the experiments; J.F.-L. performed the synthesis and characterization of the strawberry magnetic functionalized nanoparticles; M.C.S., A.S.S., J.F.-L., A.A.R., C.L., and T.C. analyzed the data. C.L., A.A.-R., and T.C. contributed reagents/materials/analysis tools; M.C.S., A.S.S., T.C., C.L., and A.A.-R. wrote and corrected the paper.

Conflicts of Interest: The authors declare no conflict of interest.

Abbreviations

The following abbreviations are used in this manuscript:

SASD	Supercritical Assisted Spray Drying
scCO$_2$	Supercritical Carbon Dioxide
FT-IR	Fourier Transform Infra Red
TEM	Transmission Electron Microscopy
SEM	Scanning Electron Microscopy
DDS	Drug Delivery System
CHT	Chitosan
NP	Nanoparticles
IBP	Ibuprofen
PDADMAC	Poly-diallyldimethylammonium chloride
PSS	Poly-sodium 4-styrenesulfonate
FPF	Fine Particle Fraction
EF	Emitted Fraction
RF	Respirable Fraction
MMAD	Mass Median Aerodynamic Diameter
GSD	Geometric Standard Deviation
ACI	Anderson Cascade Impactor

References

1. Sanders, M. Pulmonary Drug Delivery: An Historical Overview. In *Controlled Pulmonary Drug Delivery*; Springer: New York, NY, USA, 2011; pp. 51–73.

2. Provencio, M.; Isla, D.; Sánchez, A.; Cantos, B. Inoperable stage III non-small cell lung cancer: Current treatment and role of vinorelbine. *J. Thorac. Dis.* **2011**, *3*, 197–204. [PubMed]

3. Bailey, M.M.; Berkland, C.J. Nanoparticle formulations in pulmonary drug delivery. *Med. Res. Rev.* **2009**, *29*, 196–212. [CrossRef] [PubMed]

4. Buttini, F.; Colombo, P.; Rossi, A.; Sonvico, F.; Colombo, G. Particles and powders: Tools of innovation for non-invasive drug administration. *J. Control. Release* **2012**, *161*, 693–702. [CrossRef] [PubMed]

5. Odziomek, M.; Sosnowski, T.R.; Gradoń, L. Conception, preparation and properties of functional carrier particles for pulmonary drug delivery. *Int. J. Pharm.* **2012**, *433*, 51–59. [CrossRef] [PubMed]

6. Patil, J.S.; Sarasija, S. Pulmonary drug delivery strategies: A concise, systematic review. *Lung India* **2012**, *29*, 44–49. [PubMed]

7. Patel, A.; Patel, M.; Yang, X.; Mitra, A.K. Recent Advances in Protein and Peptide Drug Delivery—A Special Emphasis on Polymeric Nanoparticles. *Protein Pept. Lett.* **2014**, *21*, 1102–1120. [CrossRef] [PubMed]

8. De Jong, W.H.; Borm, P.J. Drug delivery and nanoparticles: Applications and hazards. *Int. J. Nanomed.* **2008**, *3*, 133–149. [CrossRef]

9. Han, G.; Ghosh, P.; Rotello, V.M. Functionalized gold nanoparticles for drug delivery. *Nanomedicine* **2007**, *2*, 113–123. [CrossRef] [PubMed]

10. Kim, C.; Credi, A.; Park, C.; Youn, H.; Kim, H.; Noh, T.; Kook, H.; Oh, T.; Park, J.; Kook, Y.H.; et al. Cyclodextrin-covered gold nanoparticles for targeted delivery of an anti-cancer drug. *J. Mater. Chem.* **2009**, *19*, 2261–2440.

11. Grenha, A.; Remuñán-López, C.; Carvalho, E.L.S.; Seijo, B. Microspheres containing lipid/chitosan nanoparticles complexes for pulmonary delivery of therapeutic proteins. *Eur. J. Pharm. Biopharm.* **2008**, *69*, 83–93. [CrossRef] [PubMed]

12. Andrade, F.; Rafael, D.; Videira, M.; Ferreira, D.; Sosnik, A.; Sarmento, B. Nanotechnology and pulmonary delivery to overcome resistance in infectious diseases. *Adv. Drug Deliv. Rev.* **2013**, *65*, 1816–1827. [CrossRef] [PubMed]

13. Barnaby, S.N.; Sita, T.L.; Petrosko, S.H.; Stegh, A.H.; Mirkin, C.A. Nanotechnology-Based Precision Tools for the Detection and Treatment of Cancer. *Cancer Treat. Res.* **2015**, *166*, 293–322.

14. Rana, S.; Bajaj, A.; Mout, R.; Rotello, V.M. Monolayer coated gold nanoparticles for delivery applications. *Adv. Drug Deliv. Rev.* **2012**, *64*, 200–216. [CrossRef] [PubMed]

15. Chatterjee, K.; Sarkar, S.; Rao, K.J.; Paria, S. Core/shell nanoparticles in biomedical applications. *Adv. Colloid Interface Sci.* **2014**, *209*, 8–39. [CrossRef] [PubMed]

16. Ahmad, M.Z.; Akhter, S.; Jain, G.K.; Rahman, M.; Pathan, S.A.; Ahmad, F.J.; Khar, R.K. Metallic nanoparticles: Technology overview & drug delivery applications in oncology. *Expert Opin. Drug Deliv.* **2010**, *7*, 927–942. [PubMed]

17. Takami, T.; Murakami, Y. Development of PEG-PLA/PLGA microparticles for pulmonary drug delivery prepared by a novel emulsification technique assisted with amphiphilic block copolymers. *Colloids Surf. B Biointerfaces* **2011**, *87*, 433–438. [CrossRef] [PubMed]

18. Fang, I.J.; Trewyn, B.G. *Application of Mesoporous Silica Nanoparticles in Intracellular Delivery of Molecules and Proteins*, 1st ed.; Elsevier Inc.: Amsterdam, The Netherlands, 2012.

19. Yang, Y.; Bajaj, N.; Xu, P.; Ohn, K.; Tsifansky, M.D.; Yeo, Y. Development of highly porous large PLGA microparticles for pulmonary drug delivery. *Biomaterials* **2009**, *30*, 1947–1953. [CrossRef] [PubMed]

20. Silva, A.S.; Tavares, M.T.; Aguiar-Ricardo, A. Sustainable strategies for nano-in-micro particle engineering for pulmonary delivery. *J. Nanopart. Res.* **2014**, *16*, 2602. [CrossRef]

21. Huynh, N.T.; Roger, E.; Lautram, N.; Benoît, J.-P.; Passirani, C. The rise and rise of stealth nanocarriers for cancer therapy: Passive versus active targeting. *Nanomedicine* **2010**, *5*, 1415–1433. [CrossRef] [PubMed]

22. Koo, O.M.; Rubinstein, I.; Onyuksel, H. Role of nanotechnology in targeted drug delivery and imaging: A concise review. *Nanomed. Nanotechnol. Biol. Med.* **2005**, *1*, 193–212. [CrossRef] [PubMed]

23. Lu, A.H.; Salabas, E.L.; Schuth, F. Magnetic nanoparticles: Synthesis, protection, functionalization, and application. *Angew. Chemi. Int. Ed.* **2007**, *46*, 1222–1244. [CrossRef] [PubMed]

24. Spasova, M.; Salgueiriño-Maceira, V.; Schlachter, A.; Hilgendorff, M.; Giersig, M.; Liz-Marzán, L.M.; Farle, M. Magnetic and optical tunable microspheres with a magnetite/gold nanoparticle shell. *J. Mater. Chem.* **2005**, *15*, 2095–2098. [CrossRef]

25. Araújo, J.E.; Lodeiro, C.; Capelo, J.L.; Rodríguez-González, B.; Santos, A.A.D.; Santos, H.M.; Fernandez-Lodeiro, J. Novel nanocomposites based on a strawberry-like gold-coated magnetite (Fe@Au) for protein separation in multiple myeloma serum samples. *Nano Res.* **2015**, *8*, 1189–1198. [CrossRef]

26. Silva, A.S.; Silva, M.C.; Miguel, S.P.; Bonifácio, V.D.B.; Correia, I.J.; Aguiar-Ricardo, A. Nanogold Poxylation: Towards always-on fluorescent lung cancer targeting. *RSC Adv.* **2016**, *6*, 33631–33635. [CrossRef]

27. Restani, R.B.; Conde, J.; Pires, R.F.; Martins, P.; Fernandes, A.R.; Baptista, P.V.; Bonifácio, V.D.B.; Aguiar-Ricardo, A. POxylated Polyurea Dendrimers: Smart Core-Shell Vectors with IC50 Lowering Capacity. *Macromol. Biosci.* **2015**, *15*, 1045–1051. [CrossRef] [PubMed]

28. Wilson, P.; Ke, P.C.; Davis, T.P.; Kempe, K. Poly(2-oxazoline)-based micro- and nanoparticles: A review. *Eur. Polym. J.* **2016**. [CrossRef]

29. Patton, J.S. Unlocking the opportunity of tight glycaemic control: Inovative delivery of insulin via the lung. *Diabetes Obes. Metab.* **2005**, *7*, S5–S8. [CrossRef] [PubMed]

30. Smola, M.; Vandamme, T.; Sokolowski, A. Nanocarriers as pulmonary drug delivery systems to treat and to diagnose respiratory and non respiratory diseases. *Int. J. Nanomed.* **2008**, *3*, 1–19. [CrossRef]

31. Wang, Y.-B.; Watts, A.B.; Peters, J.I.; Liu, S.; Batra, A.; Williams, R.O. In vitro and in vivo performance of dry powder inhalation formulations: Comparison of particles prepared by thin film freezing and micronization. *AAPS PharmSciTech* **2014**, *15*, 981–993. [CrossRef] [PubMed]

32. Cabral, R.P.; Sousa, A.; Silva, A.S.; Paninho, A.I.; Temtem, M.; Costa, E.; Casimiro, T. Design of experiments approach on the preparation of dry inhaler chitosan composite formulations by supercritical CO_2-assisted spray-drying. *J. Supercrit. Fluids* **2016**, *116*, 26–35. [CrossRef]

33. Illum, L. Chitosan and its use as a pharmaceutical excipient. *Pharm. Res.* **1998**, *15*, 1326–1331. [CrossRef] [PubMed]

34. Grenha, A.; Seijo, B. The potential of chitosan for pulmonary drug delivery. *J. Drug Deliv. Sci. Technol.* **2010**, *20*, 33–43. [CrossRef]

35. Kean, T.; Thanou, M. Biodegradation, biodistribution and toxicity of chitosan. *Adv. Drug Deliv. Rev.* **2010**, *62*, 3–11. [CrossRef] [PubMed]

36. Dash, M.; Chiellini, F.; Ottenbrite, R.M.; Chiellini, E. Chitosan—A versatile semi-synthetic polymer in biomedical applications. *Prog. Polym. Sci.* **2011**, *36*, 981–1014. [CrossRef]

37. Peniche, C.; Arguelles-Monal, W.; Peniche, H.; Acosta, N. Chitosan: An Attractive Biocompatible Polymer for Microencapsulation. *Macromol. Biosci.* **2003**, *3*, 511–520. [CrossRef]

38. Adami, R.; Liparoti, S.; Reverchon, E. A new supercritical assisted atomization configuration, for the micronization of thermolabile compounds. *Chem. Eng. J.* **2011**, *173*, 55–61. [CrossRef]

39. Porta, G.D.; De Vittori, C.; Reverchon, E. Supercritical assisted atomization: A novel technology for microparticles preparation of an asthma-controlling drug. *AAPS PharmSciTech* **2005**, *6*, E421–E428.

40. Shoyele, S.A.; Cawthorne, S. Particle engineering techniques for inhaled biopharmaceuticals. *Adv. Drug Deliv. Rev.* **2006**, *58*, 1009–1029. [CrossRef] [PubMed]

41. Mero, A.; Pasut, G.; Via, L.D.; Fijten, M.W.M.; Schubert, U.S.; Hoogenboom, R.; Veronese, F.M. Synthesis and characterization of poly(2-ethyl 2-oxazoline)-conjugates with proteins and drugs: Suitable alternatives to PEG-conjugates? *J. Control. Release* **2008**, *125*, 87–95. [CrossRef] [PubMed]

42. Correia, V.G.; Bonifácio, V.D.; Raje, V.P.; Casimiro, T.; Moutinho, G.; da Silva, C.L.; Pinho, M.G.; Aguiar-Ricardo, A. Oxazoline-Based Antimicrobial Oligomers: Synthesis by CROP Using Supercritical CO_2. *Macromol. Biosci.* **2011**, *11*, 1128–1137. [CrossRef] [PubMed]

43. Hirota, K.; Terada, H. Endocytosis of Particle Formulations by Macrophages and Its Application to Clinical Treatment. In *Molecular Regulation of Endocytosis*; INTECH Open Access Publisher: Rijeka, Croatia, 2012; pp. 413–428.

44. Lambert, A.R.; Shaughnessy, P.O.; Tawhai, M.H.; Hoffman, E.A.; Lin, C. Regional deposition of particles in an image-based airway model: Large-eddy simulation and left-right lung ventilation asymmetry. *Aerosol Sci. Technol.* **2011**, *45*, 11–25. [CrossRef] [PubMed]

45. Chow, A.H.L.; Tong, H.H.Y.; Chattopadhyay, P.; Shekunov, B.Y. Particle engineering for pulmonary drug delivery. *Pharm. Res.* **2007**, *24*, 411–437. [CrossRef] [PubMed]

46. Restani, R.B.; Correia, V.G.; Bonifácio, V.D.B.; Aguiar-Ricardo, A. Development of functional mesoporous microparticles for controlled drug delivery. *J. Supercrit. Fluids* **2010**, *55*, 333–339. [CrossRef]

47. Caruso, F.; Lichtenfeld, H.; Giersig, M.; Mohwald, H. Electrostatic self-assembly of silica nanoparticle-polyelectrolyte multilayers on polystyrene latex particles. *J. Am. Chem. Soc.* **1998**, *120*, 8523–8524. [CrossRef]

48. Caruso, F.; Caruso, R.A.; Möhwald, H. Nanoengineering of inorganic and hybrid hollow spheres by colloidal templating. *Science* **1998**, *282*, 1111–1114. [CrossRef] [PubMed]

49. Caruso, F. Hollow capsule processing through colloidal templating and self-assembly. *Chemistry* **2000**, *6*, 413–419. [CrossRef]

50. Salgueirino-Maceira, V.; Correa-Duarte, M.A.; Farle, M.; López-Quintela, A.; Sieradzki, K.; Diaz, R. Bifunctional Gold-Coated Magnetic Silica Spheres Bifunctional Gold-Coated Magnetic Silica Spheres. *Chem. Mater.* **2006**, *18*, 2701–2706. [CrossRef]

51. De Macedo, C.V.; da Silva, M.S.; Casimiro, T.; Cabrita, E.J.; Aguiar-Ricardo, A. Boron trifluoride catalyzed polymerisation of 2-substituted-2-oxazolines in supercritical carbon dioxide. *Green Chem.* **2007**, *9*, 948–953. [CrossRef]

52. Bosquillon, C.; Lombry, C.; Préaat, V.; Vanbever, R. Influence of formulation excipients and physical characteristics of inhalation dry powders on their aerosolization performance. *J. Control. Release* **2001**, *70*, 329–339. [CrossRef]

53. Kinnunen, H.; Hebbink, G.; Peters, H.; Shur, J.; Price, R. An investigation into the effect of fine lactose particles on the fluidization behaviour and aerosolization performance of carrier-based dry powder inhaler formulations. *AAPS PharmSciTech* **2014**, *15*, 898–909. [CrossRef] [PubMed]

54. De Boer, A.H.; Gjaltema, D.; Hagedoorn, P.; Frijlink, H.W. Can "extrafine" dry powder aerosols improve lung deposition? *Eur. J. Pharm. Biopharm.* **2015**, *96*, 143–151. [CrossRef] [PubMed]

55. Newman, S.P. Dry powder inhalers for optimal drug delivery. *Expert Opin. Biol. Ther.* **2004**, *4*, 23–33. [CrossRef] [PubMed]

56. Buttini, F.; Brambilla, G.; Copelli, D.; Sisti, V.; Balducci, A.G.; Bettini, R.; Pasquali, I. Effect of Flow Rate on In Vitro Aerodynamic Performance of NEXThaler® in Comparison with Diskus® and Turbohaler® Dry Powder Inhalers. *J. Aerosol Med. Pulm. Drug Deliv.* **2016**, *28*, 167–178. [CrossRef] [PubMed]

3

The Effect of Bisphasic Calcium Phosphate Block Bone Graft Materials with Polysaccharides on Bone Regeneration

Hyun-Sang Yoo [1,†], Ji-Hyeon Bae [1,†], Se-Eun Kim [2,†], Eun-Bin Bae [1], So-Yeun Kim [3], Kyung-Hee Choi [4], Keum-Ok Moon [4], Chang-Mo Jeong [1] and Jung-Bo Huh [1,*]

[1] Department of Prosthodontics, Dental Research Institute, Institute of Translational Dental Sciences, BK21 PLUS Project, School of Dentistry, Pusan National University, Yangsan 50612, Korea; nasis3@naver.com (H.-S.Y.); say0739@daum.net (J.-H.B.); 0228dmqls@hanmail.net (E.-B.B.); cmjeong@pusan.ac.kr (C.-M.J.)

[2] Department of Veterinary Surgery, College of Veterinary Medicine, Chonnam National University, Gwangju 61186, Korea; sen0223@gmail.com

[3] Department of Prosthodontics, Pusan National University Hospital, Pusan 49241, Korea; function3@naver.com

[4] Tissue Biotech Institute, Cowellmedi Co., Ltd., Busan 46986, Korea; ckh@cowellmedi.co.kr (K.-H.C.); moonko81@nate.com (K.-O.M.)

* Correspondence: huhjb@pusan.ac.kr or neoplasia96@hanmail.net

† These authors contributed equally to this work.

Academic Editors: Enrico Bernardo and Arne Berner

Abstract: In this study, bisphasic calcium phosphate (BCP) and two types of polysaccharide, carboxymethyl cellulose (CMC) and hyaluronic acid (HyA), were used to fabricate composite block bone grafts, and their physical and biological features and performances were compared and evaluated in vitro and in vivo. Specimens of the following were prepared as 6 mm diameter, 2 mm thick discs; BPC mixed with CMC (the BCP/CMC group), BCP mixed with crosslinked CMC (the BCP/c-CMC group) and BCP mixed with HyA (the BCP/HyA group) and a control group (specimens were prepared using particle type BCP). A scanning electron microscope study, a compressive strength analysis, and a cytotoxicity assessment were conducted. Graft materials were implanted in each of four circular defects of 6 mm diameter in calvarial bone in seven rabbits. Animals were sacrificed after four weeks for micro-CT and histomorphometric analyses, and the findings obtained were used to calculate new bone volumes (mm^3) and area percentages (%). It was found that these two values were significantly higher in the BCP/c-CMC group than in the other three groups ($p < 0.05$). Within the limitations of this study, BCP composite block bone graft material incorporating crosslinked CMC has potential utility when bone augmentation is needed.

Keywords: bone regeneration; bone substitutes; composite; biphasic calcium phosphate; carboxymethyl cellulose; crosslinking; hyaluronic acid

1. Introduction

A sufficient amount of residual bone is required for a successful outcome for dental implants, and any restoration provided should have a good long-term prognosis. However, hard tissue defects resulting from causes such as infection and trauma often require bone augmentation [1]. Various grafting materials are used for this purpose, such as, autogenic, allogenic, and xenogenic bone and synthetic calcium phosphate bone graft materials [2], and these materials should trigger osteoblast attachment, proliferation, and differentiation by binding to surrounding bone [3]. In addition, they also should degrade appropriately in concert with the speed of bone growth.

Among the bone graft materials, synthetic calcium phosphate bone graft materials, which have excellent biocompatibility, are commonly used as alternatives to autogenous bone or xenograft or allograft materials [4]. These synthetic materials are easily obtained, do not transmit disease and can be manufactured in various forms. Hydroxyapatite (HA) and β-tricalcium phosphate (β-TCP) are representative calcium phosphate graft materials [5–8]. These materials have drawn interest for bone regeneration because of their structural and chemical similarity with the inorganic component of bone [9]. Although HA is widely used in the dental field as a bone graft material for implant placement due to its excellent biocompatibility and osteoconductivity [5,6], it remains in situ for a long time due to its low in vivo solubility [7]. By comparison, β-TCP quickly dissolves in the body due to its porous structure and low mechanical strength, however this means that the space required for bone regeneration period is often not maintained when β-TCP is used alone [8]. Thus, HA and β-TCP are mixed in various ratios to form biphasic calcium phosphate (BCP) to optimize their advantages [10,11]. It is possible to adjust the degradation rate, mechanical property and the bioactivity of these materials [12].

Commercial forms of particle type synthetic calcium phosphate bone graft materials of various sizes have been marketed. These materials are grafted into defected sites and are covered with a membrane during the guided bone regeneration (GBR) procedure [1,13]. Spaces between particles promote cell invasion and angiogenesis, but also cause mechanical weakening [14]. Furthermore, when the shape of a defect is unfavorable or the size of a defect is large, the augmented site can easily collapse and the particle type graft material is often displaced or lost [14–18]. Therefore, various methods of preventing the escape of particle type bone graft materials have been suggested. Torres et al. [19] tried to prevent this from recipient sites by mixing platelet rich plasma (PRP) with particle type bone grafts, and Dung and Tu used a cap on calvarial defect sites in a rabbit model. However, the results obtained were less than satisfactory [20].

To overcome these problems, studies on the composite materials of organic and inorganic substances have been actively conducted to take advantage of the benefits of block bone and particle type bone graft materials [14,18]. Some of these studies involved introducing organic substances between bone graft particles to prevent particle loss and enhance handling properties. Due to their excellent formability, these materials can be cut or pressed into any shape to help maintain grafts at recipient sites [18,21]. In addition, the introduction of organic substances prevents structural collapse during bone graft degradation process, because their resorption rates and bone cell invasion rates are similar, and, as a result, organic substance absorption harmonizes the bone remodeling processes [21]. Collagen is a representative biodegradable material that is used as an organic scaffold for composite block bone grafts. Such composite block bone grafts are used in various clinical procedures like socket preservation and typical GBR procedures [18,22,23]. However, although they have good handling properties and produce excellent bone augmentation results, they are more expensive than xenogenic bone graft substitutes, and, as a result, several studies have been conducted on the use of degradable polymer graft materials for bone regeneration [24–26].

In this study, we used a carboxymethyl cellulose (CMC) and hyaluronic acid (HyA) to prevent particle loss and to enhance the handling properties of bone graft materials. CMC is a polysaccharide used in the food, pharmaceutical, textile, and paper industries. It is biocompatible, biodegradable, cathodic in nature, and promotes calcium phosphate mineralization. Studies on its use in bone regeneration have being actively pursued [27]. On the other hand, HyA is a water-soluble polysaccharide and a type of cathodic glycosaminoglycan, and is widely distributed in all animal tissues. HyA has affinity for calcium phosphate, a major component of extracellular matrix and joints [28,29]. When HyA is added to a particle type bone graft, viscosities are increased, and, thus, graft handling properties improved, and stability of the grafted site can be maintained [30].

CMC and HyA have been confirmed to exhibit in vivo stability, and their uses in the medical field have been extensively studied [31,32]. However, the use of their composites for bone augmentation the dental fields has not been well studied. In this study, we investigated the physical properties of BCP block bone graft materials incorporating CMC or crosslinked CMC or HyA, and compared and evaluated their biological features and performances as bone graft materials in a rabbit calvarial defect model.

2. Materials and Methods

2.1. Materials

The BCP (Bio-C, Cowellmedi Co., Ltd., Pusan, Korea) used in this study was a mixture of HA and β-TCP (3:7 ratio; Ca/P ratio 1.55). Materials were prepared as follows. To produce BCP/CMC and BCP/HyA, BCP (0.01 ± 0.002 g) was mixed with CMC (1.5%, Daejung Chem Co., Ltd., Siheung, Korea) or HyA (2.5%, Bioland Co., Ltd., Chunan, Korea) at a ratio of 1:1, and specimens were then freeze-dried on a 96-well plate at −70 °C for 24 h. To prepare BCP containing cross-linked CMC (BCP/c-CMC), 2.5% CMC was mixed with 1% ammonium persulfate (Sigma-Aldrich Corp., St. Louis, MO, USA) and 1% sodium hydrogen sulfite (Sigma-Aldrich Corp.), and then 20% of 2-hydroxyethyl methacrylate monomer ($C_6H_{10}O_3$, HEMA, Sigma-Aldrich Corp.) was added to crosslink the CMC, as previously described [33]. The prepared solution was mixed with BCP, reacted in a water bath at 40 °C for 2 h and dried at room temperature for 16 h, and the mixture so obtained was dried in oven at 60 °C for 1 h, and freeze-dried on a 96-well plate at −80 °C for 48 h. Specimens were washed with distilled water 3 times for 10 min on a sonicator (JAC-2010, Kodo Co, Ltd., Hwaseong, Korea), and then freeze-dried on a 96-well plate at −80 °C for 48 h. All specimens were prepared as 6 mm diameter, 2 mm thick block bone discs (Figure 1).

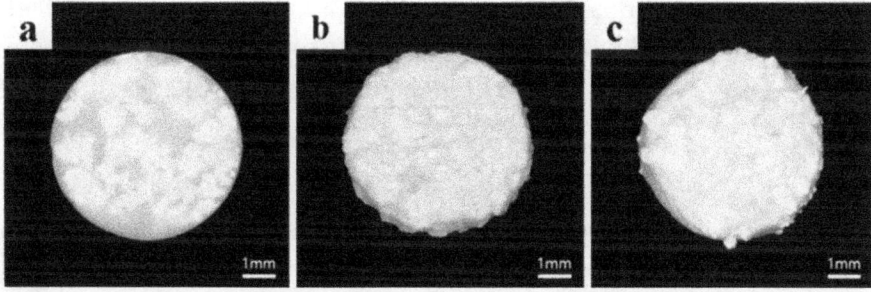

Figure 1. The specimens used in this study: (**a**) bisphasic calcium phosphate/carboxymethyl cellulose (BCP/CMC); (**b**) bisphasic calcium phosphate/cross-linked carboxymethyl cellulose (BCP/c-CMC); and (**c**) bisphasic calcium phosphate/ hyaluronic acid (BCP/HyA).

The specimens were divided based on composition into a control group (particle type BCP) and three experimental groups the BCP/CMC, BCP/c-CMC, and BCP/HyA groups.

2.2. Physical Characterization

2.2.1. Scanning Electron Microscope Surface Analysis

Specimen surfaces were observed using a scanning electron microscope (SEM, SUPRA 25, Carl Zeiss AG, Oberkochen, Germany) at a magnification of ×500 and ×3000 to assess surface microstructures. The specimens were coated with platinum using a sputter coater (Eiko IB, Tokyo, Japan) and observations were conducted at an accelerating voltage of 10 kV. For surface compositional analyses, the SEM-observed specimens were analyzed by EDX (energy dispersive X-ray spectroscopy; Apollo X, Ametek EDAX, Mahwah, NJ, USA) at an accelerating voltage of 15 kV.

2.2.2. Compressive Strength Analysis

To measure compressive strengths, specimens of BCP/CMC, BCP/c-CMC, and BCP/HyA were prepared of diameter 10 mm and thickness 2 mm ($n = 5$). Loads were applied at 0.5 ± 0.1 mm/min using a universal testing machine (3366, Instron Co., Ltd., Norwood, MA, USA). Obtained load data were divided by cross-section area, and are shown in diagram as a stress (N/cm^2, log scale) versus distance (μm, linear scale) plot. Maximum stress (N/cm^2) before fracture was recorded.

2.2.3. In Vitro Cell Test; Assessment of Cytotoxicity

Human MG-63 osteoblast-like cells were seeded into 24-well culture plates containing 0.1 g of specimens per well at a density of 5×10^4 cells/well. Plates were cultured in Dulbecco's modified eagle's medium (DMEM, Gibco BRL, Paisley, UK) containing 10% fetal bovine serum (FBS, Gibco BRL), 100 U/mL penicillin (Gibco BRL) for 24 or 72 h at 37 °C in a 5% CO_2 atmosphere. The effects of specimens on cell proliferation were evaluated using a cell counting Kit-8 (Dojindo, Tokyo, Japan). Experiments were performed five times in in triplicate.

2.2.4. Statistical Analysis

SPSS ver. 21.0 (SPSS, Chicago, IL, USA) was used for the statistical analysis. The significances of differences were determined by One-way analysis of variance (ANOVA). Statistical significance was accepted for p values of <0.05.

2.3. In Vivo Experiments

2.3.1. Experimental Animals

Seven 12- to 13-week-old male New Zealand white rabbits of average weight 3.4 kg were used in this study. Animals were individually housed in a light- and temperature-controlled environment and provided food and water ad-libitum. Animal selection, animal management, and the surgical procedure were performed in accordance with the standards issued by the Ethics Committee on Animal Experimentation at Chungbuk University (CA-15-13).

2.3.2. Surgical Procedure

General anesthesia was induced by an intramuscular injection of 0.5 mL tiletamine plus zolazepam (125 mg/mL; Zoletil, Bayer Korea, Seoul, Korea) and 0.5 mL xylazine hydrochloride (10 mg/kg body weight; Rompun, Bayer Korea). The cranium of each animal was shaved and disinfected with povidone-iodine, and the surgical site was injected with 1 mL of 2% lidocaine HCL and 1:100,000 epinephrine (Yu-Han Co., Gunpo, Korea). An about 30 mm incision was made on the skull to expose the parietal bones, and four 6-mm-diameter calvarial defects were produced on each rabbit using a dental-trephine bur as described previously [34–36]. The same amounts (0.03 ± 0.002 g) of BCP (Bio-C, Cowellmedi, Pusan, Korea), BCP/CMC, BCP/c-CMC, and BCP/HyA were randomly placed on each calvarial defect (Figure 2). No additional membrane was used. The periosteum was sutured using 4-0 Vicryl® (Johnson & Johnson, New Brunswick, NJ, USA), and skin was sutured using 3-0 silk (Ailee. Co., Ltd., Seoul, Korea).

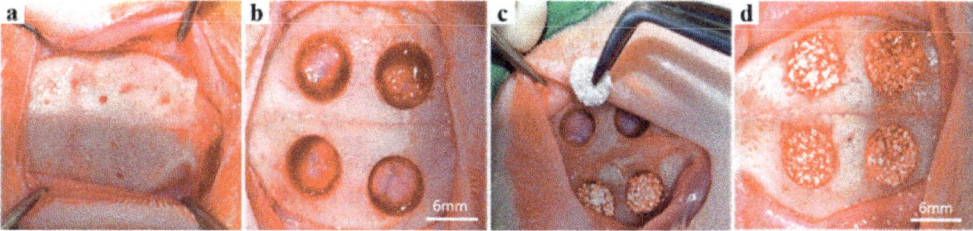

Figure 2. Surgical procedures for the in vivo study: (**a**) parietal bones were exposed by removing periosteum; (**b**) four calvarial defects were formed with a trephine bur (6 mm diameter); (**c**) the four bone grafting materials were randomly placed in defects; and (**d**) all defects were filled with graft materials.

2.3.3. Postoperative Care and Sacrifice

After surgery, rabbits received 1 mg/kg gentamicin (Kookje, Seoul, Korea) and 0.5 mL/kg Pyrin (Green Cross Veterinary Products, Seoul, Korea) intramuscularly three times daily for 3 days. Animals were allowed to recover for 4 weeks, when they were sacrificed by CO_2 inhalation. Calvarial defect sites were harvested along with surrounding bone, and harvested specimens were fixed in neutral buffered formalin (Sigma-Aldrich Corp.) for 2 weeks.

2.3.4. Micro-Computed Tomography Analysis

After fixation, micro-computed tomography three-dimensional images were obtained to determine new bone volumes in defect areas. Specimens were wrapped in Parafilm M® (Bemis Company, Inc., Neenah, WI, USA) to keep them from drying during scanning and scanned using the following settings; scan energy 130 kV, intensity 60 μA, and a pixel resolution of 7.10 μm using a bromine filter (0.25 mm) (Skyscan-1173, version 1.6, Bruker-CT, Kontich, Belgium). The NRecon reconstruction program (version 1.6.10.1, Bruker-CT, Kontich, Belgium) was performed using the same applied scan and reconstruction parameters for all specimens. New bone volumes (NBV; mm^3) were calculated within the regions of interest (Figure 3).

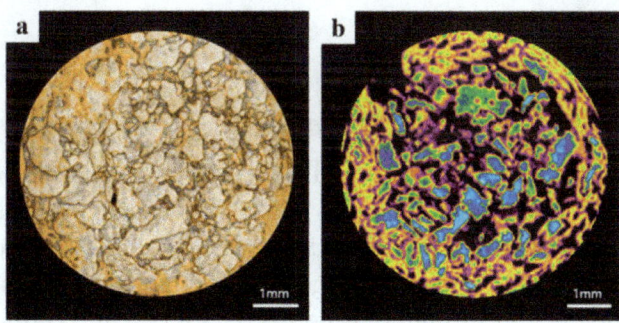

Figure 3. In micro-computed tomography analysis, images of a region of interest: (**a**) reconstructed image; and (**b**) color images of bone graft material and new bone.

2.3.5. Histomorphometric Analysis

After the micro-CT analysis, specimens were cleaned with distilled water, and calcium was removed using EDTA solution (10%, pH 7.0). After confirming calcium removal, specimens were dehydrated by increasing the ethanol concentration. The alcohol was then removed, and samples were infiltrated with paraffin (PolyFin; Triangle Biomedical Sciences, Durham, NC, USA), paraffin embedded, sectioned longitudinally at 4 μm through each defect center using a microtome (Leica RM2255, Leica Microsystems, Wetzlar, Germany), and mounted on slides. Sections were hematoxylin and eosin (H&E) and Masson's trichrome stained to visualize newly regenerated bone tissues. The central-most sections from each block were selected for histologic and histometric evaluations. Images of selected slides were captured using an optical microscope connected to a computer (BX51, OLYMPUS, Tokyo, Japan), a charged-coupled device (CCD) camera (SPOT Insight 2Mp scientific CCD digital camera system, DIAGNOSTIC Instruments Inc., Sterling Heights, MI, USA), and an adaptor (U-CMA3, OLYMPUS, Tokyo, Japan). Captured images were analyzed using i-Solution ver. 8.1 (IMT i-Solution, Inc., Coquitlam, BC, Canada). General specimen images were conducted at ×20 and histometric analyses at ×40 and ×100. The histometric analysis was conducted by one professionally trained, blinded investigator. New bone area percentages (%) (defined as defect area occupied by new bone expressed as a percentage of defect area) within defects area were recorded (Figure 4).

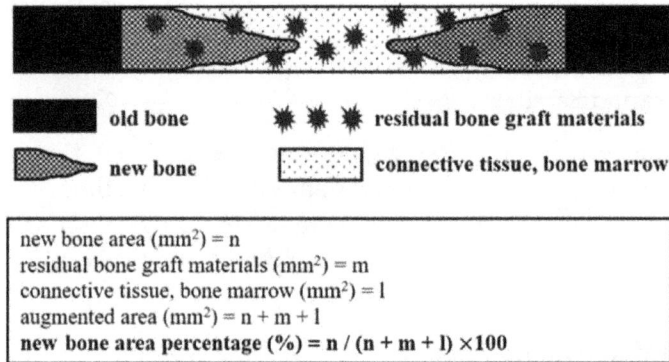

Figure 4. Schematic drawing showing the histometric analysis.

2.3.6. Statistical Analysis

Experiment results are presented as means, standard deviations, and medians. Software R (version 3.1.3, R Foundation for Statistical Computing, Vienna, Austria) was used for the statistical analyses. Brunner and Langer nonparametric analysis was to determine the significances of differences [37]. Statistical significance was accepted for p values of <0.05.

3. Results

3.1. Physical Characterization

3.1.1. Scanning Electron Microscope Surface Analysis

The surface patterns observed by SEM, and EDX surface compositional analysis results are summarized in Figure 5 and Table 1, respectively. In the BCP/CMC (Figure 5c,d) and BCP/HyA groups (Figure 5g,h), radial type polysaccharides covered the surface of BCP, whereas, in the BCP/c-CMC group, CMC was crosslinked and condensed (Figure 5e,f). EDX showed the presence of Ca and P (major components of the bone graft material) and of C, O, Na, and S (major components of the polysaccharide and the crosslinking agent) (Table 1).

Table 1. Energy dispersive X-ray spectroscopy (EDX) determined chemical compositions (wt %) in the four study groups.

Elements	Chemical Compositions (wt %)			
	BCP (Control)	BCP/CMC	BCP/c-CMC	BCP/HyA
C	4.2	25.5	30.4	26.5
O	35.7	35.1	36.2	36.3
Na	-	4.3	5.6	-
P	19.4	11.5	8.6	13.6
S	-	-	6.6	-
Ca	40.9	23.5	12.7	24.0

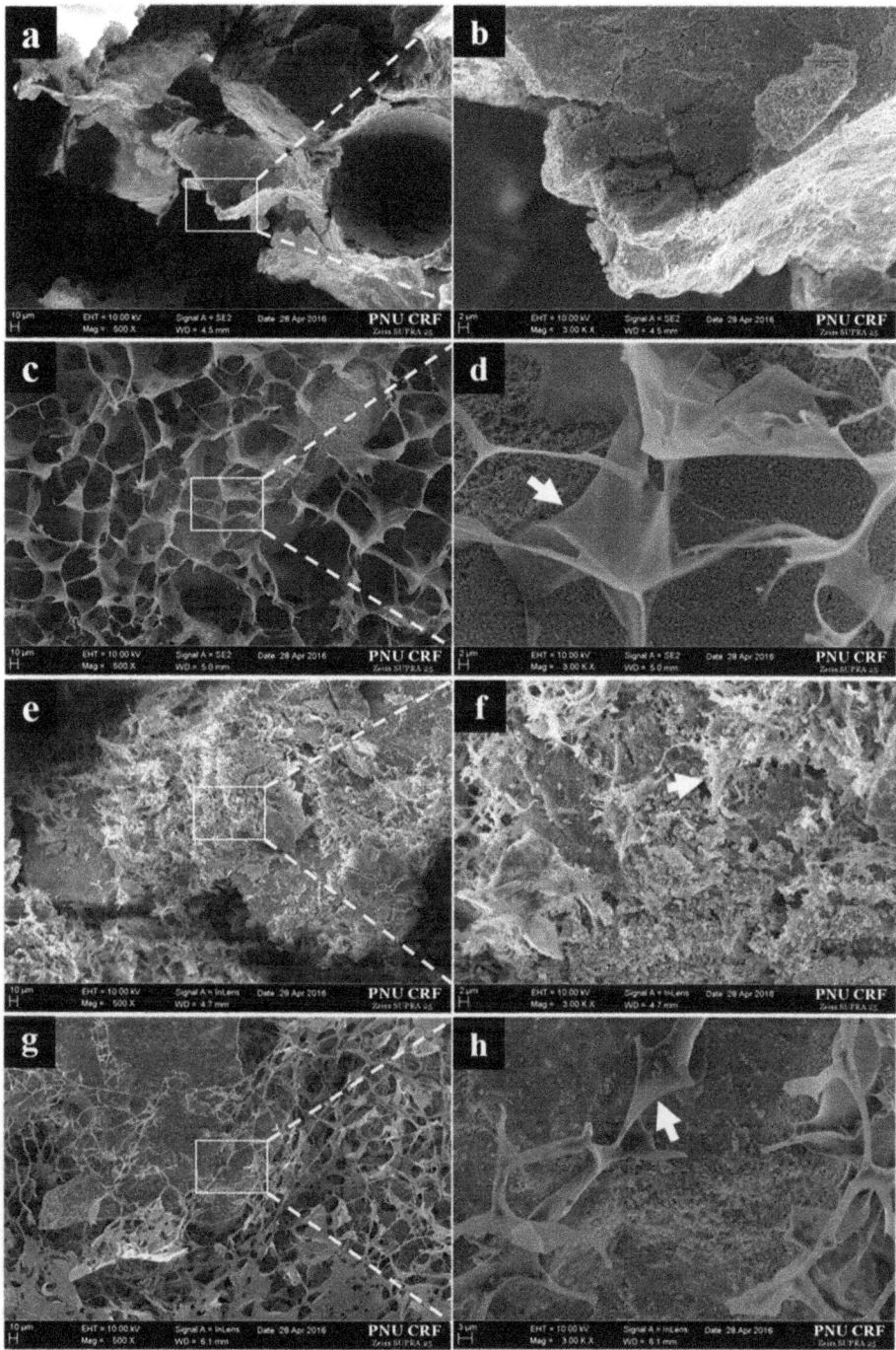

Figure 5. SEM images of bone graft materials. Surfaces of: (**a,b**) BCP; (**c,d**) BCP/CMC; (**e,f**) BCP/c-CMC; and (**g,h**) BCP/HyA samples. White arrow indicates polysaccharides (original magnifications: ×500 (**a,c,e,g**); and ×3000 (**b,d,f,h**)).

3.1.2. Compressive Strength Analysis

The compressive strength results for each group are shown in Figure 6. The BCP/CMC group produced the highest compressive strengths, followed by the BCP/HyA and BCP/c-CMC groups. However, intergroup differences were not statistically significant ($p > 0.05$).

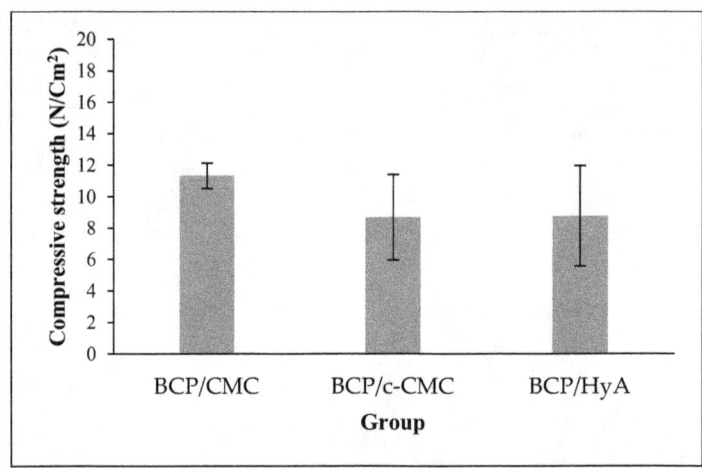

Figure 6. Compressive strengths of the experimental groups. No significant intergroup difference was observed ($p > 0.05$; $n = 5$).

3.1.3. In Vitro Assessment of Cytotoxicity

Cytotoxicity results for human MG-63 osteoblast-like cells are shown in Figure 7. No evidence of cytotoxicity was observed versus the BCP (control) group.

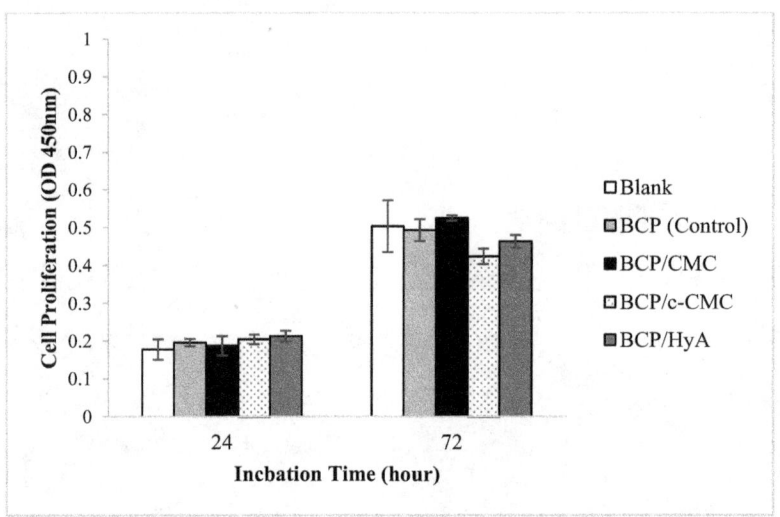

Figure 7. Cytotoxicity of the four graft materials to MG-63 osteoblast-like cells. No evidence of cytotoxicity was observed. Blank, no specimen added ($n = 5$).

3.2. *In Vivo Results*

3.2.1. Clinical Findings

All experimental animals survived the surgical procedure, and the 28 defects healed without issue. Furthermore, no infection or inflammation was observed.

3.2.2. Micro-Computed Tomography Findings

Volumetric measurements are summarized in Table 2 and Figure 8. BCP/c-CMC produced significantly more new bone (mm^3) than the other three groups at four weeks post surgery ($p < 0.05$), which were not significantly different ($p > 0.05$). Micro-CT images revealed bone graft materials in the BCP/c-CMC group had stabilized on defects and the presence of new bone regeneration, whereas in the other groups graft materials had disintegrated and scattered (Figure 9).

Table 2. New bone volumes within regions of interest ($n = 7$; mm^3).

Group	Mean ± SD	Median
BCP (control)	11.45 ± 1.87	11.69 [c]
BCP/CMC	11.95 ± 2.13	11.96 [c]
BCP/c-CMC	15.35 ± 2.39	14.80 [a,b,d]
BCP/HyA	9.83 ± 3.39	10.87 [c]
p value		<0.05

[a] Significantly different versus the BCP (control) group ($p < 0.05$); [b] Significantly different versus the BCP/CMC group ($p < 0.05$); [c] Significantly different versus the BCP/c-CMC group ($p < 0.05$); [d] Significantly different versus the BCP/HyA group ($p < 0.05$).

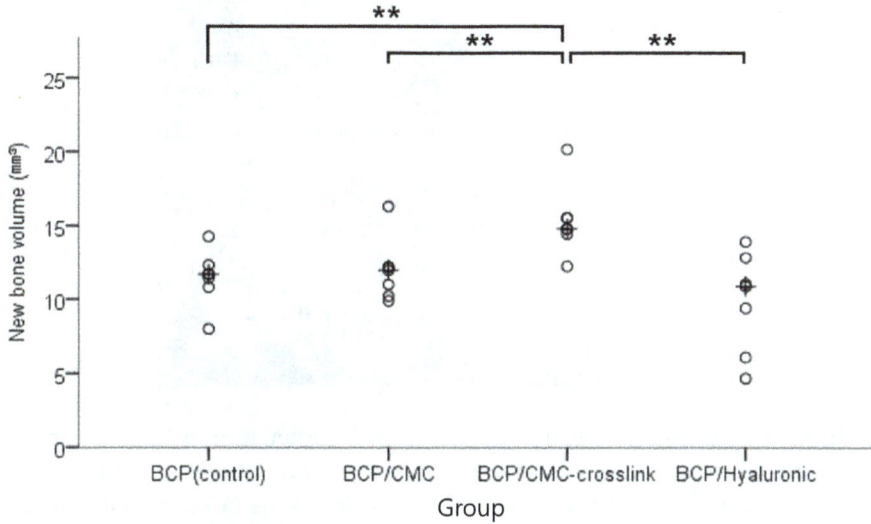

Figure 8. Micro-CT analysis results: scatter plot and median (indicated the cross) representing new bone volumes (mm^3). The BCP/c-CMC group shows the highest level of new bone production. The symbol "**" indicates statistical significance ($p < 0.05$).

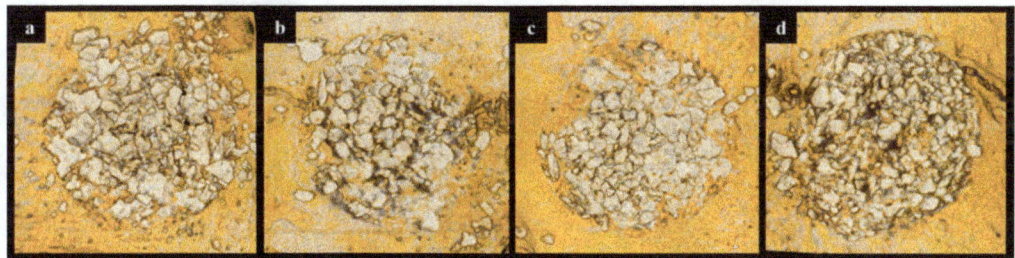

Figure 9. Micro-CT reconstructed images of each group: (**a**) BCP (control) group; (**b**) BCP/CMC group; (**c**) BCP/c-CMC group; and (**d**) BCP/HyA group.

3.2.3. Histologic Findings

In the BCP group (Figure 10), we observed new bone formation derived from existing old bone and around grafted materials. No tissue inflammation was observed.

In the BCP/CMC group (Figure 11), we observed a small amount of new bone formation derived from existing bone and around the grafted materials. The grafted material was evenly distributed, and large amounts of fibrous tissues without inflammation were observed throughout defect sites.

In the BCP/c-CMC group (Figure 12), we observed a large amount of new bone formation derived from existing old bone and grafted material. The boundary of the periosteum was clearly observed, and the bone graft material well occupied defects. No tissue inflammation was observed.

In the BCP/HyA group (Figure 13), we observed new bone formation derived from existing bone and around grafted material. As was observed in the BCP/CMC group, large amounts of fibrous tissues without inflammation were observed throughout defect sites.

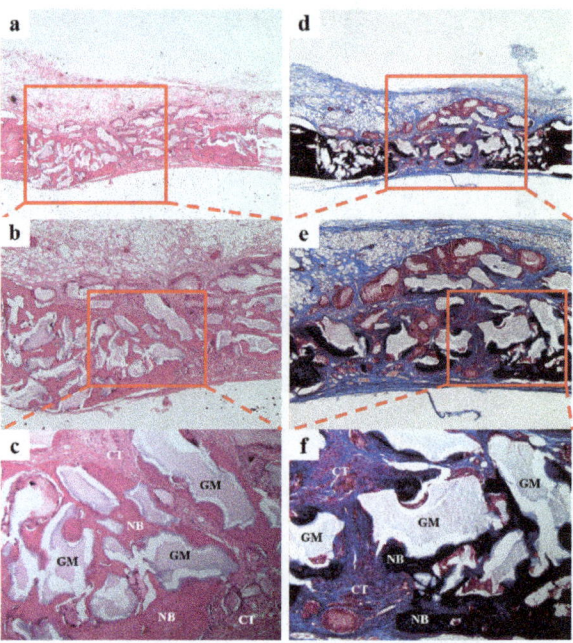

Figure 10. Histological sections of the BCP group at 4 weeks after surgery. A small amount of new bone formation and fibrous connective tissue were observed. NB: new bone; GM: grafted material; CT: connective tissue. Hematoxylin and eosin (H&E) stain results (**a**–**c**); and Masson's trichrome staining results (**d**–**f**) (original magnifications: ×20 (**a**,**d**); ×40 (**b**,**e**); and ×100 (**c**,**f**)).

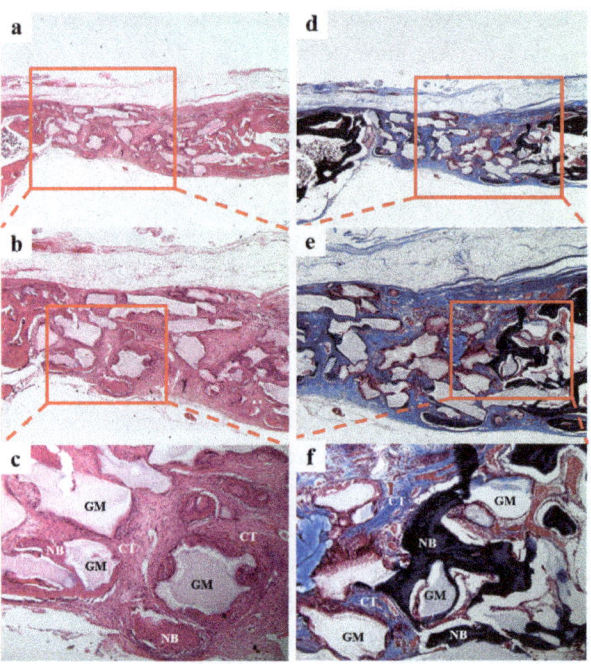

Figure 11. Histological sections of the BCP/CMC group at 4 weeks after surgery. A small amount of new bone formation and a substantial amount of fibrous connective tissue were observed. NB, new bone; GM, grafted material; CT, connective tissue. H&E stain results (**a**–**c**); and Masson's trichrome stain results (**d**–**f**) (original magnifications: ×20 (**a**,**d**); ×40 (**b**,**e**); and ×100 (**c**,**f**)).

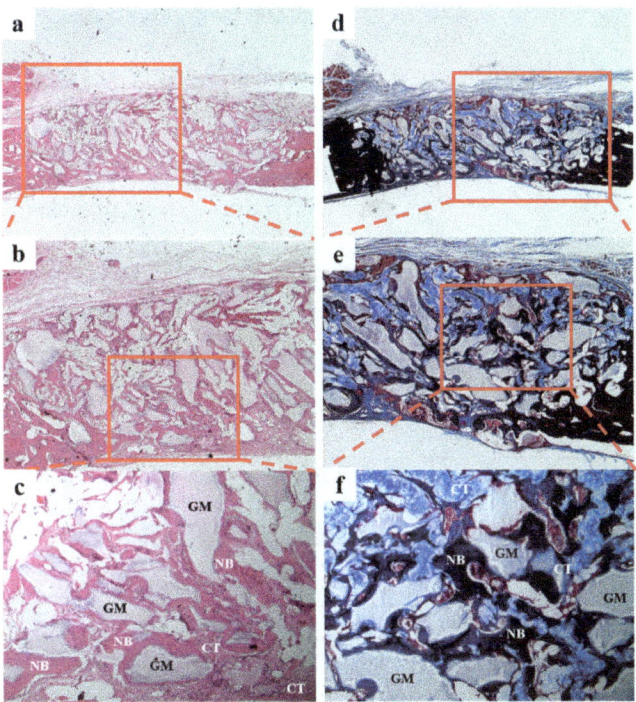

Figure 12. Histological sections of the BCP/c-CMC group at four weeks after surgery. A great amount of new bone formation and a small amount of fibrous connective tissue were observed. NB, new bone; GM, grafted material; CT, connective tissue. H&E stain results (**a–c**); and Masson's trichrome stain results (**d–f**) (original magnification: ×20 (**a,d**); ×40 (**b,e**); and ×100 (**c,f**)).

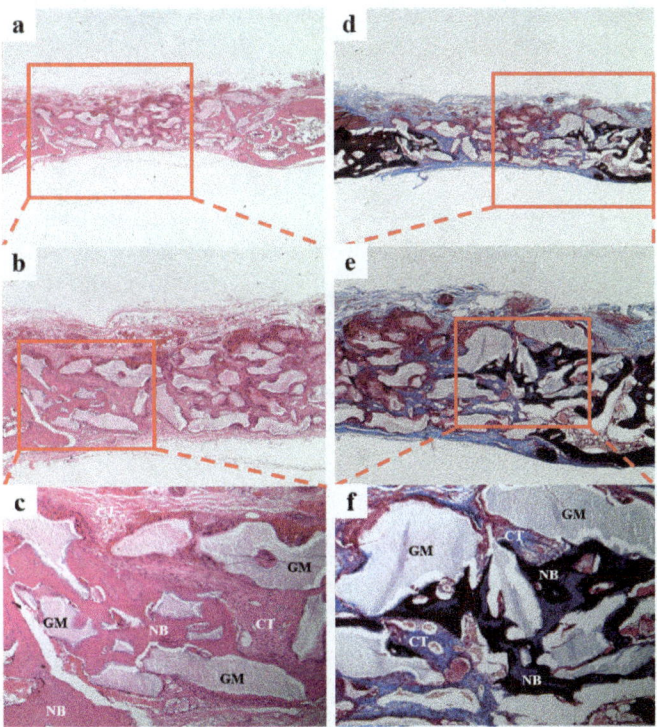

Figure 13. Histological sections of the BCP/HyA group at 4 weeks after surgery. A small amount of new bone formation and fibrous connective tissue were observed. NB, new bone; GM, grafted material; CT, connective tissue. H&E stain results (**a–c**); and Masson's trichrome stain results (**d–f**) (original magnification: ×20 (**a,d**); ×40 (**b,e**); and ×100 (**c,f**)).

3.2.4. Histometric Findings

Histometric measurements are summarized in Table 3 and Figure 14. The BCP/c-CMC group exhibited a significantly higher new bone area percentage (%) ($p < 0.05$). No significant difference was observed between the other three groups ($p > 0.05$).

Table 3. New bone area percentage within the region of interest ($n = 7$; %).

Group	Mean ± SD	Median
BCP (control)	14.27 ± 2.92	15.09 [c]
BCP/CMC	13.52 ± 3.56	12.17 [c]
BCP/c-CMC	17.43 ± 2.59	16.61 [a,b,d]
BCP/HyA	12.68 ± 5.49	13.42 [c]
p value		0.033

[a] Significantly different versus the BCP(control) group ($p < 0.05$); [b] Significantly different versus the BCP/CMC group ($p < 0.05$); [c] Significantly different versus the BCP/c-CMC group ($p < 0.05$); [d] Significantly different versus the BCP/HyA group ($p < 0.05$).

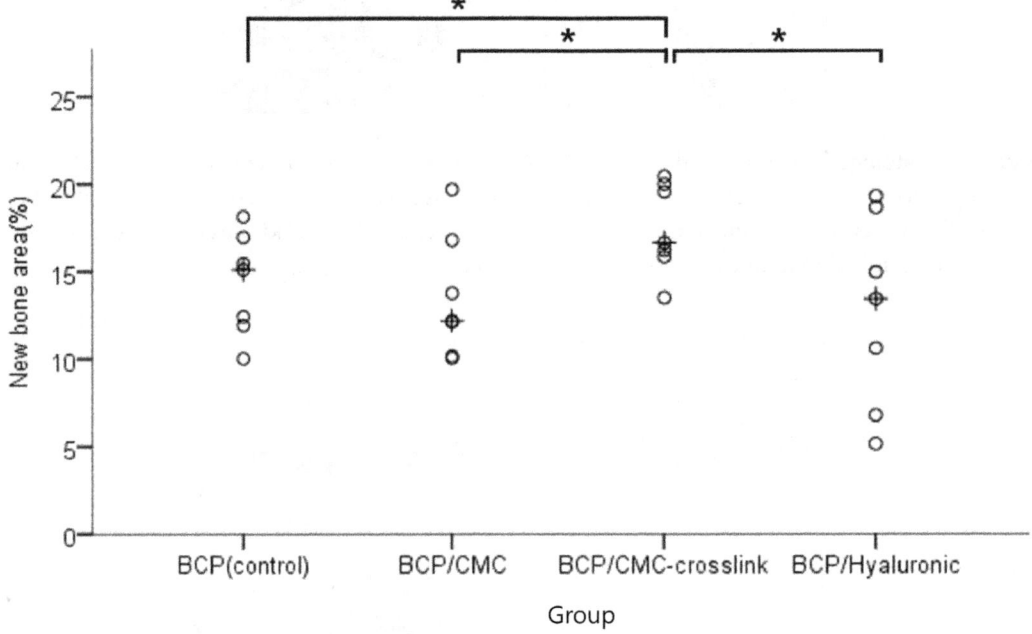

Figure 14. Scatter plot and median (indicated by the cross) of new bone areas (%). The BCP/c-CMC group exhibited more new bone area than the other three groups. The symbol "*" indicates statistical significance ($p < 0.05$).

4. Discussion

Many methods can be used to improve the properties of polymers [38–40], but crosslinking is the most popular method of improving material properties. For example, as compared with native collagen, crosslinked collagen is resorbed more slowly in vivo and acts a barrier membrane to maintain long-term functional stability [40,41]. Crosslinking also makes it possible incorporate polysaccharides to improve properties [42]. In the case of CMC, the crosslinked block bone was compared with non-crosslinked block bone a to examine differences in physical properties and new bone formation ability. In the present study, crosslinking was conducted using the method previously described for chitosan, which has a chemical structure is similar to that of CMC [33]. We also considered crosslinking HyA for comparison purposes using glutaraldehyde or divinyl sulfone, but it because the use of glutaraldehyde might resulted in toxic residuals and divinyl sulfone is not free of safety concerns [43], we did not include crosslinked HyA in the study.

According to micro-CT results, the BCP/c-CMC group showed about 15 mm^3 of new bone formation per defect, whereas the other three groups achieved 9–11 mm^3. Furthermore, micro-CT showed bone grafts in the BCP, BCP/CMC, and BCP/HyA groups had disintegrated and scattered, while in the BCP/c-CMC group, bone grafts were intact and well stabilized in defect sites, which was attributed to crosslinking. However, no significant differences in compressive strength were observed between specimens. These observations concur with those of a previous on the use of crosslinked collagen membranes, which took longer to degrade than corresponding non-crosslinked membranes [41].

In the BCP/CMC and BCP/HyA groups, no difference in new bone formation was observed when particle type BCP was grafted. Because the calvarial defects made were closed, that is, surrounded by a bony wall, dura mater, and periosteum graft materials could not escape as is observed for typical one- or two-wall defects. Thus, the differences among the groups may be significant in the clinical dehiscence defects which cannot maintain the graft materials easily, further study designs may be necessary in the future for confirmation.

Our histometric analysis results of tissue specimens showed the same tendencies as micro-CT results. In the BCP/c-CMC group, new bones accounted for about 17% of defect area, while in the other groups new bones levels ranged from 12% to 14%, which was a significant difference. In tissue specimens, the BCP/c-CMC group showed well maintained graft volumes at defect sites, whereas the graft volumes were reduced in the other groups, particularly in the center areas of defect sites. In addition to maintaining graft volume, it would appear the presence CMC promoted bone metabolism, and that this resulted in the formation of thicker, new bone around grafts in the BCP/c-CMC group than was clearly observed in the other experimental groups.

In histometric analysis of the BCP/CMC and BCP/HyA groups, no significant differences in new bone formation were observed due to the use of particle type BCP (Figure 10b,e). However, the BCP/CMC (Figure 11b,e), BCP/c-CMC (Figure 12b,e), and BCP/HyA (Figure 13b,e) groups showed an obvious boundary between graft and periosteum, whereas in the BCP group the boundary was rough. During GBR (guided bone regeneration) procedures, a membrane is required to prevent soft tissue invasion, but no invasion was observed despite the lack of a membrane in the BCP/CMC, BCP/c-CMC, or BCP/HyA groups, which potentially could simplify the GBR procedure. Zecha et al. reported [44] the application of an absorptive membrane superior to the graft did not result in additional bone formation when bone augmentation was performed using HA/collagen composite block bone. Rothamel et al. [45] found no difference in the bone augmentation results according to the presence/absence of a membrane in allogeneic block bone grafts in a dog model.

The 6 mm diameter defect made in rabbit calvarial bone in the present study was smaller than the 10 mm critical defect size used to evaluate re-ossification in rabbits [46]. However, in the present study, control defects were filled with grafts like the experimental groups. Therefore, there were no limitations in the comparison of new bone formation ability of the graft materials. If a specimen is placed on dura mater, about 10 mmHg intracranial pressure is introduced [47]. This pressure can be assumed to be typical of the pressure applied to a graft by surrounding tissues during GBR. The calvarial bone is mainly composed of cortical bones with relatively smaller bone marrow cells. Thus, graft stability during new bone formation is more critical than in other cell rich areas, which is in-line with the objectives of the present study.

The matrices of the commercially available organic substances used in composite block bone grafts usually consist of collagen. Collagen maintains inorganic particles and blood clots, and has excellent formability, which allows it to be shaped to fit defects. These properties and other make collagen the most useful material for preparing composite materials [18,48]; furthermore, its safety has been well proven in clinical practice. However, no material has ideal properties for bone grafts, and, thus, efforts to identify better new materials are ongoing. In the present study, the BCP/HyA and BCP/CMC groups showed the same ability to promote new bone formation as the BCP control group, and no soft tissue invasion was observed even without a barrier membrane. The BCP/c-CMC group showed most new bone formation, which suggests is has potential use as composite block bone graft material.

5. Conclusions

The micro-CT and histomorphometric analysis showed the BCP/c-CMC group promoted most new bone formation and higher new bone area percentages (%). No significant difference was observed between the other three groups in this respect. The present study shows, BCP/crosslinked CMC composite block bone graft material has potential utility as a means of bone augmentation.

Acknowledgments: This study was supported by a grant from the Korean Health Technology R&D Project through the Korean Health Industry Development Institute (KHIDI), funded by the Korean Ministry of Health & Welfare (grant number: HI14C3309).

Author Contributions: Jung-Bo Huh and Chang-Mo Jeong conceived and designed the experiments; Ji-Hyeon Bae, Se-Eun Kim, Eun-Bin Bae, So-Yeun Kim performed the animal experiments; Kyung-Hee Choi and Keum-Ok Moon performed in-vitro studies and fabricated specimen. Hyun-Sang Yoo and Ji-Hyeon Bae analyzed the data; Se-Eun Kim contributed reagents; every authors contributed to write the paper.

Conflicts of Interest: The authors have no conflict of interest to declare.

References

1. Liu, J.; Kerns, D.G. Mechanisms of guided bone regeneration: A review. *Open Dent. J.* **2014**, *8*, 56–65. [CrossRef] [PubMed]
2. Bernhardt, A.; Lode, A.; Peters, F.; Gelinsky, M. Comparative evaluation of different calcium phosphate-based bone graft granules—An in vitro study with osteoblast-like cells. *Clin. Oral Implants Res.* **2013**, *24*, 441–449. [CrossRef] [PubMed]
3. Yip, I.; Ma, L.; Mattheos, N.; Dard, M.; Lang, N.P. Defect healing with various bone substitutes. *Clin. Oral Implants Res.* **2015**, *26*, 606–614. [CrossRef] [PubMed]
4. Elliot, J.C. *Structure and Chemistry of the Apatites and Other Calcium Orthophosphates*, 1st ed.; Elsevier: Amsterdam, The Netherlands, 1994.
5. Misch, C.E.; Dietsh, F. Bone-grafting materials in implant dentistry. *Implant Dent.* **1993**, *2*, 158–167. [CrossRef] [PubMed]
6. Holmes, R.E.; Bucholz, R.W.; Mooney, V. Porous hydroxyapatite as a bone graft substitute in diaphyseal defects: A histometric study. *J. Orthop. Res.* **1987**, *5*, 114–121. [CrossRef] [PubMed]
7. Barralet, J.; Akao, M.; Aoki, H.; Aoki, H. Dissolution of dense carbonate apatite subcutaneously implanted in Wistar rats. *J. Biomed. Mater. Res* **2000**, *49*, 176–182. [CrossRef]
8. Jensen, S.S.; Broggini, N.; Hjørting-Hansen, E.; Schenk, R.; Buser, D. Bone healing and graft resorption of autograft, anorganic bovine bone and beta-tricalcium phosphate. A histologic and histomorphometric study in the mandibles of minipigs. *Clin. Oral Implants Res.* **2006**, *17*, 237–243. [CrossRef] [PubMed]
9. Knowles, J.C. Phosphate based glasses for biomedical applications. *J. Mater. Chem.* **2003**, *13*, 2395–2401. [CrossRef]
10. Greenwald, A.S.; Boden, S.D.; Goldberg, V.M.; Khan, Y.; Laurencin, C.T.; Rosier, R.N. Bone-graft substitutes: Facts, fictions, and applications. *J. Bone Jt. Surg. Am.* **2001**, *83*, 98–103. [CrossRef]
11. Nevins, M.; Nevins, M.L.; Schupbach, P.; Kim, S.W.; Lin, Z.; Kim, D.M. A prospective, randomized controlled preclinical trial to evaluate different formulations of biphasic calcium phosphate in combination with a hydroxyapatite collagen membrane to reconstruct deficient alveolar ridges. *J. Oral Implantol.* **2013**, *39*, 133–139. [CrossRef] [PubMed]
12. LeGeros, R.Z.; Lin, S.; Rohanizadeh, R.; Mijares, D.; LeGeros, J.P. Biphasic calcium phosphate bioceramics: Preparation, properties and applications. *J. Mater. Sci. Mater. Med.* **2003**, *14*, 201–209. [CrossRef] [PubMed]
13. Buser, D.; Dula, K.; Belser, U.; Hirt, H.P.; Berthold, H. Localized ridge augmentation using guided bone regeneration. 1. Surgical procedure in the maxilla. *Int. J. Periodontics Restor. Dent.* **1993**, *13*, 29–45.
14. Gu, S.J.; Sohn, J.Y.; Lim, H.C.; Um, Y.J.; Jung, U.W.; Kim, C.S.; Choi, S.H. The effects of bone regeneration in rabbit calvarial defect with particulated and block type of hydroxyapatite. *J. Korean Acad. Periodontol.* **2009**, *39*, 321–329. [CrossRef]

15. Le, B.T.; Borzabadi-Farahani, A. Simultaneous implant placement and bone grafting with particulate mineralized allograft in sites with buccal wall defects, a three-year follow-up and review of literature. *J. Craniomaxillofac. Surg.* **2014**, *42*, 552–559. [CrossRef] [PubMed]

16. Seibert, J.S.; Salama, H. Alveolar ridge preservation and reconstruction. *Periodontol 2000* **1996**, *11*, 69–84. [CrossRef] [PubMed]

17. Choi, B.H.; Im, C.J.; Huh, J.Y.; Suh, J.J.; Lee, S.H. Effect of platelet-rich plasma on bone regeneration in autogenous bone graft. *Int. J. Oral Maxillofac. Surg.* **2004**, *33*, 56–59. [CrossRef] [PubMed]

18. Wong, R.W.; Rabie, A.B. Effect of bio-oss collagen and collagen matrix on bone formation. *Open Biomed. Eng. J.* **2010**, *4*, 71–76. [CrossRef] [PubMed]

19. Torres, J.; Tamimi, F.; Tresguerres, I.F.; Alkhraisat, M.H.; Khraisat, A.; Blanco, L.; Lopez-Cabarcos, E. Effect of combining platelet-rich plasma with anorganic bovine bone on vertical bone regeneration: Early healing assessment in rabbit calvariae. *Int. J. Oral Maxillofac. Implants* **2010**, *25*, 123–129. [PubMed]

20. Dung, S.Z.; Tu, Y.K. Effect of different alloplast materials on the stability of vertically augmented new tissue. *Int. J. Oral Maxillofac. Implants* **2012**, *27*, 1375–1381. [PubMed]

21. Kato, E.; Lemler, J.; Sakurai, K.; Yamada, M. Biodegradation property of beta-tricalcium phosphate-collagen composite in accordance with bone formation: A comparative study with Bio-Oss Collagen® in a rat critical-size defect model. *Clin. Implant Dent. Relat. Res.* **2014**, *16*, 202–211. [CrossRef] [PubMed]

22. Araújo, M.; Linder, E.; Wennström, J.; Lindhe, J. The influence of Bio-Oss Collagen on healing of an extraction socket: An experimental study in the dog. *Int. J. Periodontics Restor. Dent.* **2008**, *28*, 123–135.

23. Araújo, M.G.; Linder, E.; Lindhe, J. Bio-Oss collagen in the buccal gap at immediate implants: A 6-month study in the dog. *Clin. Oral Implants Res.* **2011**, *22*, 1–8. [CrossRef] [PubMed]

24. Shim, J.H.; Moon, T.S.; Yun, M.J.; Jeon, Y.C.; Jeong, C.M.; Cho, D.W.; Huh, J.B. Stimulation of healing within a rabbit calvarial defect by a PCL/PLGA scaffold blended with TCP using solid freeform fabrication technology. *J. Mater. Sci. Mater. Med.* **2012**, *23*, 2993–3002. [CrossRef] [PubMed]

25. Shim, J.H.; Won, J.W.; Sung, S.J.; Lim, D.H.; Yun, W.S.; Jeon, Y.C.; Huh, J.B. Comparative efficacies of a 3D-printed PCL/PLGA/β-TCP membrane and a titanium membrane for guided bone regeneration in beagle dogs. *Polymers* **2015**, *7*, 2067–2077. [CrossRef]

26. Won, J.Y.; Park, C.Y.; Bae, J.H.; Ahn, G.; Kim, C.; Lim, D.H.; Cho, D.W.; Yun, W.S.; Shim, J.H.; Huh, J.B. Evaluation of 3D printed PCL/PLGA/β-TCP versus collagen membranes for guided bone regeneration in a beagle implant model. *Biomed. Mater.* **2016**, *7*, 1–15. [CrossRef] [PubMed]

27. Salama, A.; Abou-Zeid, R.E.; El-Sakhawy, M.; El-Gendy, A. Carboxymethyl cellulose/silica hybrids as templates for calcium phosphate biomimetic mineralization. *Int. J. Biol. Macromol.* **2015**, *74*, 155–161. [CrossRef] [PubMed]

28. Collins, M.N.; Birkinshaw, C. Hyaluronic acid based scaffolds for tissue engineering-a review. *Carbohydr. Polym.* **2013**, *92*, 1262–1279. [CrossRef] [PubMed]

29. Lisignoli, G.; Fini, M.; Giavaresi, G.; Nicoli, A.N.; Toneguzzi, S.; Facchini, A. Osteogenesis of large segmental radius defects enhanced by basic fibroblast growth factor activated bone marrow stromal cells grown on non-woven hyaluronic acid-based polymer scaffold. *Biomaterials* **2002**, *23*, 1043–1051. [CrossRef]

30. Schwartz, Z.; Goldstein, M.; Raviv, E.; Hirsch, A.; Ranly, D.M.; Boyan, B.D. Clinical evaluation of demineralized bone allograft in a hyaluronic acid carrier for sinus lift augmentation in humans: A computed tomography and histomorphometric study. *Clin. Oral Implants Res.* **2007**, *18*, 204–211. [CrossRef] [PubMed]

31. Lee, J.H.; Jeong, B.O. The effect of hyaluronate-carboxymethyl cellulose on bone graft substitute healing in a rat spinal fusion model. *J. Korean Neurosurg. Soc.* **2011**, *50*, 409–414. [CrossRef] [PubMed]

32. Hesaraki, S.; Nezafati, N. In vitro biocompatibility of chitosan/hyaluronic acid-containing calcium phosphate bone cements. *Bioprocess Biosyst. Eng.* **2014**, *37*, 1507–1516. [CrossRef] [PubMed]

33. El Kady, A.M.; Mohamed, K.R.; El-Bassyouni, G.T. Fabrication, characterization and bioactivity evaluation of calcium pyrophosphate/polymeric biocomposites. *Ceram. Int.* **2009**, *35*, 2933–2942. [CrossRef]

34. Paknejad, M.; Rokn, A.R.; Yaghobee, S.; Moradinejad, P.; Heidari, M.; Mehrfard, A. Effects of two types of anorganic bovine bone on bone regeneration: A histological and histomorphometric study of rabbit calvaria. *J. Dent. (Tehran)* **2014**, *11*, 687–695.

35. Fekrazad, R.; Sadeghi Ghuchani, M.; Eslaminejad, M.B.; Taghiyar, L.; Kalhori, K.A.; Pedram, M.S.; Shayan, A.M.; Aghdami, N.; Abrahamse, H. The effects of combined low level laser therapy and mesenchymal stem cells on bone regeneration in rabbit calvarial defects. *J. Photochem. Photobiol. B* **2015**, *151*, 180–185. [CrossRef] [PubMed]

36. Schmidlin, P.R.; Nicholls, F.; Kruse, A.; Zwahlen, R.A.; Weber, F.E. Evaluation of moldable, in situ hardening calcium phosphate bone graft substitutes. *Clin. Oral Implants Res.* **2013**, *24*, 149–157. [CrossRef] [PubMed]

37. Brunner, E.; Langer, F. Nonparametric analysis of ordered categorical data in designs with longitudinal observations and small sample sizes. *Biom. J.* **2000**, *42*, 663–675. [CrossRef]

38. Moses, O.; Shemesh, A.; Aboodi, G.; Tal, H.; Weinreb, M.; Nemcovsky, C.E. Systemic tetracycline delays degradation of three different collagen membranes in rat calvaria. *Clin. Oral Implants Res.* **2009**, *20*, 189–195. [CrossRef] [PubMed]

39. Kozlovsky, A.; Aboodi, G.; Moses, O.; Tal, H.; Artzi, Z.; Weinreb, M.; Nemcovsky, C.E. Bio-degradation of a resorbable collagen membrane (Bio-Gide) applied in a double-layer technique in rats. *Clin. Oral Implants Res.* **2009**, *20*, 1116–1123. [CrossRef] [PubMed]

40. Sela, M.N.; Babitski, E.; Steinberg, D.; Kohavi, D.; Rosen, G. Degradation of collagen-guided tissue regeneration membranes by proteolytic enzymes of Porphyromonas gingivalis and its inhibition by antibacterial agents. *Clin. Oral Implants Res.* **2009**, *20*, 496–502. [CrossRef] [PubMed]

41. Rothamel, D.; Schwarz, F.; Sager, M.; Herten, M.; Sculean, A.; Becker, J. Biodegradation of differently cross-linked collagen membranes: An experimental study in the rat. *Clin. Oral Implants Res.* **2005**, *16*, 369–378. [CrossRef] [PubMed]

42. Cui, J.; Liang, J.; Wen, Y.; Sun, X.; Li, T.; Zhang, G.; Sun, K.; Xu, X. In vitro and in vivo evaluation of chitosan/β-glycerol phosphate composite membrane for guided bone regeneration. *J. Biomed. Mater. Res. A* **2014**, *102*, 2911–2917. [CrossRef] [PubMed]

43. Schanté, C.E.; Zuber, G.; Herlin, C.; Vandamme, T.F. Chemical modifications of hyaluronic acid for the synthesis of derivatives for a broad range of biomedical applications. *Carbohydr. Polym.* **2011**, *85*, 469–489. [CrossRef]

44. Zecha, P.J.; Schortinghuis, J.; van der Wal, J.E.; Nagursky, H.; van den Broek, K.C.; Sauerbier, S.; Vissink, A.; Raghoebar, G.M. Applicability of equine hydroxyapatite collagen (eHAC) bone blocks for lateral augmentation of the alveolar crest. A histological and histomorphometric analysis in rats. *Int. J. Oral Maxillofac. Surg.* **2011**, *40*, 533–542. [CrossRef] [PubMed]

45. Rothamel, D.; Schwarz, F.; Herten, M.; Ferrari, D.; Mischkowski, R.A.; Sager, M.; Becker, J. Vertical ridge augmentation using xenogenous bone blocks: A histomorphometric study in dogs. *Int. J. Oral Maxillofac. Implants* **2009**, *24*, 243–250. [PubMed]

46. Lundgren, D.; Nyman, S.; Mathisen, T.; Isaksson, S.; Klinge, B. Guided bone regeneration of cranial defects, using biodegradable barriers: An experimental pilot study in the rabbit. *J. Craniomaxillofac. Surg.* **1992**, *20*, 257–260. [CrossRef]

47. Zornow, M.H.; Scheller, M.S.; Sheehan, P.B.; Strnat, M.A.; Matsumoto, M. Intracranial pressure effects of dexmedetomidine in rabbits. *Anesth. Analg.* **1992**, *75*, 232–237. [CrossRef] [PubMed]

48. Sculean, A.; Chiantella, G.C.; Windisch, P.; Arweiler, N.B.; Brecx, M.; Gera, I. Healing of intra-bony defects following treatment with a composite bovine-derived xenograft (Bio-Oss Collagen) in combination with a collagen membrane (Bio-Gide PERIO). *J. Clin. Periodontol.* **2005**, *32*, 720–724. [CrossRef] [PubMed]

4

Microstructural Evolution of Dy$_2$O$_3$-TiO$_2$ Powder Mixtures during Ball Milling and Post-Milled Annealing

Jinhua Huang, Guang Ran *, Jianxin Lin, Qiang Shen, Penghui Lei, Xina Wang and Ning Li

College of Energy, Xiamen University, Xiamen 361102, China; jhhuang@stu.xmu.edu.cn (J.H.); jxlin@stu.xmu.edu.cn (J.L.); shenqiang1989@126.com (Q.S.); p.h.lei@foxmail.com (P.L.); xina_wang@126.com (X.W.); Ningli@xmu.edu.cn (N.L.)
* Correspondence: gran@xmu.edu.cn

Academic Editor: Jan Ingo Flege

Abstract: The microstructural evolution of Dy$_2$O$_3$-TiO$_2$ powder mixtures during ball milling and post-milled annealing was investigated using XRD, SEM, TEM, and DSC. At high ball-milling rotation speeds, the mixtures were fined, homogenized, nanocrystallized, and later completely amorphized, and the transformation of Dy$_2$O$_3$ from the cubic to the monoclinic crystal structure was observed. The amorphous transformation resulted from monoclinic Dy$_2$O$_3$, not from cubic Dy$_2$O$_3$. However, at low ball-milling rotation speeds, the mixtures were only fined and homogenized. An intermediate phase with a similar crystal structure to that of cubic Dy$_2$TiO$_5$ was detected in the amorphous mixtures annealed from 800 to 1000 °C, which was a metastable phase that transformed to orthorhombic Dy$_2$TiO$_5$ when the annealing temperature was above 1050 °C. However, at the same annealing temperatures, pyrochlore Dy$_2$Ti$_2$O$_7$ initially formed and subsequently reacted with the remaining Dy$_2$O$_3$ to form orthorhombic Dy$_2$TiO$_5$ in the homogenous mixtures. The evolutionary mechanism of powder mixtures during ball milling and subsequent annealing was analyzed.

Keywords: microstructure; ball milling; dysprosium oxide; neutron absorber; phase evolution

1. Introduction

High-energy ball milling has been widely used to prepare various types of materials, such as supersaturated solid solutions, metastable crystalline materials [1], quasicrystal phases [2], nanostructured materials [3], and amorphous alloys [4]. The technology was initially used in place of blending and sintering at elevated temperatures to prepare ceramic-strengthened alloys [5,6]. A large amount of mechanical energy is transformed into intrinsic energy in the target materials, which induces the formation of numerous defects in the crystal structure, such as vacancies, interstitials, cavities and dislocations, which are always in a non-equilibrium state [7–9]. The defects and structural disorders will increase the mobility of atomic diffusion and induce chemical reactions amongst components that are not present under equilibrium conditions [10]. Therefore, based on its excellent characteristics, ball milling was used to prepare bulk Dy$_2$TiO$_5$, which can be used as a neutron absorber in control rods in nuclear power plants. Control rods are very important in both operating and accident conditions because the nucleon reactivity must be controlled in order to safely operate a nuclear reactor [11]. In fact, bulk Dy$_2$TiO$_5$ prepared by ball milling and sintering has been used in Russian power plant water reactors, such as MIR and VVER-1000 RCCAs [12,13], because of the excellent nucleon characteristics of the element dysprosium, as natural dysprosium consists of five stable isotopes with high thermal neutron absorption cross sections. The decay products are Ho and Er, which are also able to absorb neutrons. All of the radionuclides have low gamma activity and short half-life periods. The absorption cross sections of dysprosium isotopes range from 130 barn to 2600 barn. The region of resonance absorption is 1.6–25 eV, in which the absorption cross-section can reach approximately 1000 barn [14].

According to its equilibrium phase diagram, Dy_2TiO_5 has three crystal structural types depending on the temperature, orthorhombic $\overset{1350\ °C}{\leftrightarrow}$ hexagonal $\overset{1680\ °C}{\leftrightarrow}$ cubic [15], which have different physical properties and radiation resistance abilities. In fact, bulk Dy_2TiO_5 in the cubic crystal structure has the lowest neutron irradiation swelling and highest irradiation resistance. Therefore, it is necessary to synthesize bulk Dy_2TiO_5 in the cubic crystal structure. Jung [16] synthesized bulk Dy_2TiO_5 with high purity and density using a polymer carrier chemical synthesis process, in which ethylene glycol was used as an organic carrier for metal cations. An amorphous phase was detected below 800 °C, and orthorhombic Dy_2TiO_5 was observed after sintering for 1 h at 1300 °C, while little else was observed while sintering in the range of 800 to 1300 °C. Panneerselvam [17] used both solid-state synthesis and wet chemical synthesis to prepare Dy_2TiO_5. However, the effect of sintering temperature on the phase evolution needs further investigation. The sinterability of Dy_2O_3 and TiO_2 with different molar ratios was determined for various ball-milling and sintering conditions [18]. Amit Sinha [19] reported the synthesis of bulk Dy_2TiO_5 from mixtures of equimolar Dy_2O_3 and TiO_2 powders in a two-step process: (I) pyrochlore $Dy_2Ti_2O_7$ was initially formed and (II) $Dy_2Ti_2O_7$ then reacted with the remaining Dy_2O_3 to form orthorhombic Dy_2TiO_5. The powder mixtures used in sintering were simply mixed during ball milling. Garcia-Martinez [7] observed the experimental phenomena of the transformation of Dy_2O_3 from cubic to monoclinic and the synthesis of a hexagonal high-temperature phase, reported as Dy_2TiO_5, in an equimolar Dy_2O_3-TiO_2 mixture during ball milling. Therefore, further investigation is needed into the evolutionary behavior of the microstructure under different ball-milling conditions and the effect of the state of the ball-milled powder on the sintering behavior. In the present work, the microstructural evolutionary behavior and corresponding reaction mechanism of Dy_2O_3-TiO_2 powder mixtures under two types of ball-milling parameters were investigated. The annealing behavior of the ball-milled mixtures was also examined.

2. Experiments

Powders of Dy_2O_3 (cubic crystal structure) and TiO_2 (rutile crystal structure) with an average particle diameter of 5 μm and 50 nm, respectively, were used as raw materials. The raw powders of Dy_2O_3 and TiO_2 were purchased from Beijing HWRK Chem Co., Ltd. (Beijing, China). The purity of the raw powders of both Dy_2O_3 and TiO_2 was 99.9%. Ball milling of the molar fraction Dy_2O_3-50% TiO_2 (Dy_2O_3:TiO_2 = 1:1) powder mixtures was carried out on an SFM-1 high-energy planetary ball mill at room temperature. Stainless steel balls that were 5 mm in diameter were used as the milling media. The ball-to-powder mass ratio was 10:1, and the rotational speed was 200 rpm and 500 rpm. No more than one weight percent stearic acid was added to the powder mixtures as a process control agent to prevent excessive cold welding and aggregation amongst the powder particles. During ball milling, a 5-min stopping interval was used after milling for 55 min to prevent excess heat generation, which has an obvious effect on the ball-milling procedure. The powder mixtures used for microstructural analysis were extracted from the loose powders in the steel can, not from powders adhered to the surface of the stainless balls or the steel can wall, after ball milling for 4, 12, 24, 48, and 96 h.

After various milling times, a small amount of ball-milled powders taken from the container were characterized and analyzed by X-ray diffraction (XRD) on a Rigaku Ultima IV X-ray diffractometer (Rigaku, Tokyo, Japan) with Cu Kα radiation (λ = 0.1540598 nm) and transmission electron microscopy (TEM) on a JEM-2100 instrument (JEOL, Tokyo, Japan). Analysis was also carried out for ball-milled powder mixtures annealed at different temperatures.

The grain size was calculated using Suryanarayana and Grant Norton's formula [20].

$$B_r \cos\theta = \frac{K\lambda}{L} + \eta \sin\theta \tag{1}$$

where, K is a constant (with a value of 0.9); λ is the wavelength of the X-ray radiation; L and η are the grain size and internal strain, respectively; and θ is the Bragg angle. B_r is the full width at half-maximum (FWHM) of the diffraction peak after instrumental correction and can be calculated from the following equation:

$$B = B_r + B_s \tag{2}$$

where, B and B_S are the FWHM of the broadened Bragg peaks and the standard sample's Bragg peaks, respectively.

The ball-milled mixtures and annealed mixtures were first put in ethyl alcohol, and then adequately dispersed by ultrasonic vibration. A carbon-coated copper grid was used to collect the dispersed powders in the ethyl alcohol and then dried by ultraviolet lamp. After that, the prepared samples were observed by TEM. Differential scanning calorimetry (DSC) was used to analyze the thermal behavior of ball-milled powders at a 5 °C/min heating rate in argon atmosphere using a SAT 449C instrument (NETZSCH, Bavarian State, Germany). The powder mixtures milled for 96 h were annealed at temperatures ranging from 700 to 1150 °C in a tube furnace under atmospheric conditions. The heating and cooling rates were both 5 °C/min.

3. Results and Discussion

The XRD patterns of Dy_2O_3-TiO_2 powder mixtures milled at 500 rpm and 200 rpm for different times are shown in Figure 1. The XRD results show that the crystal structure of the original Dy_2O_3 phase and TiO_2 phase are cubic and rutile, respectively. At the condition of 500 rpm, the diffraction peaks of cubic Dy_2O_3 and TiO_2 broadened significantly and reduced in intensity with increased milling time. The broadening of the X-ray diffraction peaks is associated with the refinement in grain size and lattice distortions. Meanwhile, the diffraction peaks of monoclinic Dy_2O_3 can be observed in the X-ray patterns as indicated by the black inverted triangles in Figure 1a. Ball milling induces a Dy_2O_3 phase transformation from the cubic to the monoclinic crystal structure. A broad, singular diffraction peak is also present that indicates the formation of the amorphous phase during ball milling. Interestingly, the amorphous peak is present at the location of the diffraction peak for the monoclinic Dy_2O_3 phase, not at the location of the diffraction peak for the cubic Dy_2O_3 phase, which indicates that the formed amorphous phase is derived from the monoclinic Dy_2O_3 phase, not from the cubic Dy_2O_3 phase. The transformation from cubic to monoclinic increases with increased milling time. After milling for 96 h, only the amorphous phase can be observed, which indicates that the monoclinic Dy_2O_3 phase was fully converted to the amorphous phase. Additionally, this behavior indicates that no new compounds are synthesized during ball milling. Even if new compounds were formed in the milled powders, the amount is very low and does not reach the sensitivity range of the X-ray measurement. Therefore, in the present work, the evolution of Dy_2O_3-TiO_2 powder mixtures is as follows: ball milling first induces the transformation of Dy_2O_3 from the cubic to the monoclinic crystal phase, then monoclinic Dy_2O_3 undergoes amorphization, and finally the powder mixtures completely transform to the amorphous phase.

However, the ball-milling behavior of powder mixtures at 200 rpm is distinctly different from that at 500 rpm. The change of the diffraction peaks with increased milling time at 200 rpm is shown in Figure 1b. Although the diffraction peaks of cubic Dy_2O_3 and TiO_2 are also broadened and reduced in intensity with increased milling time, the diffraction peaks of TiO_2 can be observed in the XRD spectrums and are not disappeared. After ball milling for 96 h, the intensity of diffraction peaks of Dy_2O_3 and TiO_2 are also high. The powder mixtures are not changed completely to amorphization. According to the shape of XRD diffraction spectrums, the effect of ball milling on powder mixtures after milling for 96 h at 200 rpm is only similar to that after milling for 4 h at 500 rpm. Therefore, at low ball-milling rotation speeds, the powder mixtures are only fined and homogenized.

Our experimental results are different from the results of G. Garcia-Martinez [7]. In their research, ball milling induced a phase transformation in Dy_2O_3 from cubic to monoclinic. However, a Dy_2TiO_5 compound with a hexagonal crystal structure was formed simultaneously. The ball-milled powders consisted of mixed phases of hexagonal Dy_2TiO_5 and monoclinic Dy_2O_3. The Dy_2O_3-TiO_2 powder mixtures did not completely transform to the amorphous phase, but instead produced the hexagonal Dy_2TiO_5 phase. This difference can be attributed to the different ball-milling conditions used in this research, with special attention to the different ball-milling facilities. During ball milling, experimental parameters such as rotation speed, ball-milling time, ball-milling media, and the ball-to-powder mass ratio have an important influence on the ball-milled products even when using the same proportion and type of oxides. For example, Gajović reported that nanosized $ZrTiO_4$ formed in ZrO_2-TiO_2 powder mixtures, whereas only amorphous mixtures were obtained during ball milling in Stubičar's work [21,22]. In addition, the polymorphic transformation of Ln_2O_3 was also observed in Gd_2O_3-TiO_2 and Y_2O_3-$2TiO_2$ powder systems [23].

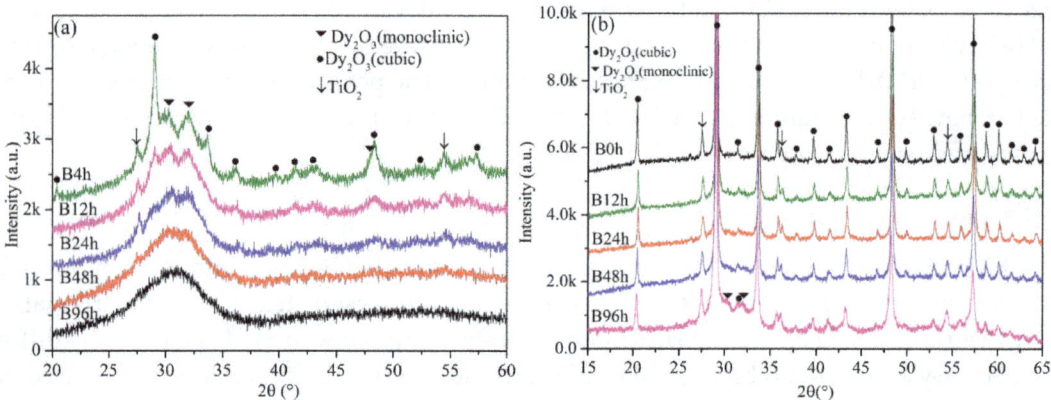

Figure 1. X-ray diffraction patterns of the powder mixtures milled at (**a**) 500 rpm and (**b**) 200 rpm for various times, respectively.

The variation of Dy_2O_3 grain size with ball-milling time at the rotational speeds of 500 rpm and 200 rpm is shown in Figure 2. Actually, the size of Dy_2O_3 grain was calculated for Dy_2O_3 with cubic structure, not for Dy_2O_3 with monoclinic structure, because ball milling induced a phase transformation in Dy_2O_3 from cubic to monoclinic and simultaneously from monoclinic to amorphous. It is difficult to calculate the grain size of Dy_2O_3 with monoclinic structure. It can be seen that ball milling results in a fast decrease of Dy_2O_3 grain size in the initial stage at both 500 rpm and 200 rpm. The refinement rate of crystallite size is roughly logarithmic with ball-milling time at 200 rpm. After 96 h of ball milling, the size of Dy_2O_3 grain is up to approximately 60 nm. However, at 500 rpm, the diffraction peaks of Dy_2O_3 with cubic structure are hardly observed in the XRD spectrum after milling for 24 h as shown in Figure 1a, especially, when the ball-milling time is over 48 h. Therefore, the size of Dy_2O_3 with cubic structure is calculated only before 12 h of ball-milling time. It can be seen that the grain size is quickly decreased. In addition, after same ball-milling time, the grain size of Dy_2O_3 phase at the 500 rpm is obviously smaller than that at the 200 rpm. The effect of ball milling on the grain refinement of powder mixtures at 500 rpm is significantly more intense than that at 200 rpm. The size of Dy_2O_3 grain in the powder mixtures after milling for 4 h at 500 rpm is about 52 nm, which is smaller than that after milling for 96 h at 200 rpm (approximately 60 nm).

The morphology evolution of Dy_2O_3-TiO_2 powder mixtures with increasing ball-milling time at 500 rpm is shown in Figure 3. Both TiO_2 and Dy_2O_3 are brittle components, which are fragmented during ball milling and particle size reduces continuously as a consequence of the energy provided during ball milling. The morphology of large particles is changed significantly due to fracture, agglomeration, and deagglomeration processes. The morphology of the original powder mixtures consists of large-sized Dy_2O_3 particles in micrometer size and small-sized TiO_2 particles in nanometer. The shape of the powder particles is irregular. The line-scanning results of elemental Dy, Ti, and O in the characteristic position in Figure 3a are shown in Figure 3b and also inserted in Figure 3a. It can be seen that the small-sized particles are TiO_2 component and the large-sized particles are Dy_2O_3

components from the variation of the elemental diffraction intensity. The brittle Dy_2O_3 particles are fragmented by ball-powder-ball collisions, leading to a considerable reduction in the powder particle size and subsequent amorphization as milling time increases. After ball milling for 4 h, the size of particles decreases significantly. A large number of small size of Dy_2O_3 particles in nanometer can be observed in the milled powders as shown in Figure 3c. The morphology of the powder mixtures is transformed to uniform, as shown in Figure 3c–g, where the ball-milling time ranges from 4 to 96 h. The morphologies demonstrate that the refining effects of the powder particles are proportional to the ball-milling time for the same rotational speed. After 96 h of ball milling, a large number of nanoparticles agglomerate to form a large-sized particle, as shown in Figure 3g. In addition, TiO_2 particles disappear after ball milling for 96 h, as shown in Figure 1a. The surfaces of the Dy_2O_3 particles in Figure 3g are clean compared with those in Figure 3a. It can be concluded that the particle size in the powder mixtures is refined to the nanoscale after ball milling for 96 h. The line-scanning results of elemental Dy, Ti, and O in the characteristic position in Figure 3g is shown in Figure 3h and also inserted in Figure 3g, which indicates these elements are uniformly distributed in the ball-milled particles according the variation of the elemental diffraction intensity. In addition, the morphology evolution of powder mixtures at an 200 rpm dose are not provided in the present work because the powder mixtures are only fined and homogenized according to the XRD results as shown in Figure 1b. The morphology of mixtures after ball milling for 96 h at 200 rpm is similar with that after ball milling for 4 h at 500 rpm.

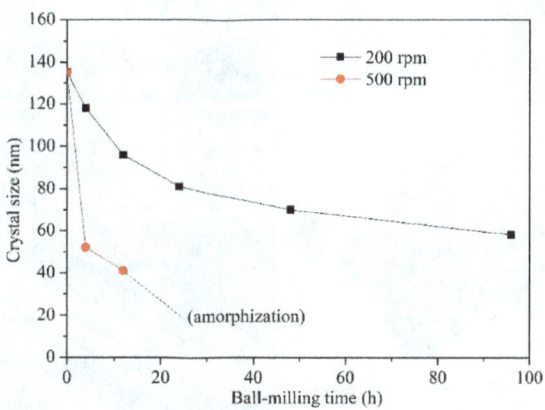

Figure 2. Curves of Dy_2O_3 grain size vs. ball-milling time.

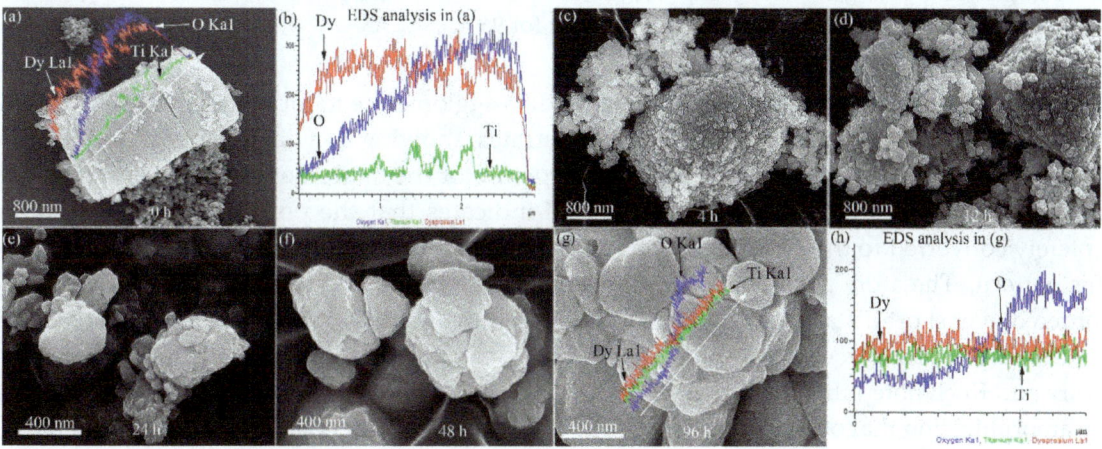

Figure 3. SEM analysis results of the powder mixtures milled at 500 rpm for different ball-milling times. The images showing the morphology of ball-milled mixtures for (**a**) 0 h; (**c**) 4 h; (**d**)12 h; (**e**) 24 h; (**f**) 48 h; and (**g**) 96 h; (**b,h**) are EDS analysis results of ball-milled particles in (**a,g**), respectively.

Figure 4 shows TEM images and corresponding selected area electron diffraction (SAED) patterns of Dy_2O_3-TiO_2 powder mixtures milled for 4 h and 96 h. After milling for 4 h, nano-sized ball-milled powder particles aggregate to form large-sized particles, as shown in Figure 4a, due to the high active surface energy created upon ball milling. The size of the original TiO_2 particles is approximately 50 nm. From the XRD results, it can be observed that ball milling leads to TiO_2 particle refinement, dissolution, and finally disappearance after 96 h. Therefore, the main particles presented in the TEM image are Dy_2O_3. The bright zones near the edge of the powder particles are the thin areas where the electron beam penetrates. The dark areas in the images of the powder particles are the thick areas where the electron beam rarely penetrates. Indexing and analyzing the ring-shaped SAED pattern taken from the area denoted by the letter "A" indicates that the Dy_2O_3 grains are already nanocrystalline. The diffraction spots coming from cubic Dy_2O_3, monoclinic Dy_2O_3, and TiO_2 grains are present in the SAED pattern in Figure 4b. The diffraction halo in the SAED pattern also indicates the formation of an amorphous phase during high-energy ball milling.

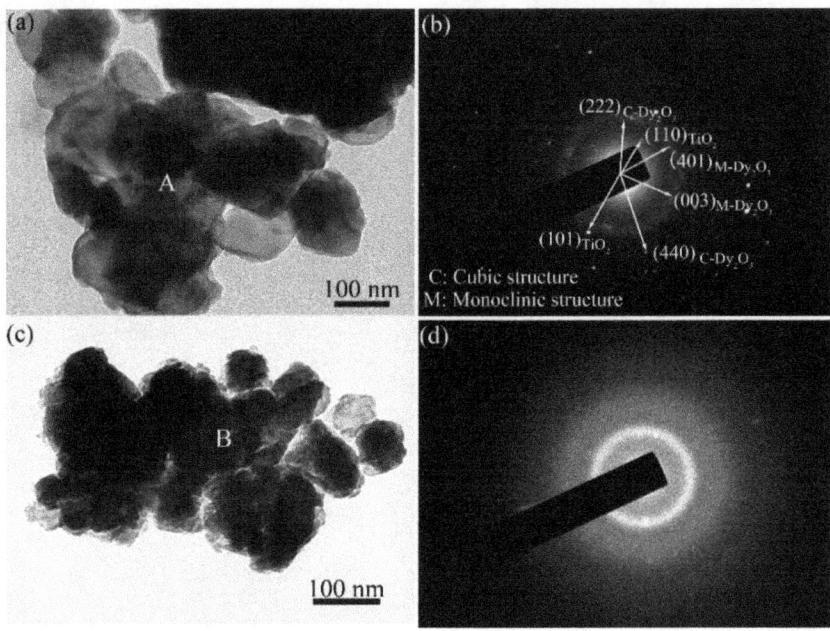

Figure 4. TEM analysis results of the mixtures ball milled at 500 rpm: (**a**) the bright field TEM image and (**b**) corresponding SAED pattern of mixtures milled for 4 h; (**c**) Bright field TEM image and (**d**) corresponding SAED pattern of mixtures milled for 96 h.

After ball milling for 96 h, the ball-milled powders agglomerate to form large particles with large thicknesses such that the electron beam rarely penetrates, showing as a dark color in the TEM image in Figure 4c. It is difficult to observe the microstructure of the agglomerated particles. The SAED pattern from the area marked with the letter "B" indicates that the powder mixtures are almost completely converted to the amorphous phase, although sporadic diffraction spots are also present in this pattern. The atom arrangement in the amorphous phase is disordered over long distances, but ordered over short distances. As grain size decreases, the number of atoms at the grain boundaries increases. The proportion of atoms in the crystal volume relative to the crystal boundary decreases. Schwarz and Koch noted that the formation of an amorphous phase in as-milled powders was similar to the amorphization that occurs during the isothermal annealing of crystalline metallic thin films [24]. In high-energy ball milling, the intense deformation accelerates interdiffusion, and the large defect density increases the free energy of the components in the mixture to form an amorphous product.

Monoclinic Ln_2O_3 is initially formed in the mechanical alloying of lanthanum titanate or dititanate. In Moreno's research, the formation of $Gd_2(Ti_{(1-y)}Zr_y)_2O_7$ pyrochlores occurred in the final step of ball milling starting from an amorphous matrix of Gd_2O_3, TiO_2, and ZrO_2 [23]. However, in the present work, after 96 h of ball milling, the monoclinic Dy_2O_3 phase could not be detected and was completely transformed to the amorphous phase. Moreover, dysprosium titanate also could not be detected. To investigate the sintering behavior of the ball-milled powder mixtures, DSC was carried out on the powder mixtures milled for 96 h at test temperatures ranging from 200 to 1200 °C. The DSC curve in Figure 5 shows one exothermic peak close to 880 °C and one endothermic peak close to 1145 °C. An exothermic peak in a DSC curve can generally be attributed to a transition from disordered to ordered, the recrystallization of original components from the amorphous phase or the formation of a new compound from the ball-milled amorphous powders. Therefore, subsequent X-ray analysis of the ball-milled powders annealed at different temperatures is used to further analyze occurrences in the heating process in more detail. The powder mixtures transform completely to the amorphous state after ball milling for 96 h. Therefore, it seems feasible that the transition from disordered to ordered produced the exothermic peak in the DSC curve. In fact, only the amorphous peak is observed in the XRD pattern of the 96 h ball-milled powder after annealing for 24 h at 700 °C. No diffraction peaks for Dy_2O_3 or TiO_2 are detected. Even with prolonged annealing time, the XRD results are the same. Therefore, the exothermic peak in the DSC curve should be related to the new phase generated from the amorphous mixtures.

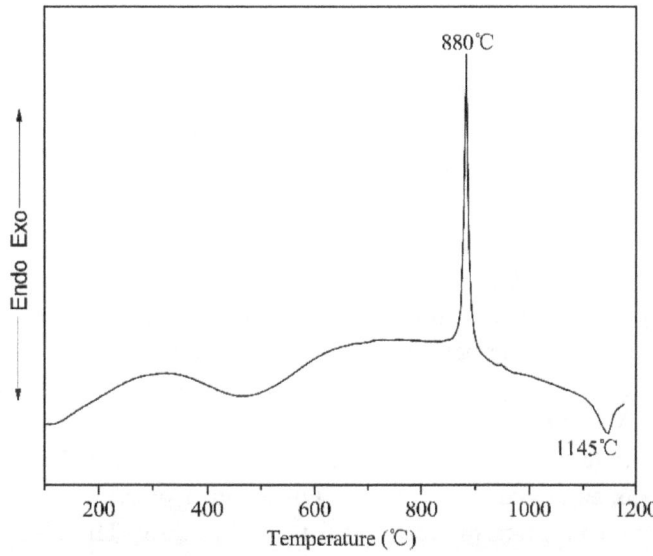

Figure 5. DSC curve of Dy_2O_3-TiO_2 powder mixtures milled for 96 h.

The powder mixtures milled for 96 h were annealed for 3 h at 800, 900, 1000, 1050, 1100, and 1150 °C. The XRD results of the annealed powder mixtures are shown in Figure 6a,b. Several diffraction peaks different from the diffraction peaks of Dy_2O_3 (cubic and monoclinic crystal structure) and TiO_2 (rutile structure) are observed, which indicates new components with crystal structures generated from the amorphous mixtures. This experimental phenomenon of synthesizing new compounds is similar to Stubičar's research in which the orthorhombic $ZrTiO_4$ phase was generated from the high-temperature annealing of amorphous mixtures formed from the ball milling of a ZrO_2-TiO_2 powder system [21] and consistent with Khor's results that zirconia was produced from annealing amorphous mixtures formed from the ball milling of an equimolar $ZrSiO_4$ and Al_2O_3 powder system [25].

According to the XRD standard database, cubic Dy_2TiO_5 {111} presents at $2\theta = 30.043°$, hexagonal Dy_2TiO_5 {102} presents at $2\theta = 32.411°$, orthorhombic Dy_2TiO_5 {201} presents at $2\theta = 29.554°$, and pyrochlore $Dy_2Ti_2O_7$ {111} presents at $2\theta = 30.698°$; the difference in the above diffraction angles is not very large. Three diffraction peaks are observed in the XRD patterns of the powder mixtures annealed for 3 h at 800, 900, and 1000 °C. The main diffraction peak representing the crystalline phase is at $2\theta = 30.0°$, as shown in Figure 6b. Therefore, the newly formed product in the powder system annealed between 800 °C and 1000 °C is not pyrochlore $Dy_2Ti_2O_7$. This result is different than that found in Amit Sinha's research in which pyrochlore $Dy_2Ti_2O_7$ was initially created in the formation of Dy_2TiO_5 [19]. In their research using an equimolar Dy_2O_3-TiO_2 system, the chemical reaction of Dy_2O_3 and TiO_2 initially formed pyrochlore $Dy_2Ti_2O_7$, and then $Dy_2Ti_2O_7$ reacted with the remaining Dy_2O_3 to form orthorhombic Dy_2TiO_5.

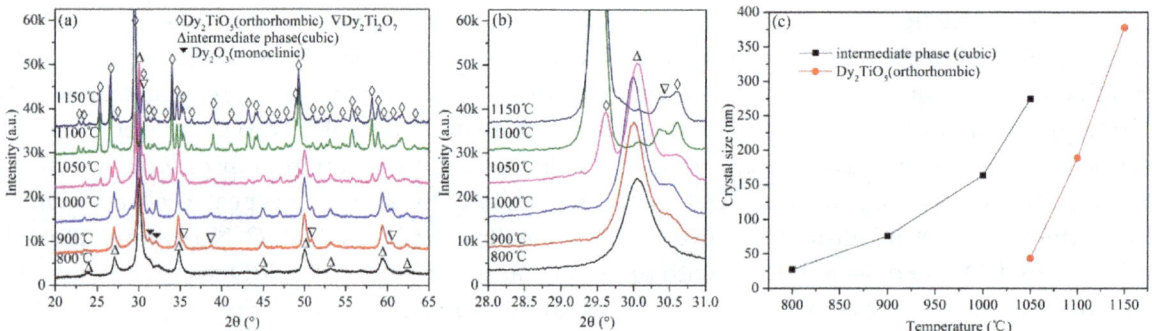

Figure 6. XRD patterns of the ball-milled powder mixtures annealed for 3 h at various temperatures at diffraction angles 2θ ranging from (**a**) $20°$ to $65°$ and (**b**) $28°$ to $31°$; (**c**) Curves of the grain size of main characteristic phase vs. annealing temperature. The powder mixtures were previously milled for 96 h at 500 rpm.

According to the equilibrium phase diagram, the Dy_2TiO_5 phase has three crystal structure types depending on the temperature: orthorhombic $\overset{1350\ °C}{\leftrightarrow}$ hexagonal $\overset{1680\ °C}{\leftrightarrow}$ cubic [15]. Transformation from the high-temperature phase to the low-temperature phase is an exothermic process. For example, the polymorphic transformation of Gd_2TiO_5 from hexagonal to orthorhombic produced an exothermic peak in the DSC curve [7]. However, in the present work, there is only one exothermic peak in the DSC curve. Although the diffraction peaks in the XRD patterns match well with the diffraction peaks in the standard pattern of cubic Dy_2TiO_5, the generated phase should not be cubic Dy_2TiO_5 because low-temperature phases of orthorhombic and hexagonal Dy_2TiO_5 were not detected after annealing over temperatures ranging from 700 to 1000 °C for annealing times ranging from several minutes to 3 h; additionally, as mentioned above, the formation temperature of cubic Dy_2TiO_5 is over 1680 °C. It is not possible to achieve such a high temperature during ball milling. Therefore, the high-temperature phase of cubic Dy_2TiO_5 should not be produced. Instead, the generated phase is an intermediate phase that has a similar crystal structure to cubic Dy_2TiO_5 and is a metastable state that phase transforms to orthorhombic Dy_2TiO_5 when the annealing temperature is above 1050 °C. After annealing for 3 h at 1100 °C, orthorhombic Dy_2TiO_5 and a small amount of pyrochlore $Dy_2Ti_2O_7$ and cubic Dy_2O_3 are detected; notably, cubic Dy_2TiO_5 is not observed in the ball-milled powder mixtures.

The change of the grain size of the main characteristic phase in the annealed powder mixtures with annealing temperature is shown in Figure 6c. According to the above results, the grain size of the intermediate phase and orthorhombic Dy_2TiO_5 are calculated using Suryanarayana and Grant Norton's formula. It can be seen that the grain size increases with increasing annealing temperature. The grain size of the intermediate phase is about 27 nm in the powder mixtures annealed for 3 h at 800 °C and is about 275 nm in the powder mixtures annealed for 3 h at 1050 °C. Because the intermediate phase transforms to orthorhombic Dy_2TiO_5 when the annealing temperature is above 1050 °C, the grain size of orthorhombic Dy_2TiO_5 is calculated at the annealing temperature ranging from 1050 °C to 1150 °C. The grain size of orthorhombic Dy_2TiO_5 is about 43 and 380 nm after annealing for 3 h at 1050 °C and 1150 °C, respectively.

In addition, the pressure created during ball milling is not high enough to transform the crystal structure of Dy_2TiO_5 from orthorhombic to hexagonal or from hexagonal to cubic. The average pressure on the contact surface of two colliding mill balls is approximately 8.5 GPa [26], which is far below the 100 GPa needed to induce a pressure wave to cause the Gd_2TiO_5 phase transformation from the low-temperature orthorhombic phase to the high-temperature hexagonal phase [27]. Therefore, the intermediate phase is not produced by collision pressure. Further investigation is needed into the cause of the formation of the intermediate phase.

To further investigate the annealing behavior of the ball-milled powder mixtures, the powder mixtures milled for 96 h at 200 rpm were sintered for 3 h at 800, 900, 1000, 1050, 1100, and 1150 °C. The phase evolution of the annealed powder mixtures identified by XRD analysis is shown in Figure 7. Under these ball-milling conditions, the Dy_2O_3-TiO_2 powder mixtures are homogenized, and the polymorphic transformation of Dy_2O_3 from cubic to monoclinic is not observed in Figure 7a. In the XRD pattern of the powder mixtures annealed for 3 h at 1000 °C, diffraction peaks for Dy_2O_3 and $Dy_2Ti_2O_7$ phase are observed. However, the main diffraction peak for orthorhombic Dy_2TiO_5 is not detected. Under these conditions, the annealed powder mixtures are composed of cubic Dy_2O_3 and pyrochlore $Dy_2Ti_2O_7$. The powder mixtures generate the orthorhombic Dy_2TiO_5 phase at 1050 °C and are composed of cubic Dy_2O_3, pyrochlore $Dy_2Ti_2O_7$, and orthorhombic Dy_2TiO_5. The powder mixtures are almost completely transformed to orthorhombic Dy_2TiO_5 after annealing at 1150 °C for 3 h. The diffraction peak intensity of orthorhombic Dy_2TiO_5 gradually increases with increasing annealing temperature, and simultaneously, the diffraction peak intensity of Dy_2O_3 and $Dy_2Ti_2O_7$ decreases with increasing annealing temperature. For powder mixtures ball milled for 96 h at 200 rpm, the evolutionary behavior at various annealing temperatures is consistent with the data presented in Ref. [19], in which Amit Sinha reported that the chemical reaction of Dy_2O_3 and TiO_2 initially formed pyrochlore $Dy_2Ti_2O_7$, and then $Dy_2Ti_2O_7$ reacted with the remaining Dy_2O_3 to form orthorhombic Dy_2TiO_5. However, this experimental phenomenon is different from that observed in the annealed powder mixtures milled for 96 h at 500 rpm, as shown in Figure 6, due to the initial conditions of the ball-milling mixtures.

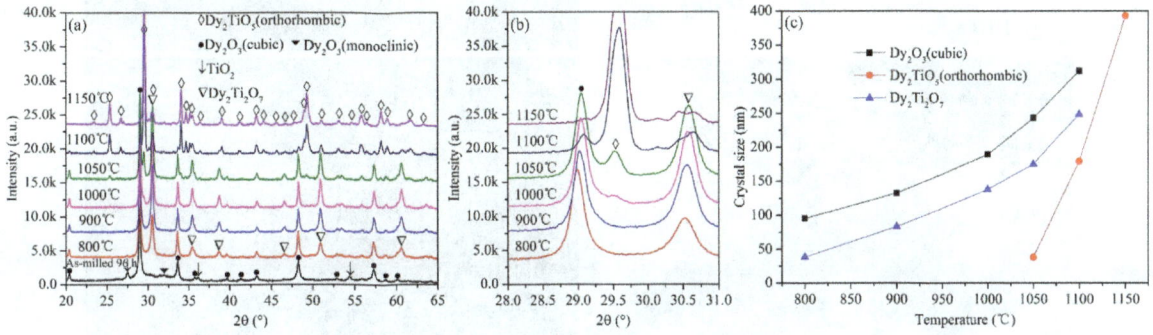

Figure 7. XRD patterns of the ball-milled powder mixtures annealed for 3 h at various temperatures at diffraction angles 2θ ranging from (**a**) 20° to 65° and (**b**) 28° to 31.0°; (**c**) Curves of the grain size of characteristic phase vs. annealing temperature. The powder mixtures were previously milled for 96 h at 200 rpm.

The change of the grain size of main characteristic phase in the annealed powder mixtures with annealing temperature is shown in Figure 7c. The grain size of the cubic Dy_2O_3, pyrochlore $Dy_2Ti_2O_7$ and orthorhombic Dy_2TiO_5 are calculated using Suryanarayana and Grant Norton's formula. The grain size of cubic Dy_2O_3 and pyrochlore $Dy_2Ti_2O_7$ increases with increasing annealing temperature. Because the orthorhombic Dy_2TiO_5 is detected in the powder mixtures annealed at 1050 °C for 3 h, the grain size of orthorhombic Dy_2TiO_5 is calculated when the annealing temperature is over 1050 °C. The grain size of orthorhombic Dy_2TiO_5 is about 38 nm and 395 nm after annealing 3 h at 1050 °C and 1150 °C, respectively.

Figure 8a,b shows the bright field TEM image and corresponding SAED pattern of the ball-milled Dy_2O_3-TiO_2 powder mixtures annealed at 1000 °C for 3 h, respectively. The powder mixtures are previously milled for 96 h at 500 rpm. After annealing for 3 h, the grain size of powder mixtures is kept in nanometer scale, which can be supported by the corresponding SAED pattern taken from the region marked letter "A" in Figure 8a. The diffraction ring is a typical SAED pattern of nanocrystal materials. After analyzing and indexing the ring-shaped SAED pattern, it is indicated that this SAED pattern belongs to the intermediate Dy_2TiO_5 phase that has a similar crystal structure to cubic Dy_2TiO_5, which is in accord with the XRD results as shown in Figure 6. The small-sized powders agglomerate to form large particles with large thicknesses as shown in Figure 8a. The bright zones near the edge of the powder particles is the thin area where the electron beam penetrates. The dark area in the image of the powder particles is the thick area where the electron beam rarely penetrates. After annealing, the amorphous ball-milled powder mixtures are changed to the intermediate Dy_2TiO_5 phase with crystal structure. Figure 8c,d is the bright field TEM image and corresponding SAED pattern of the annealed Dy_2O_3-TiO_2 powder mixtures that were previously milled for 96 h at 200 rpm. Indexing and analyzing the ring-shaped SAED pattern taken from the area denoted by the letter "B" indicates that the particles are composed of cubic Dy_2O_3 and pyrochlore $Dy_2Ti_2O_7$, which is accord with the XRD results of the powder mixtures annealed at 1000 °C for 3 h as shown in Figure 7. This experimental result is different from that observed in the annealed powder mixtures milled for 96 h at 500 rpm.

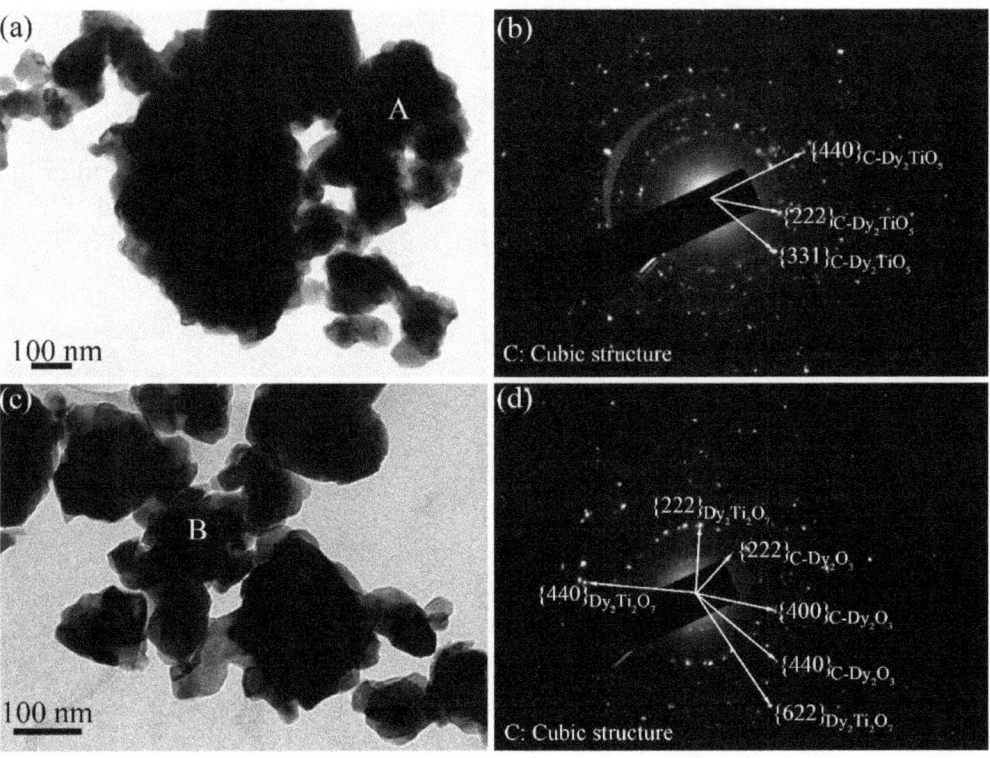

Figure 8. The bright field TEM images and corresponding SAED patterns of the ball-milled Dy_2O_3-TiO_2 powder mixtures annealed at 1000 °C for 3 h, (**a,b**) the powder mixtures are previously milled for 96 h at 500 rpm; (**c,d**) the powder mixtures are previously milled for 96 h at 200 rpm.

4. Conclusions

The microstructural evolution of Dy_2O_3-TiO_2 powder mixtures during ball milling and post-milled annealing was investigated using TEM, SEM, XRD, and DSC. The conclusions can be made as follows:

1. The ball-milling parameters had a great effect on ball milling and the subsequent annealing process.

2. At 500 rpm rotation speeds, the mixtures were fined, homogenized, nanocrystallized, and then completely amorphized, and the crystal structure of Dy_2O_3 was transformed from cubic to monoclinic. The amorphous transformation resulted from monoclinic Dy_2O_3, not from cubic Dy_2O_3. However, at 200 rpm rotation speeds, the Dy_2O_3-TiO_2 powder mixtures were only homogenized, and the polymorphic transformation of Dy_2O_3 from cubic to monoclinic was not observed. Meanwhile, the powder mixtures did not transform to the amorphous phase.

3. The powder mixtures milled for 96 h at 500 rpm were annealed for 3 h at a temperature range of 800 to 1000 °C. An intermediate phase with a crystal structure similar to that of cubic Dy_2TiO_5 was synthesized, which was a metastable phase that transformed to orthorhombic Dy_2TiO_5 when the annealing temperature was above 1050 °C. However, the powder mixtures milled for 96 h at 200 rpm did not transform to the amorphous phase. The annealing behavior showed that the chemical reaction of Dy_2O_3 with TiO_2 initially formed pyrochlore $Dy_2Ti_2O_7$, and then $Dy_2Ti_2O_7$ reacted with the remaining Dy_2O_3 to form orthorhombic Dy_2TiO_5.

Acknowledgments: The work was supported by the National Natural Science Foundation of China through Grant No. 11305136.

Author Contributions: Guang Ran conceived and designed the experiments; Jinhua Huang performed the experiments and analyzed the data; Qiang Shen conducted the TEM experiment; Jinhua Huang, Jianxin Lin, Penghui Lei, and Xina Wang wrote the paper under the supervision of Guang Ran. All authors contributed to the scientific discussion of the results and reviewed the manuscript.

Conflicts of Interest: The authors declare no conflicts of interest.

References

1. Ran, G.; Zhou, J.E.; Xi, S.Q.; Li, P.L. Formation of nanocrystalline and amorphous phase of Al-Pb-Si-Sn-Cu powder during mechanical alloying. *Mater. Sci. Eng. A* **2006**, *416*, 45–50.

2. Eckert, J.; Schultz, L.; Urban, K. Formation of quasicrystals by mechanical alloying. *Appl. Phys. Lett.* **1989**, *55*, 117–119. [CrossRef]

3. Luo, Y.; Ran, G.; Chen, N.J.; Shen, Q.; Zhang, Y.L. Microstructural evolution, thermodynamics and kinetics of Mo-Tm2O3 powder mixtures during ball milling. *Materials* **2016**, *9*, 834. [CrossRef]

4. Kimura, H.; Kimura, M.; Takada, F. Development of an extremely high energy ball mill for solid state amorphizing transformations. *J. Less Common Met.* **1988**, *40*, 113–118. [CrossRef]

5. Gilman, P.S.; Benjamin, J.S. Mechanical alloying. *Annu. Rev. Mater. Sci.* **1983**, *13*, 279–300. [CrossRef]

6. Kong, L.B.; Ma, J.; Zhu, W.; Tan, O.K. Preparation of PMN-PT ceramics via a high-energy ball milling process. *J. Alloys Compd.* **2002**, *335*, 290–296. [CrossRef]

7. G-Martinez, G.; MGonzalez, L.G.; Escalante-García, J.I.; Fuentes, A.F. Phase evolution induced by mechanical milling in Ln2O3:TiO2 mixtures (Ln = Gd and Dy). *Powder Technol.* **2005**, *152*, 72–78. [CrossRef]

8. Fuentes, A.F.; Takacs, L. Preparation of multicomponent oxides by mechanochemical methods. *J. Mater. Sci.* **2013**, *48*, 598–611. [CrossRef]

9. Suryanarayana, C. Mechanical alloying and milling. *Prog. Mater. Sci.* **2001**, *46*, 1–184. [CrossRef]

10. Khakpour, Z.; Youzbashi, A.A.; Maghsoudipour, A.; Ahmadi, K. Synthesis of nanosized gadolinium doped ceria solid solution by high energy ball milling. *Powder Technol.* **2011**, *214*, 117–121. [CrossRef]

11. Huang, J.H.; Ran, G.; Liu, T.J.; Shen, Q.; Li, N. Microstructure and Physical Properties of Tb2TiO5 Neutron Absorber Synthesized by Ball Milling and Sintering. *J. Mater. Eng. Perform.* **2016**, *25*, 4266–4273. [CrossRef]

12. Risovany, V.D.; Varlashova, E.E.; Suslov, D.N. Dysprosium titanate as an absorber material for control rods. *J. Nucl. Mat.* **2000**, *281*, 84–89. [CrossRef]

13. Risovany, V.D.; Klochkov, E.P.; Varlashova, E.E. Hafnium and dysprosium titanate based control rods for thermal water-cooled reactors. *At. Energy* **1996**, *81*, 764–769. [CrossRef]

14. Kermit, W.; Theilacker, J.S. *Neutron Absorber Materials for Reactor Control*; Naval Reactors Division of Reactor Development United States Atomic Energy Commission: Washington, DC, USA, 1962.

15. Petrova, M.A.; Novikova, A.S.; Grebenshchikov, R.G. Polymorphism of rare earth titanates of Ln2TiO5 composition. *Izv. Akad. Nauk SSSR. Neorg. Mater.* **1982**, *18*, 287–291.

16. Jung, C.H.; Kim, C.J.; Lee, S.J. Synthesis and sintering studies on Dy_2TiO_5 prepared by polymercarrier chemical process. *J. Nucl. Mater.* **2006**, *354*, 137–142. [CrossRef]

17. Panneerselvam, G.; Venkata Krishnan, R.; Antony, M.P.; Nagarajam, K.; Vasudevan, T.; Vasuadeva Rao, P.R. Thermophysical measurements on dysprosium and gadolinium titanates. *J. Nucl. Mater.* **2004**, *327*, 220–225. [CrossRef]

18. Kim, J.S.; Kim, H.S.; Jeong, C.Y. Phase study on the $Dy_xTi_yO_z$ pellets. In Proceedings of the Transactions of the Korean Nuclear Society Autumn Meeting, Busan, Korea, 27–28 October 2005.

19. Sinha, A.; Prakash, B. Development of Dysprosium Titanate Based Ceramics. *J. Am. Ceram. Soc.* **2005**, *88*, 1064–1066. [CrossRef]

20. Suryanarayana, C.; Norton, M.G. *X-ray Diffraction: A Practical Approach*; Plenum Press: New York, NY, USA, 1998.

21. Stubičar, M.; Bermanec, V.; Stubičar, N.; Dudrnovski, D.; Drumes, D. Microstructure evolution of an equimolar powder mixture of ZrO_2-TiO_2 during high-energy ball-milling and post-annealing. *J. Alloys Compd.* **2001**, *316*, 316–320. [CrossRef]

22. Gajović, A.; Djerdj, I.; Furić, K.; Schlögl, R.; Su, D.S. Preparation of nanostructured $ZrTiO_4$ by solid state reaction in equimolar mixture of TiO_2 and ZrO_2. *J. Am. Ceram. Soc.* **2006**, *89*, 2196–2205. [CrossRef]

23. Moreno, K.J.; Rodrigo, R.S.; Fuentes, A.F. Direct synthesis of $A_2(Ti_{(1-y)}Zr_y)_2O_7$ ($A = Gd^{3+}$, Y^{3+}) solid solutions by ball milling constituent oxides. *J. Alloys Compd.* **2005**, *390*, 230–235. [CrossRef]

24. Koch, C.C.; Cavin, O.B. Preparation of "amorphous"Ni60Nb40 by mechanical alloying. *Appl. Phys. Lett.* **1986**, *49*, 146–148. [CrossRef]

25. Khor, K.A.; Li, Y. Effect of mechanical alloying on the reaction sintering of $ZrSiO_4$ and Al_2O_3. *Mater. Sci. Eng. A* **1998**, *256*, 271–279. [CrossRef]

26. Huang, J.H.; Lu, J.Q.; Ran, G.; Chen, N.J.; Qu, P.D. Formation of nanocrystalline and amorphization phase of Fe-Dy_2O_3 powder mixtures induced by ball milling. *J. Mater. Res.* **2016**. [CrossRef]

27. Shcherbakova, L.G.; Kolesnikov, A.V.; Breusov, O.N. Investigation of TiO_2-Ln_2O_3 systems under shock wave action. *Inorg. Mater.* **1979**, *15*, 1724–1729.

5

Thermal Properties of the Mixed *n*-Octadecane/Cu Nanoparticle Nanofluids during Phase Transition: A Molecular Dynamics Study

Qibin Li [1,2,*], Yinsheng Yu [2], Yilun Liu [3,*], Chao Liu [2] and Liyang Lin [1]

[1]   Chongqing Key Laboratory of Heterogeneous Material Mechanics, College of Aerospace Engineering, Chongqing University, Chongqing 400044, China; jack_linliyang@cqu.edu.cn

[2]   Key Laboratory of Low-grade Energy Utilization Technology & System, Ministry of Education, College of Power Engineering, Chongqing University, Chongqing 400044, China; 20123858@cqu.edu.cn (Y.Y.); liuchao@cqu.edu.cn (C.L.)

[3]   State Key Laboratory for Strength and Vibration of Mechanical Structures, School of Aerospace Engineering, Xi'an Jiaotong University, Xi'an 710049, China

*   Correspondence: qibinli@cqu.edu.cn (Q.L.); yilunliu@mail.xjtu.edu.cn (Y.L.)

Academic Editor: Ming Hu

Abstract: Paraffin based nanofluids are widely used as thermal energy storage materials and hold many applications in the energy industry. In this work, equilibrium and nonequilibrium molecular dynamics simulations are employed to study the thermal properties of the mixed nanofluids of *n*-octadecane and Cu nanoparticles during phase transition. Four different nanofluids systems with different mass ratios between the *n*-octadecane and Cu nanoparticles have been studied and the results show that Cu nanoparticles can improve the thermal properties of *n*-octadecane. The melting point, heat capacity and thermal conductivity of the mixed systems are decreased with the increasing of the mass ratio of *n*-octadecane.

Keywords: *n*-octadecane; Cu nanoparticle; phase transition; thermal conductivity; molecular dynamics simulation

1. Introduction

Today, severe challenges are encountered during the development of modern society in the whole world, such as energy crisis, environmental pollution, global climate warming, etc. Use of renewable and clean energy and improving the energy efficiency are the main solutions for the above challenges. One of the most attractive approaches is the thermal energy storage in phase change materials (PCMs) [1], which can not only provide sufficient renewable energy, but also can optimize the energy utilization. Actually, the PCMs have many potential applications in a lot of fields, such as building energy saving, thermal insulation of woven textile fabrics, electronic component cooling, solar thermal power generation, waste heat recovery, and so on [2].

Paraffin is one of the typical PCMs for its large latent heat, low cost, chemical stability and non-toxicity. However, the thermal conductivity of paraffin is low, which limits its application in thermal energy storage. However, the thermophysical properties of working fluid can be significantly improved by adding nanoparticles into it, which is the so-called nanofluid/nanocomposite [3]. Therefore, paraffin based nanofluid/nanocomposite have been extensively studied since the last decade. Zhu et al. [4,5] experimentally studied the thermal properties of paraffin based PCMs by adding Cu, Al, carbon nanotube and graphite nanoparticles. The thermal conductivities of PCMs with Cu and graphite were improved significantly. Besides, the thermal properties of PCMs with nanoparticles are stable after a long-time service of thermal cycling. Ho and Gao [6] prepared Al_2O_3/paraffin emulsion and found that both the thermal conductivity and dynamic viscosity of the emulsion nonlinearly increase with the mass fraction of the nanoparticles. However, up to now, a fundamental understanding of the enhancing mechanism of the thermal properties for the paraffin based nanofluids/nanocomposite is still lacking. The systematical studies of the interactions between nanoparticles and paraffin are helpful to the understanding of the enhancing mechanism and the design of high performance PCMs.

Due to the nanoscale dimension of nanoparticles, it is difficult to investigate the interactions between nanoparticles and paraffin by conventional experimental methods. Therefore, several alternative approaches are proposed to investigate the thermal properties of paraffin based nanofluid/nanocomposite, among which molecular dynamics (MD) simulation [7] has been proved to be a powerful tool in studying the nanoscale thermal and mechanical behaviors [8,9]. Rao et al. [10,11] used the MD simulation to study the thermal properties of a nano-capsule, that is the n-octadecane, n-nonadecane, n-eicosane, n-heneicosane or n-docosane as a core and SiO_2 as a shell. Their results agreed well with experiments and verified the validity of studying the thermal properties of the nanofliud/nanocomposite PCMs via MD simulations from the nanoscale. Wang et al. [12] also studied the octadecane–water PCM via MD simulations. They found that the heat capacity of octadecane slurry decreased with the mass ratio of the octadecane. Since the nanofluid/nanocomposite PCMs usually consist of complicated and various components, MD simulation is an ideal method to investigate the thermal properties of the nanofluid/nanocomposite PCMs. As a common additive of nanofluids, Cu nanoparticles can improve the thermal properties of working fluid [4,13]. Therefore, in this work, MD simulations are employed to investigate the thermal properties of n-octadecane/Cu nanoparticle nanofluids during phase transition.

2. Model and Computational Method

In MD simulation, the motions of the atoms are governed by Newton's second law, while the force applied on every atom is determined by positions of the atoms and the corresponding molecular force fields.

2.1. Simulation Model and Molecular Force Fields

In general, n-octadecane is the common ingredient in paraffin. The molecular structure of n-octadecane is a straight chain consisting of 18-alkanes (CH_3–$(CH_2)_{16}$–CH_3), as shown in Figure 1. The Cu nanoparticle studied in this work consists of $2 \times 2 \times 2$ unit cell (64 Cu atoms) as shown in Figure 2. By considering the complicated interaction of alkane, the condensed-phase optimized molecular potentials for atomistic simulation studies (COMPASS) force field [14] is used to describe the interactions of the investigated 18-alkanes and Cu atoms.

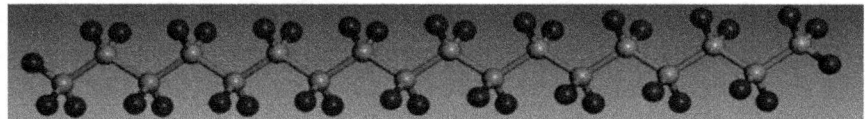

Figure 1. Molecular structure of n-octadecane. (Grey spheres: Carbon atoms, black spheres: Hydrogen atoms.)

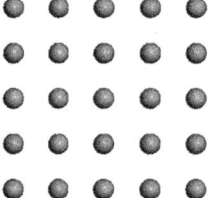

Figure 2. Atoms configuration of Cu nanoparticles.

Four systems are created to study the effect of Cu nanoparticles on the thermal properties of the *n*-octadecane/Cu nanoparticle nanofluids, which are (a) 128 *n*-octadecane molecules (2304 carbon atoms and 4864 hydrogen atoms) in a simulation box of $4.2 \times 4.2 \times 4.2$ nm^3; (b) 1 Cu nanoparticle and 512 *n*-octadecane molecules (9216 carbon atoms, 19,456 hydrogen atoms and 64 Cu atoms) in a simulation box of $6.5 \times 6.5 \times 6.5$ nm^3; (c) 1 Cu nanoparticle and 256 *n*-octadecane molecules (4608 carbon atoms, 9728 hydrogen atoms and 64 Cu atoms) in a simulation box of $5.8 \times 5.8 \times 5.8$ nm^3; and (d) 1 Cu nanoparticle and 128 *n*-octadecane molecules (2304 carbon atoms, 4864 hydrogen atoms and 64 Cu atoms) in a simulation box of $4.7 \times 4.7 \times 4.7$ nm^3. The initial configurations of the studied systems, as shown in Figure 3, are obtained by the AMORPHOUS CELL module of Materials Studio (Accelrys Software Inc., San Diego, CA, USA) [15], in which we set the initial temperature of the systems at 320 K.

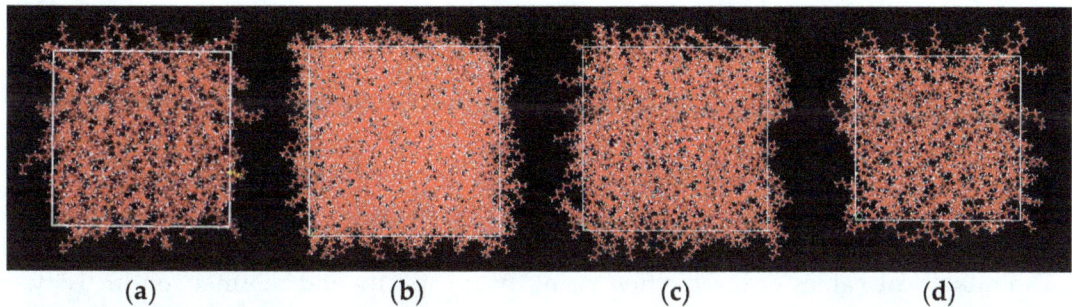

(a)   (b)   (c)   (d)

Figure 3. Initial configuration of systems. (**a**) 128 *n*-octadecane molecules; (**b**) 1 Cu nanoparticle and 512 *n*-octadecane molecules; (**c**) 1 Cu nanoparticle and 256 *n*-octadecane molecules; (**d**) 1 Cu nanoparticle and 128 *n*-octadecane molecules.

2.2. Computational Method

The simulations are performed by Materials Studio. Periodic boundary conditions are applied in X, Y, Z directions. The timestep is set as 1 fs in the simulations. Since, in this work, we focus on the thermal properties of the *n*-octadecane/Cu nanoparticle nanofluids during the phase transition, the four systems are equilibrated at 285 K, 295 K, 300 K, 305 K, 310 K, 320 K, 325 K and 330 K in NVT (canonical) ensemble for 50 ps in the FORCITE module of Materials Studio, respectively, to calculate their thermal properties near melting point. Note that the melting point for pure *n*-octadecane is about 301 K [16]. These simulations are the so-called equilibrium MD (EMD) simulations. Some of the thermal properties are analyzed by our homemade PERL script. The Berendsen method [17] is used to control the temperature of the simulation systems and the Velocity–Verlet algorithm is applied to update the atomic motions.

2.3. Thermal Conductivity Calculation

Then, the above equilibrium systems are simulated by nonequilibrium MD (NEMD) based on our homemade PERL script to compute their thermal conductivities. The system is divided into 20 slabs along the X direction, as shown in Figure 4. The Muller–Plathe algorithm [18,19] is used to exchange the kinetic energy of particles in the first and eleventh slabs, every 100 steps. The stable temperature gradient is generated after performing the exchange process for 50,000 timesteps. Then, the thermal conductivity of the *n*-octadecane/Cu nanoparticle nanofluids is calculated in the following 100,000 timesteps. Here, the total exchanged kinetic energy during the MD simulation by using the Muller–Plathe method is recorded and output. Therefore, the average heat flux of the *n*-octadecane/Cu nanoparticle nanofluids in the X direction is calculated through dividing the total exchanged kinetic energy by time and the cross section area, i.e., Y dimension × Z dimension of the simulation box. Thus, according to the Fourier law of thermal conduction, the thermal conductivity of the *n*-octadecane/Cu nanoparticle nanofluids can be calculated through dividing the heat flux by the temperature gradient in the X direction. Note that due to the nanoscale dimension of the *n*-octadecane/Cu nanoparticle nanofluids, we ignore the convection heat transfer in this work. The details of this method are described in References [18,19]. The other computational parameters for the NEMD are the same as that in Section 2.2.

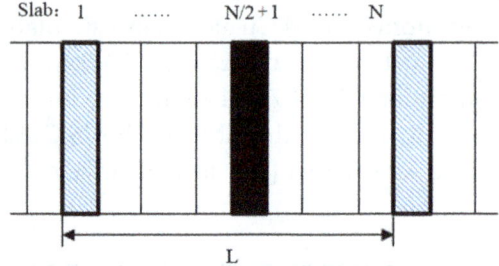

Figure 4. NEMD for calculating thermal conductivity.

3. Results and Discussion

Indeed, the thermal properties obtained from MD are usually dependent on the atomic force field used in the simulations. Therefore, the thermal properties of the *n*-octadecane/Cu nanoparticle nanofluids during phase transition discussed in this work are based on the COMPASS force field. The numerical values of the melting point, heat capacity and thermal conductivity of the *n*-octadecane/Cu nanoparticle nanofluids may be different for different atomic force fields. However, we believe that the trends of these thermal properties of the mass ratio of *n*-octadecane are similar.

3.1. Diffusion Coefficient

Generally, for a given material, the diffusion coefficient of the liquid state is larger than that of its solid state. Therefore, the phase transition of the *n*-octadecane/Cu nanoparticle nanofluids is calibrated by the transition of the relationship between the diffusion coefficient and temperature. The diffusion coefficient of the system is determined by:

$$D = \frac{1}{6} \lim_{t \to \infty} \frac{d}{d_t} \sum_{i=1}^{N} MSD \qquad (1)$$

where the *MSD* is the mean square displacement of atoms in the system [20], *t* is the simulation time. The *MSD* is a typical dynamic parameter defined as:

$$MSD = \left\langle \left| \vec{r}(t) - \vec{r}(0) \right|^2 \right\rangle \qquad (2)$$

$\vec{r}(t)$ is the position of the atom at the time *t*.

The *MSD* of system (c) (1 Cu nanoparticle and 256 *n*-octadecane molecules) for different temperatures, from 295 K to 325 K during the MD simulations, is plotted in Figure 5. The *MSD* increases with the simulation time and the relationships between the *MSD* and simulation time are almost linear after sufficient simulation time, e.g., 30 ps in our simulations. Thus, the diffusion coefficient is defined as the slope of the *MSD*-simulation time curves for the simulation time larger than 30 ps. Besides, the *MSD* increases as the temperature increases, which is because the atoms have a higher degree of mobility for higher temperature.

Next, the diffusion coefficient of the studied systems with different temperatures is calculated and presented in Figure 6. It is shown that the diffusion coefficients of the four systems are almost constant for the system temperature lower than 295 K, which means that the four systems are at the solid state. Then, the transition of the diffusion coefficient–temperature curves for the four systems occurs at about 300 K. The slope of the diffusion coefficient–temperature curves after transition decreases from system

(a) to system (d), as shown in Figure 6. Therefore, it can be concluded that the sequence of the phase transition is from system (a) to system (d) as the temperature increases. Furthermore, the melting point of the *n*-octadecane/Cu nanoparticle nanofluids increases as the mass ratio of the Cu nanoparticle increases. Further study has shown that the increasing of the melting point is proportional to the mass ratio of the Cu nanoparticle in *n*-octadecane.

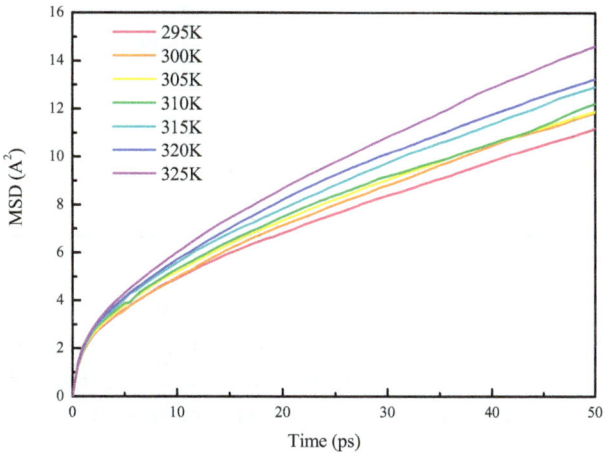

Figure 5. MSD of system (c) (1 Cu nanoparticle and 256 *n*-octadecane molecules) for different system temperatures.

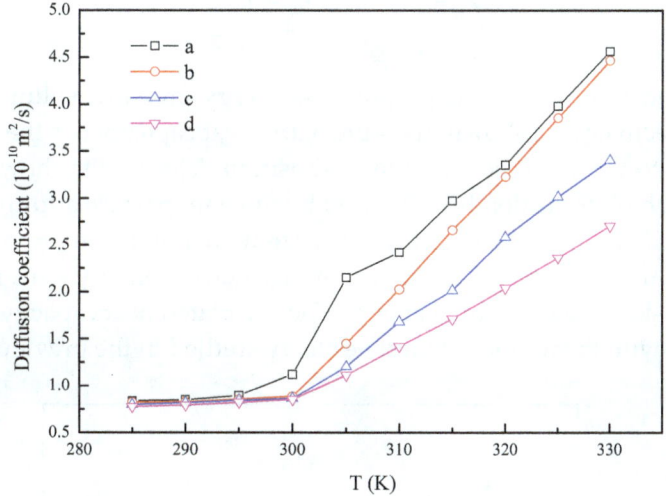

Figure 6. Diffusion coefficient of the different *n*-octadecane/Cu nanoparticle nanofluids with different system temperatures. (**a**) 128 *n*-octadecane molecules; (**b**) 1 Cu nanoparticle and 512 *n*-octadecane molecules; (**c**) 1 Cu nanoparticle and 256 *n*-octadecane molecules; (**d**) 1 Cu nanoparticle and 128 *n*-octadecane molecules.

3.2. Heat Capacity

The potential energy of the systems also increases with the system temperature increasing, as shown in Figure 7. The transitions are also observed in the potential energy–temperature curves near the melting point. Hence, the latent heat of phase transition is partly represented by the variation of the potential energy. Here, the variation of the potential energy near the melting point is 18.9 kJ/mol, 22.3 kJ/mol, 28.2 kJ/mol and 27.4 kJ/mol for system (a), system (b), system (c) and system (d), respectively. So, the *n*-octadecane/Cu nanoparticle nanofluids have larger latent heat of phase transition than that of pure *n*-octadecane, which is beneficial for the thermal energy storage.

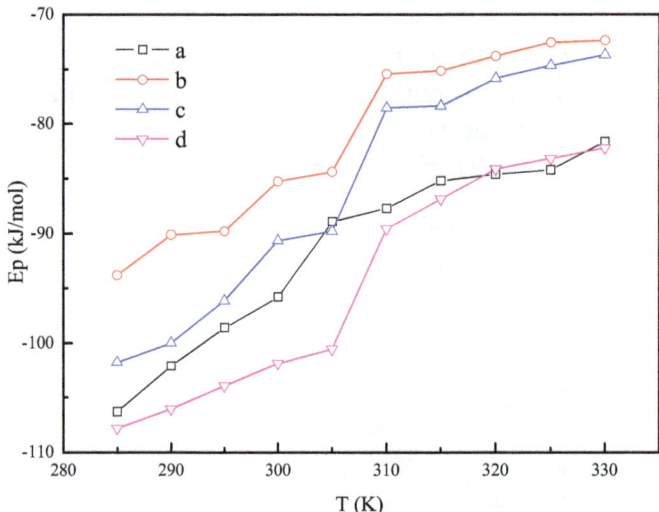

Figure 7. Potential energy of the *n*-octadecane/Cu nanoparticle nanofluids with different temperatures. (**a**) 128 *n*-octadecane molecules; (**b**) 1 Cu nanoparticle and 512 *n*-octadecane molecules; (**c**) 1 Cu nanoparticle and 256 *n*-octadecane molecules; (**d**) 1 Cu nanoparticle and 128 *n*-octadecane molecules.

Heat capacity (C_V) is an important parameter for the thermodynamic system. In MD simulation, C_V is defined as [10]:

$$C_V = \left(\frac{\partial E}{\partial T}\right)_v = \frac{\left\langle \delta (E_k + U + PV)^2 \right\rangle}{k_B T^2} \tag{3}$$

where E_k, U, P, V, k_B and T are kinetic energy, potential energy, pressure, volume, Boltzmann constant and temperature, respectively. As shown in Figure 8, the heat capacity has the maximum value near the melting point due to the latent heat of phase transition. The results show that heat capacity of working fluid during phase transition is influenced by the nanoparticle additives. The heat capacity of the *n*-octadecane/Cu nanoparticle nanofluid is large with a high content of the Cu nanoparticle, which is because the Cu nanoparticles can absorb more thermal energy. This result agrees with the aforementioned discussion of the potential energy. The calculated heat capacity of pure *n*-octadecane, system (a), also agrees with that of the octadecane slurry studied in the previous literature [12].

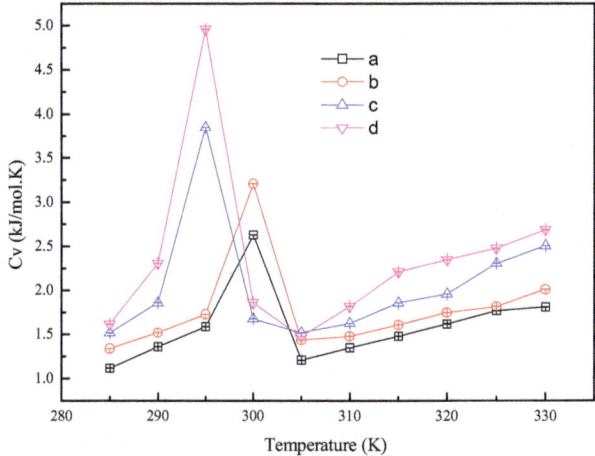

Figure 8. Heat capacity of the *n*-octadecane/Cu nanoparticle nanofluids with different temperatures. (**a**) 128 *n*-octadecane molecules; (**b**) 1 Cu nanoparticle and 512 *n*-octadecane molecules; (**c**) 1 Cu nanoparticle and 256 *n*-octadecane molecules; (**d**) 1 Cu nanoparticle and 128 *n*-octadecane molecules.

3.3. Thermal Conductivity

The thermal conductivity of nanofluid has been extensively studied for many years. First, the thermal conductivity of *n*-octadecane/Cu nanoparticle nanofluids is calculated through a theoretical approach to verify the MD simulation results. In general, the thermal conductivity of nanofluid (k_{eff}) can be calculated by the relations between the thermal conductivity of nanoparticles (k_p) and the thermal conductivity of working fluid (k_f), which has the following general formula [3]:

$$k_{eff} = \frac{k_p \alpha_p (dT/dx)_p + k_f \alpha_f (dT/dx)_f}{\alpha_p (dT/dx)_p + \alpha_f (dT/dx)_f} \tag{4}$$

However, Equation (4), which is limited by many factors, is usually inaccurate to predict the thermal conductivity of nanofluid. Thus, some modified models are proposed, such as the Maxwell model:

$$\frac{k_{eff}}{k_f} = \frac{k_p + 2k_f - 2\phi\left(k_f - k_p\right)}{k_p + 2k_f + \phi\left(k_f - k_p\right)} \tag{5}$$

where ϕ is the volume fraction of nanoparticles. Here, Equation (5) is used to calculate the thermal conductivity of system (d) at 320 K.

The results obtained by NEMD and the Maxwell model are listed in Table 1. Furthermore, the maximum deviation between the present NEMD result and theoretical prediction is about 5.1%, which further verifies the validity of the MD simulation results.

Table 1. Thermal conductivity (W/(m·K)) of system (d) at 320 K.

NEMD	Maxwell
0.2142	0.2038

As mentioned in Section 2.3, the thermal conductivity is calculated based on the Fourier law of thermal conduction, so that the temperature gradient is an important parameter in the calculation of thermal conductivity. The temperature of every slab for system (a) is shown in Figure 9. The system temperature linearly decreases from the heat source (first slab) to the cold source (eleventh slab) which guarantees the validity of the Fourier law of thermal conduction. The temperature of the intermediate slab (sixth or sixteenth slab) is about 305 K and the calculated thermal conductivity represents the thermal conductivity of the system at this temperature. The temperature gradient is the temperature difference between the heat source and cold source divided by the distance of the two slabs.

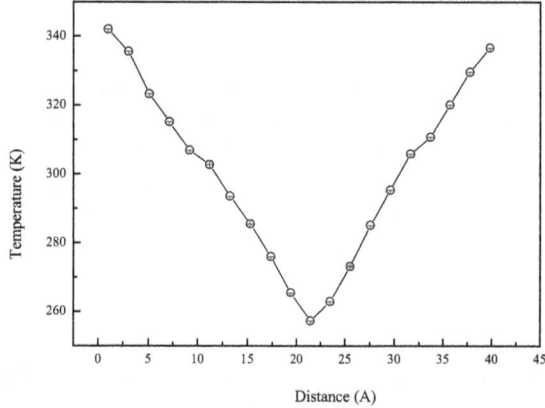

Figure 9. Temperature profile of system (a) (128 *n*-octadecane molecules) in NEMD simulation with an average system temperature of 305 K.

The thermal conductivities of the four systems are presented in Figure 10 which shows that the thermal conductivity increases with the temperature increase when the systems are in the solid or liquid state, while the thermal conductivity decreases sharply near the phase transition point. However, it indicates that the thermal conductivity of *n*-octadecane is enhanced by adding Cu nanoparticle into it and the thermal conductivity is large for high content of Cu nanoparticle. In the present study, the thermal conductivity of system (d) is about 14% larger than that of system (a) which consists of the basic fluid.

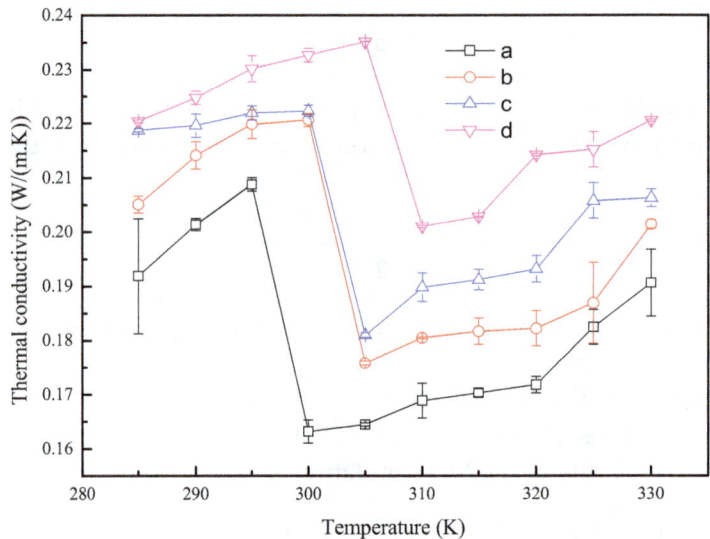

Figure 10. Thermal conductivities of the *n*-octadecane/Cu nanoparticle nanofluids with different temperatures. (**a**) 128 *n*-octadecane molecules; (**b**) 1 Cu nanoparticle and 512 *n*-octadecane molecules; (**c**) 1 Cu nanoparticle and 256 *n*-octadecane molecules; (**d**) 1 Cu nanoparticle and 128 *n*-octadecane molecules.

4. Conclusions

In this work, EMD is used to investigate the thermodynamic properties of *n*-octadecane/Cu nanoparticle nanofluids during phase transition, while NEMD is employed to study their thermal conductivity at the phase transition state. The results could draw the following conclusions.

The Cu nanoparticle could improve the thermal properties of *n*-octadecane. The diffusion coefficient at different temperatures indicates that the melting point of *n*-octadecane/Cu nanoparticle nanofluid increases with the increase of the mass ratio of the Cu nanoparticle. A similar trend for the heat capacity and thermal conductivity is also found. The reason is that the *n*-octadecane with Cu nanoparticles could store more energy than the pure *n*-octadecane system.

Acknowledgments: This work is supported by the National Natural Science Foundation of China (Nos. 51506013, 51576019 and 11332013), Chongqing Research Program of Basic Research and Frontier Technology (No. cstc2015jcyjA50008), Fundamental Research Funds for the Central Universities (No. 104167693).

Author Contributions: Qibin Li and Yilun Liu conceived and designed the simulations; Yinsheng Yu performed the simulations; Qibin Li, Chao Liu and Liyang Lin analyzed the data; Qibin Li and Yilun Liu wrote the paper.

Conflicts of Interest: The authors declare no conflict of interest.

References

1.   Zalba, B.; Marin, J.M.; Cabeza, L.F.; Mehling, H. Review on thermal energy storage with phase change: Materials, heat transfer analysis and applications. *Appl. Therm. Eng.* **2003**, *23*, 251–283. [CrossRef]
2.   Sharma, A.; Tyagi, V.V.; Chen, C.R.; Buddhi, D. Review on thermal energy storage with phase change materials and applications. *Renew. Sustain. Energy Rev.* **2009**, *13*, 318–345. [CrossRef]

3. Xuan, Y.; Li, Q. Heat transfer enhancement of nanofluids. *Int. J. Heat Fluid Flow* **2000**, *21*, 58–64. [CrossRef]

4. Wu, S.; Zhu, D.; Zhang, X.; Huang, J. Preparation and melting/freezing characteristics of Cu/paraffin nanofluid as phase-change material (PCM). *Energy Fuel* **2010**, *24*, 1894–1898. [CrossRef]

5. Wang, N.; Zhang, X.R.; Zhu, D.D.; Gao, J.W. The investigation of thermal conductivity and energy storage properties of graphite/paraffin composites. *J. Therm. Anal. Calorim.* **2012**, *107*, 949–954. [CrossRef]

6. Ho, C.J.; Gao, J.Y. Preparation and thermophysical properties of nanoparticle-in-paraffin emulsion as phase change material. *Int. Commun. Heat Mass* **2009**, *36*, 467–470. [CrossRef]

7. Frenkel, D.; Smit, B. *Understanding Molecular Simulation: From Algorithms to Applications*, 2nd ed.; Academic Press: New York, NY, USA, 2002.

8. Li, Q.; Liu, C. Molecular dynamics simulation of heat transfer with effects of fluid-lattice interactions. *Int. J. Heat Mass Transf.* **2012**, *55*, 8088–8092. [CrossRef]

9. Peng, T.; Firouzi, M.; Li, Q.; Peng, K. Surface force at the nano-scale: Observation of non-monotonic surface tension and disjoining pressure. *Phys. Chem. Chem. Phys.* **2015**, *17*, 20502–20507. [CrossRef] [PubMed]

10. Rao, Z.; Wang, S.; Peng, F. Self diffusion and heat capacity of *n*-alkanes based phase change materials: A molecular dynamics study. *Int. J. Heat Mass Transf.* **2013**, *64*, 581–589. [CrossRef]

11. Rao, Z.; Wang, S.; Peng, F. Molecular dynamics simulations of nano-encapsulated and nanoparticle-enhanced thermal energy storage phase change materials. *Int. J. Heat Mass Transf.* **2013**, *66*, 575–584. [CrossRef]

12. Wang, Y.; Chen, Z.; Ling, X. A molecular dynamics study of nano-encapsulated phase change material slurry. *Appl. Therm. Eng.* **2016**, *98*, 835–840. [CrossRef]

13. Eastman, J.A.; Choi, S.U.S.; Li, S.; Yu, W.; Thompson, L.J. Anomalously increased effective thermal conductivities of ethylene glycol-based nanofluids containing copper nanoparticles. *Appl. Phys. Lett.* **2001**, *78*, 718–720. [CrossRef]

14. Sun, H. COMPASS: An ab initio force-field optimized for condensed-phase applications—Overview with details on alkane and benzene compounds. *J. Phys. Chem. B* **1998**, *102*, 7338–7364. [CrossRef]

15. BIOVIA. *Materials Studio*; Accelrys Software Inc.: San Diego, CA, USA, 2010.

16. National Institute of Standards and Technology (NIST). Thermophysical Properties of Fluid Systems. Available online: http://webbook.nist.gov/chemistry/fluid/ (accessed on 18 November 2016).

17. Berendsen, H.J.C.; Postma, J.P.M.; van Gunsteren, W.F.; DiNola, A.; Haak, J.R. Molecular dynamics with coupling to an external bath. *J. Chem. Phys.* **1984**, *81*, 3684–3690. [CrossRef]

18. Muller-Plathe, F. A simple nonequilibrium molecular dynamics method for calculating the thermal conductivity. *J. Chem. Phys.* **1997**, *106*, 6082–6085. [CrossRef]

19. Zhang, M.; Lussetti, E.; de Souza, L.E.S.; Muller-Plathe, F. Thermal conductivities of molecular liquids by reverse nonequilibrium molecular dynamics. *J. Phys. Chem. B* **2005**, *109*, 15060–15067. [CrossRef] [PubMed]

20. Li, Q.; Peng, X.; Peng, T.; Tang, Q.; Zhang, X.; Huang, C. Molecular dynamics simulation of Cu/Au thin films under temperature gradient. *Appl. Surf. Sci.* **2015**, *357*, 1823–1829. [CrossRef]

6

Optically Clear and Resilient Free-Form μ-Optics 3D-Printed via Ultrafast Laser Lithography

Linas Jonušauskas [1,*], **Darius Gailevičius** [1], **Lina Mikoliūnaitė** [2], **Danas Sakalauskas** [2], **Simas Šakirzanovas** [2], **Saulius Juodkazis** [3,4,*] and **Mangirdas Malinauskas** [1,*]

[1] Department of Quantum Electronics, Faculty of Physics, Vilnius University, Saulėtekio Ave. 10, Vilnius LT-10223, Lithuania; darius.gailevicius@ff.vu.lt

[2] Department of Applied Chemistry, Vilnius University, Naugarduko Str. 24, Vilnius LT-03225, Lithuania; lina.mikoliunaite@chf.vu.lt (L.M.); danas.sakalauskas@chf.vu.lt (D.S.); simas.sakirzanovas@chf.vu.lt (S.S.)

[3] Center for Micro-Photonics, Faculty of Engineering and Industrial Sciences, Swinburne University of Technology, Hawthorn 3122, Australia

[4] Melbourne Center for Nanofabrication, Australian National Fabrication Facility, Clayton 3168, Australia

* Correspondence: linas.jon@gmail.com (L.J.); sjuodkazis@swin.edu.au (S.J.); mangirdas.malinauskas@ff.vu.lt (M.M.)

Academic Editor: Martin Byung-Guk Jun

Abstract: We introduce optically clear and resilient free-form micro-optical components of pure (non-photosensitized) organic-inorganic SZ2080 material made by femtosecond 3D laser lithography (3DLL). This is advantageous for rapid printing of 3D micro-/nano-optics, including their integration directly onto optical fibers. A systematic study of the fabrication peculiarities and quality of resultant structures is performed. Comparison of microlens resiliency to continuous wave (CW) and femtosecond pulsed exposure is determined. Experimental results prove that pure SZ2080 is ~20 fold more resistant to high irradiance as compared with standard lithographic material (SU8) and can sustain up to 1.91 GW/cm^2 intensity. 3DLL is a promising manufacturing approach for high-intensity micro-optics for emerging fields in astro-photonics and atto-second pulse generation. Additionally, pyrolysis is employed to homogeneously shrink structures up to 40% by removing organic SZ2080 constituents. This opens a promising route towards downscaling photonic lattices and the creation of mechanically robust glass-ceramic microstructures.

Keywords: direct laser writing; ultrafast laser; 3D laser lithography; 3D printing; hybrid polymer; integrated micro-optics; optical damage; photonics; pyrolysis; ceramic 3D structures

1. Introduction

Hybrid organic-inorganic polymers have emerged as great materials for fabricating objects in both 2D and 3D configurations [1–3]. They are the material of choice for lithographic 3D femtosecond laser structuring due to several convenient features, which include optical transparency in the visible part of the spectrum [4] and the use of photoinitiators (PI) absorbing the UV radiation [5–7]. The latter makes them perfectly suitable for multiphoton polymerization [5,8] achieved by an ultrafast laser and employed in true free-form structuring by 3D laser lithography (3DLL) [9]. Additionally, their refractive index and mechanical properties can be tuned by changing the proportion between the organic and inorganic components [4]. This led to extensive research in this area and, to date, new hybrid materials containing Si [1], Zr [4] and Ge [10] were made for 3DLL.

The Zr containing hybrid photopolymer, mostly referred to as SZ2080, is especially interesting. It combines all of the best properties offered by these materials, such as low shrinkage, a hard gel form during fabrication, and transparency for visible light [4], and thus is widely used in creating structures to be employed in various applications in medicine [11], micro-optics [12] and photonics [13]. In the standard 3DLL case, photopolymerization of this material is initiated by nonlinear absorption in the PI molecule [14]. However, recent works showed that photopolymerization in SZ2080 can be achieved without PI using both tight focusing with high numerical aperture (NA) objective ($NA > 1$) and loose ($NA < 1$) focusing [15,16]. It is considered that this reaction is induced when nonlinear absorption takes place, which initiates the breaking of chemical bonds. Generated free electrons are subsequently accelerated by the intense electric field and provide bond cleavage via avalanche ionization [16,17]. This combination of multiphoton and avalanche ionization is responsible for subsequent crosslinking, which allows 3D microstructures to be formed out of pure material. Currently, it is known that SZ2080 in its pure form is biocompatible [11] and has a high optical damage threshold [18].

This paper aims at expanding knowledge of 3DLL with pure SZ2080. Special attention is given to its possible application in the field of micro-optics. Thus, experiments for determining the properties of pure SZ2080 relevant to this field are carried out. Results are compared to those obtained with photosensitized SZ2080. Functional micro-optical elements are manufactured and their resilience to continuous wave (CW) and femtosecond light exposure is tested. Furthermore, by applying pyrolysis, we remove the organic component of the hybrid material, leaving structures composed mainly of the glass-ceramic component. Finally, pyrolysis-induced shrinkage is employed in a controlled manner to create periodic lattices consisting of thin (~170-nm-wide) sintered rods.

2. Results

First, the fidelity of 3D structuring of SZ2080 with and without PI was studied and microlenses were fabricated. Then, a comparative study of microlens performance with high irradiance was carried out down to the structural degradation level. Finally, sintering via pyrolysis aimed at retrieving glass-ceramic 3D structures with significant 40% size reduction was studied.

2.1. Comparison of Structuring Properties

In 3DLL, a well chosen PI can improve fabrication throughput and structure qualities for the material used [6,7,19–21]. The set of parameters needed for structuring the material is generally referred to as the *fabrication window*. In essence, when all other experimental parameters are fixed, it can be considered to be an empirically determined intensity range ΔI between the irreversible polymerization threshold I_t and I_d at which the material is optically damaged: $\Delta I = I_d - I_t$. We chose I (calculated using Equation (1)) as the main parameter to quantify the *fabrication window* instead of translation velocity/writing speed v because changes caused by the former are substantially less noticeable during experimentation than the deviations in structuring properties induced by even modest variations in the I. Conventional thinking would lead us to believe that forgoing photosensitization would lead to the absence of absorption, completely preventing photopolymerization or hindering it to the point of heavily inefficient crosslinking and narrow ΔI. In order to determine if that is the case, we designed an experiment in which an array of identical structures was fabricated by varying the v and P, as these are the two parameters that can be changed most practically during manufacturing. The structure chosen for this experiment was a cube with integrated single suspended lines. With this configuration, it is possible to determine several important factors. First, this array provides information about the size of ΔI by showing a structure survival rate dependent on the set of parameters used. The cube shows if it is possible to produce true 3D structures. Single lines give the possibility to measure fabricated feature sizes in transverse (d) and longitudinal (l) directions. For this reason, it is called a *resolution array*. The result outlined showing ΔI achieved with this experiment is provided in Figure 1.

Data provided in Figure 1 shows several important features of pure SZ2080 in comparison to that containing PI. I_t, required to polymerize pure material, is higher by $\Delta I_t = I_{(t\ pure)} - I_{(t\ IRG)} = 0.34$ TW/cm^2. The width of ΔI_{IRG} is only 15.5% wider than ΔI_{pure}. By counting the sectors in which structures are of the best quality, we conclude that pure material provides only a 12.5% lesser survival rate compared to that of photosensitized polymer. In addition, the PI containing polymer provides structures that maintain their initial structural features even if parts of the object are greatly affected by the defects caused by overexposure, while in the case of objects formed from pure material, they completely collapse if non-optimal parameters are used (Figure 2a). This suggests that, without PI, the crosslinking process is not as efficient and provides a final polymer matrix that is considerably weaker. This result coincides well with other works showing that the degree of crosslinking during 3DLL is essential for the mechanical and optical properties of finished structures [22–24]. However, despite this, if fabrication parameters are within the ΔI, even advanced micro-optical elements, like suspended microlenses on the tip of an optical fiber, can be fabricated out of pure material (Figure 2b).

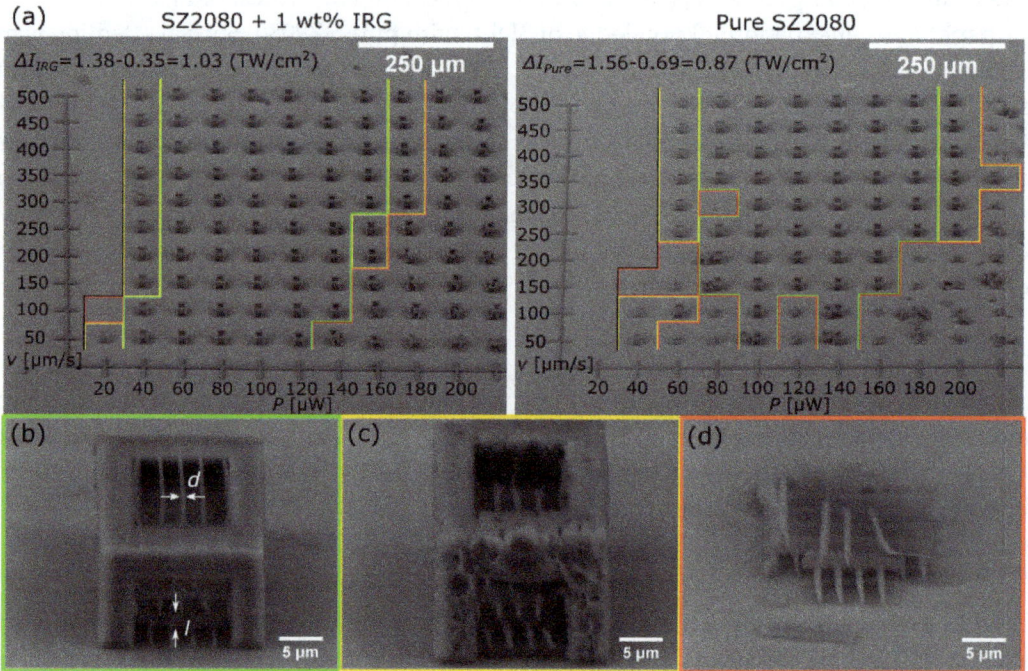

Figure 1. (**a**) SEM images of resolution arrays of photosensitized (**left**) and non-photosensitized (**right**) SZ2080. Structures with severe structural damage (**red**), with poor (**yellow**) and good (**green**) quality are outlined. The structure is considered good if internal single lines are observable and the shape of the cube is as designed. Average laser powers of the bottom and the top of the ΔI are recalculated to the peak intensity I_p (shown at the top); (**b**) one of the good quality structures in the array is shown in a greater detail; l and d marks the longitudinal and transverse sizes of the lines; and (**c**) an example of a poor quality structure and (**d**) the failed one.

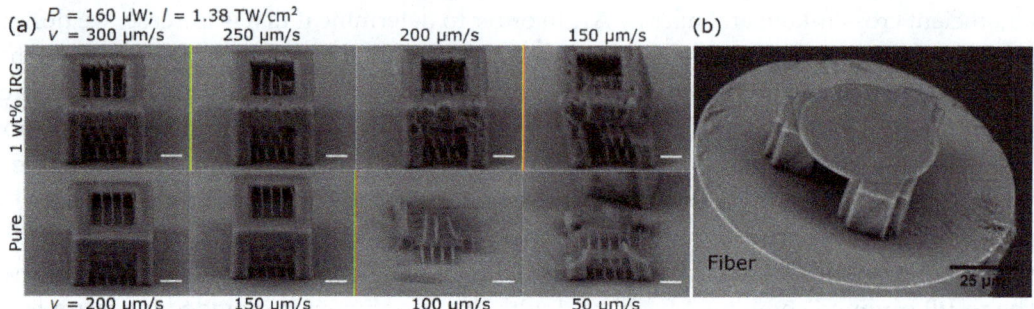

Figure 2. (**a**) reduction of structural quality in the case of photosensitized and pure SZ2080. Photoinitiator Irgacure 369 (IRG) containing cubes degrade slower and in a more progressive fashion. Conversely, the structures out of non-photosensitized material completely break up as soon as the fabrication parameters are not in the ΔI. All scales are 5 μm; and (**b**) monolithic micro-optical element on the tip of an optical fiber fabricated out of non-photosensitized SZ2080.

Dimensions of the lines inside cubes were measured. The case of $v = 250$ µm/s was chosen as it is in the middle of the tested range in both photosensitized and non-photosensitized materials. It revealed that both transverse and longitudinal line dimensions are smaller in pure SZ2080 (Figure 3a). This correlates well with earlier findings [17]. It also reveals that features produced out of photosensitized material easily exceed the calculated spot size. On the other hand, lines produced out of pure SZ2080 are all about the same size as the focus spot. This can be explained by the fact that photochemical chain reactions, in the case of photosensitized material, can expand more easily out of the volume in which nonlinear absorption took place. In the case of pure resist, such a process is less prominent. In addition, the aspect ratios of the formed voxels are very similarly (Figure 3b), which shows that non-photosensitized material does not provide any benefit related to the control of the aspect ratio of a voxel. To better illustrate this, we provide data for the line aspect ratio for 500 µm/s writing speed in Figure 3b as well.

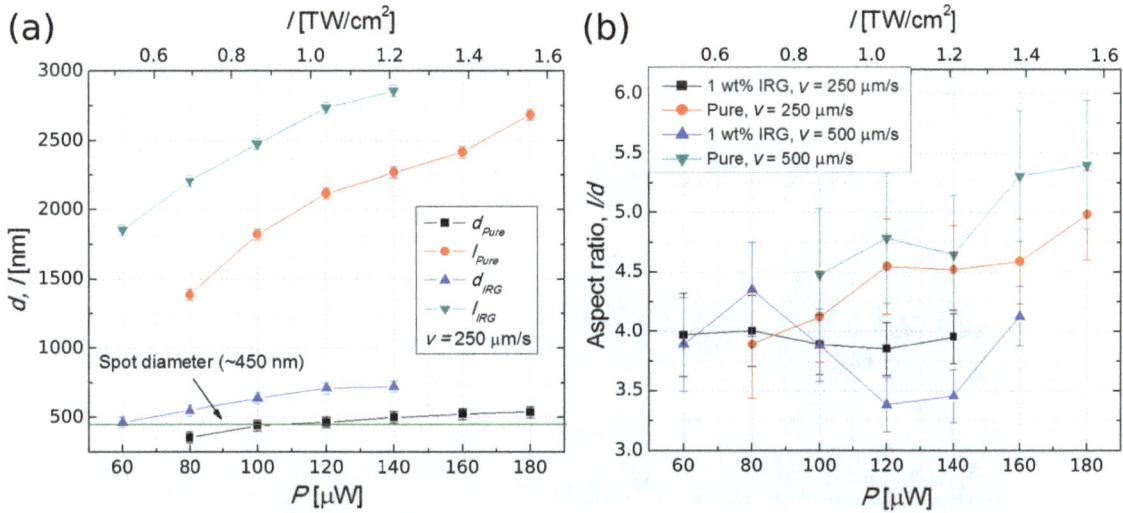

Figure 3. (**a**) feature width d and height l measured in the resolution array at writing speed of 250 µm/s; and (**b**) aspect ratio of lines produced for cases of 250 µm/s and 500 µm/s speeds.

2.2. Surface Roughness

Considering the application of SZ2080 in micro-optics, another important parameter is the surface quality of the final structures. There are several ways to quantify this property using data from precision measurement tools, such as an atomic force microscope (AFM) or very high magnification (>50 k) scanning electron microscope (SEM). The most common way to evaluate measured surface elevations is the standard root mean square (RMS), which was chosen for this study. It is common knowledge that if the surface roughness of a material is higher than $\lambda/8$, it is considered that the surface quality is insufficient for use in optics. On the other hand, if the roughness is smaller than $\lambda/20$, the material is considered suitable for optical applications. We are assuming micro-optical elements to be designed for use in the visible part of the spectrum, thus the lowest operational λ was chosen as 400 nm. AFM was employed to measure the surface profile of both pure and photosensitized SZ2080 samples. The geometry was of a flat square slab with side length of 100 µm (Figure 4a). Several different values of transverse voxel overlap (dx) were used to establish at which condition it was sufficient to achieve optical grade quality of the finished structure. It is important to note that our goal was to determine if optical grade surface quality can be achieved at all with parameters similar to those applied in microlens fabrication and, if so, whether it is easier to obtain it with photosensitized or pure polymers. For a control/comparison, we used a slab produced with one photon polymerization via homogeneous radiation of IV harmonic of an Nd:YAG laser ($\lambda = 266$ nm) similar to the one used in laser-induced damage threshold (LIDT) experiments [18,25]. After UV exposure, the samples were also submerged in the developer following the same protocol as laser-produced samples. This ensured that any difference in surface profile resulted from the polymerization method and not from the sample preparation.

Both polymers had a surface acceptable for optical applications. With the IRG containing material, surface roughness of $RMS < 20$ nm can be achieved with a smaller voxel overlap ($dx = 300$ nm, $RMS = 17.1$ nm), while in the case of pure polymer, the required dx is 100 nm ($RMS = 13.5$ nm) (Figure 4b). The femtosecond laser structured pure SZ2080 is inherently rougher when manufacturing parameters are taken from the middle of the ΔI. Even at $dx = 50$ nm, when the surface details of photosensitized SZ2080 become smooth, a non-photosensitized resist still exhibits clear nanofringes (Figure 4c). This could be explained by the fact that, with the non-photosensitized SZ2080, the polymerization mechanism is more chaotic, due to different and random process initiation pathways, compared to the photosensitized sample. It would also explain why such microstructures lose mechanical integrity much more quickly when the applied parameters are outside ΔI. It should be noted that a difference in roughness is induced by employing 3DLL, as the control samples of both IRG containing and pure SZ20820 polymerized by UV radiation showed the same flatness with RMS being less than 1 nm.

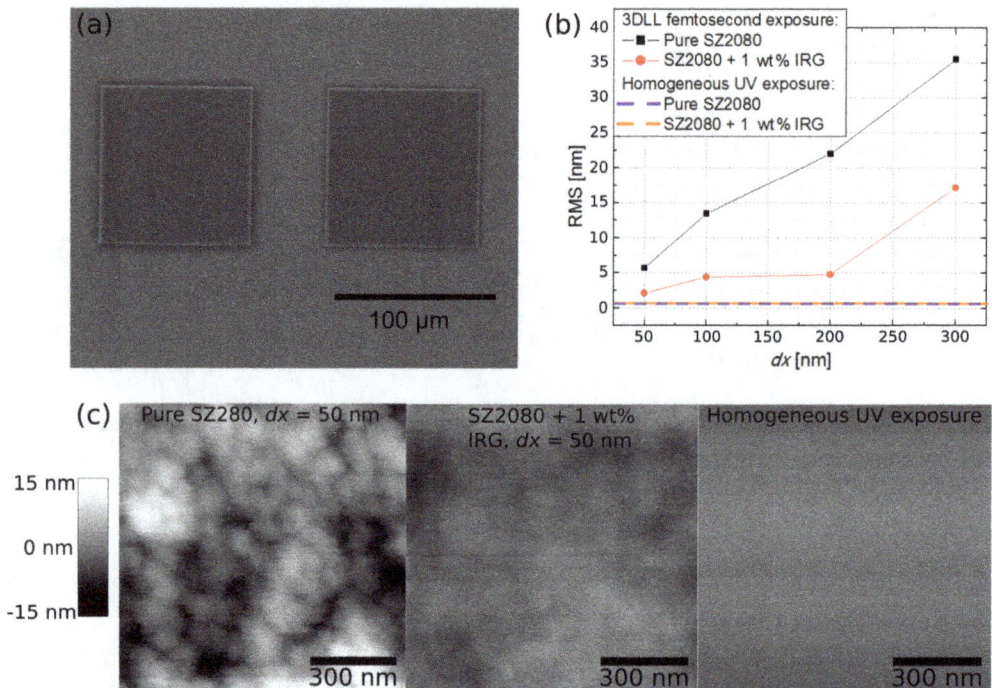

Figure 4. (a) SEM micrograph of the measured square polymerized structures; (b) RMS calculated for surfaces fabricated with different dx for SZ2080 containing PI and without it; and (c) AFM images of surfaces of pure and photosensitized polymer obtained with highest voxel overlap ($dx = 50$ nm) as well as one which was produced via homogeneous UV exposure.

2.3. Resilience of Micro-Optics to High Irradiance

We next tested how functional micro-optical elements would perform under different light radiation conditions. It is known that pure SZ2080 should have a higher optical damage threshold when thin films characterized by standard LIDT [18,25]. However, it is still unclear how this translates into the operation of standard microlenses both qualitatively and quantitatively.

First, an experiment was carried out with a CW $\lambda = 405$ nm laser operating at $P \sim 17$ mW. The laser beam was focused to a $w \sim 250$ µm radius spot onto 50 µm diameter microlenses, resulting in average intensity $I_A = P/\pi w^2 \sim 8.66$ W/cm^2. The microlenses were produced following the procedure described earlier [26]. Intense laser radiation could induce changes both in the volume and on the surface of the micro-optical element. For this reason, focusing properties prior and after exposure to potentially damaging light irradiation were examined using a CCD camera to determine whether the microlenses were affected. Microlenses were left in 405 nm light for 30 h. The light source used to measure the focusing properties of microlenses in this experiment was an HeNe laser. As shown in Figure 5, with a CW UV laser operating at 405 nm wavelength and exposing the lenses for 30 h, no effects on the microlens focusing were observed.

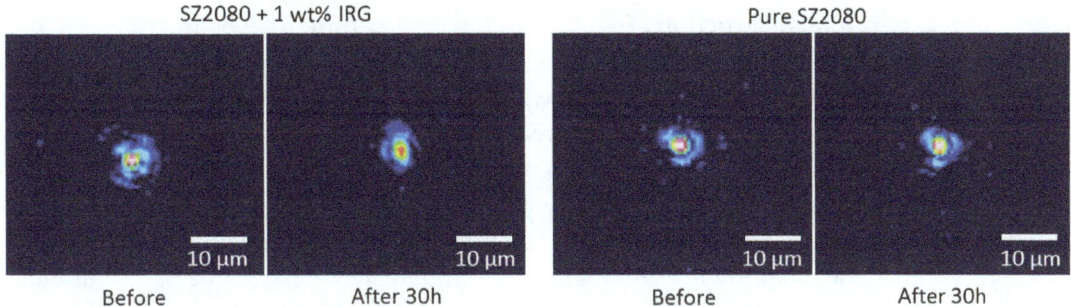

Figure 5. Images of the focal plane of a microlens before and after 30 h of exposure to 405 nm CW laser radiation. No significant change in the image at the focal point can be discerned.

Next, microlenses were left for 20 h exposed to a pulsed laser beam operating at 300 fs, 200 kHz and 515 nm wavelength. The microfabrication setup was applied for this experiment because it offered the possibility of both controlling the femtosecond laser beam irradiation parameters and simultaneously monitoring the focusing of the microlenses via the built-in microscope. A $40\times$ $NA = 0.95$ objective was employed for imaging the microlenses as well as to provide focusing for 515 nm radiation. An LED was imaged through the lenses as an illumination source. The objective was retracted 85 µm from being directly focused on the micro-optical elements, thus allowing the laser beam to expand and to form a laser beam spot of $\sim$250 µm radius and $I_p = 0.85\,\mathrm{GW/cm^2}$. Furthermore, it was deliberately offset in a transverse direction by 55 µm from the center of the microlenses in order to see if damage to the microlenses would depend on the I of the laser beam. Such prolonged exposure to femtosecond laser pulses resulted in severely damaged microlenses, which showed changes in focusing properties and the overall integrity of the structure (Figure 6). This investigation of microlens focusing shows that a micro-optical element made out of pure SZ2080 suffered less damage.

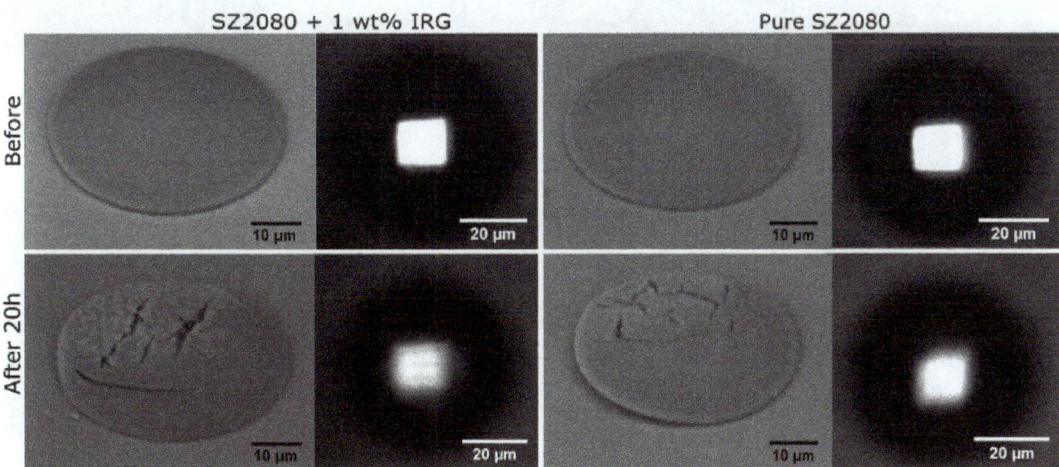

Figure 6. SEM images of microlenses before and after 20 h exposition to a loose focusing of 515 nm 300 fs laser radiation and an image of an LED made by the lens. Degradation of a lateral light distribution in the focal plane can be seen both in the structural quality of the lenses and the degraded projected image. The PI containing microlenses were more degraded.

In order to determine whether micro-optical elements of pure SZ2080 are indeed more resilient to intense laser radiation, the following time-dependent experiment was carried out. We used the $100\times NA = 0.9$ objective to in situ monitor changes in the focal plane of the microlenses. As the objective was retracted 50 µm from the microlenses, a ~100 µm radius laser beam spot was formed, with $I_p = 1.91$ GW/cm^2. This exposure resulted in fast degradation of the image (Figure 7a). Two steps were discernible: when lens degradation becomes observable and when the LED image is completely obscured. In the case of lenses containing standard 1 IRG wt %, deterioration started after 20 s of irradiation and caused total destruction after 30 s. In the case of pure SZ2080 microlenses, the time period up to the beginning of deterioration was 60 s and 100 s to a fully obscured LED image. Hence, a microlens made out of pure SZ2080 can withstand about a three times larger exposure dose. Furthermore, the damage to the microlenses differed. The entire microlens was structurally damaged in the case of SZ2080 with 1 wt % IRG (Figure 7b) resembling a thermomechanical failure, while pure SZ2080 lenses were damaged only in the central region by homogeneous melting (Figure 7c). The experiments were repeated several times and showed that the result deviated by no more than 7 s from one experiment to another, which is in the range of tens-of-percent from the measured values. For better comparison, more tests were performed varying IRG concentrations (0.5 wt % and 2 wt %) and standard lithographic material SU8 (Figure 7d). Results proved that an increase in PI concentration leads to less time being needed before the laser light damages the microlens, namely 9 s with 2 wt % IRG. The non-hybrid SU8 started to deteriorate just ~3 s after the shutter was opened with no noticeable deviations in time between the experiments. This shows that, even with relatively high (2 wt %) IRG concentrations, SZ2080 can withstand intense laser radiation at least three times longer than SU8. Removal of PI increases this superiority by more than one order (~20 times), which agrees well with earlier findings in thin films [18]. Thus, in the case of intense continuous exposure, the hybrid material without PI is the best for microlens fabrication in terms of optical resiliency.

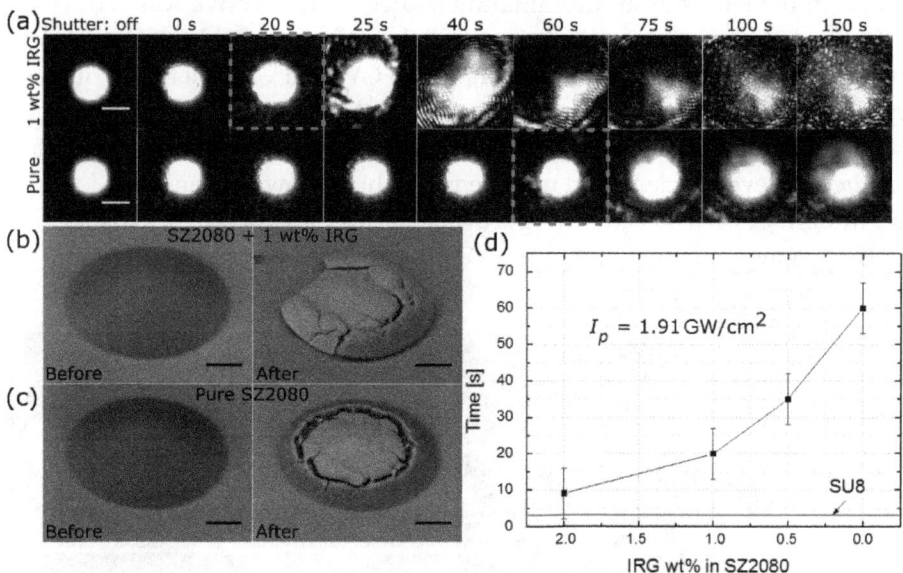

Figure 7. (**a**) real-time monitoring of a lateral intensity distribution of LED through microlenses produced using photosensitized and pure SZ2080 during irradiation with 515 nm 300 fs pulses at 200 kHz with $I_p = 1.91$ GW/cm^2 (spot radius of 100 µm). Faster deterioration of SZ2080 containing 1 wt % IRG as compared with pure SZ2080 is evident, as the image at the focus starts to degrade after 60 and 20 s (marked by **red** dashed squares), respectively; (**b,c**) SEM micrographs of the tested lenses before and after exposure. The photosensitized element is entirely destroyed, while the one produced out of pure SZ2080 exhibits relatively low damage; and (**d**) start of the microlens degradation for different concentrations of PI in SZ2080, as well as time needed to damage SU8. All scale bars are 10 µm.

The analysis presented shows degradation of micro-lenses during exposure. Whether it is caused by the accumulated dose or when critical temperature is reached was not established. Next, one set of lenses was continuously exposed for 15 min $I_p = 1.27\ GW/cm^2$ radiation while the other set was irradiated for a combined 15 min exposure delivered in 10 s light bursts followed by 10 s pauses (30 min total). This resulted in the microlenses being severely damaged for both pure and photosensitized SZ2080 when exposed in a continuous manner (Figure 8a). However, the multi-burst exposure of non-photosensitized microlenses showed no noticeable distortions, while the photosensitized lenses were unmistakably degraded (Figure 8b). To further explore this effect, lenses were exposed to 5 min of continuous $1.91\ GW/cm^2$ radiation with pulse repetition rates of 10, 100 and 200 kHz (Figure 8b). With 10 kHz illumination, both pure and photosensitized SZ2080 samples showed no noticeable changes in lateral light distribution at the focal plane, while both types of microlenses were completely destroyed by 200 kHz radiation. For this reason, repetition rates below or over these values were not tested. The response to 100 kHz varied between non-photosensitized and IRG containing SZ2080 microlenses. The effect on the pure SZ2080 microlens was inconspicuous, while the lens containing IRG displayed appreciable distortions. All these results clearly demonstrate that, in the case of pure SZ2080, the cause of damage is the heat accumulated in the volume of the lens. A better mitigation of the heat load could solve the thermal degradation of micro-optical elements.

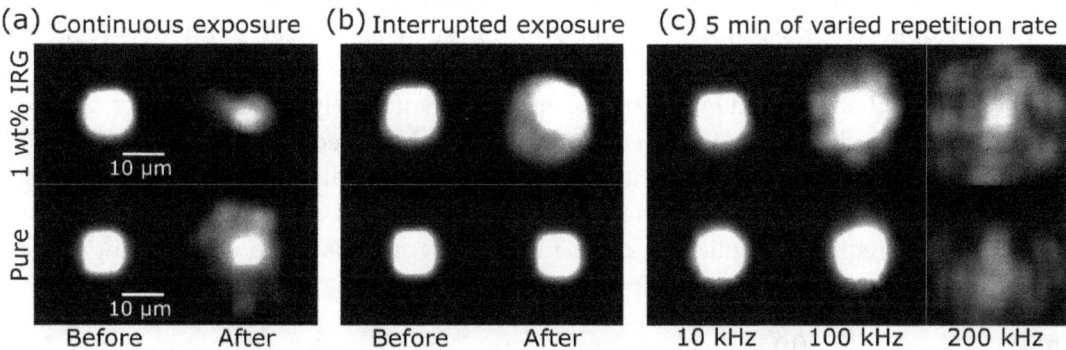

Figure 8. Focusing performance of microlenses after exposure to $I_p = 1.27\ GW/cm^2$ 515 nm 300 fs radiation in continuous (**a**) and multi-burst mode: 10 s exposure followed by a 10 s pause (**b**); (**c**) lateral distribution at the focus of microlenses exposed to 5 min of continuous radiation at $I_p = 1.91\ GW/cm^2$ (515 nm 300 fs) achieved with repetition rates of 10, 100 and 200 kHz. The 10 kHz case also serves as a before image, as there were no changes in focusing properties after this experiment.

2.4. Freeform Ceramic Structures out of Sintered SZ2080

Despite being more rigid and optically resilient in comparison to standard photopolymers [18], hybrid materials still have the limitations imposed by their organic component. On the other hand, the organic component is the reason why freeform laser nano-structuring is possible. Recent developments in the field of material processing showed that hybrid materials can be processed via pyrolysis by removing the organic component and leaving only the ceramic structure. This was successfully demonstrated using stereolithography at a millimeter-scale [27,28] or at a smaller scale using commercially available photoresists [29,30]. Here, we show a similar result in the micro- and nano-scales for SZ2080.

The initial sample structures chosen for the experiment were thick supporting walls with free hanging chain between them (Figure 9a). The sintered structure shrunk to about 65% in respect to all initial dimensions (Figure 9b). This indicates homogeneous reduction in sizes and volume during pyrolysis. Thus, in further experiments, we considered that the measurement of change in one dimension is sufficient to determine the overall change in the volume. In addition, the free hanging ring shrunk and deformed unevenly during pyrolysis. Peeling of supporting structures from the substrate was observed.

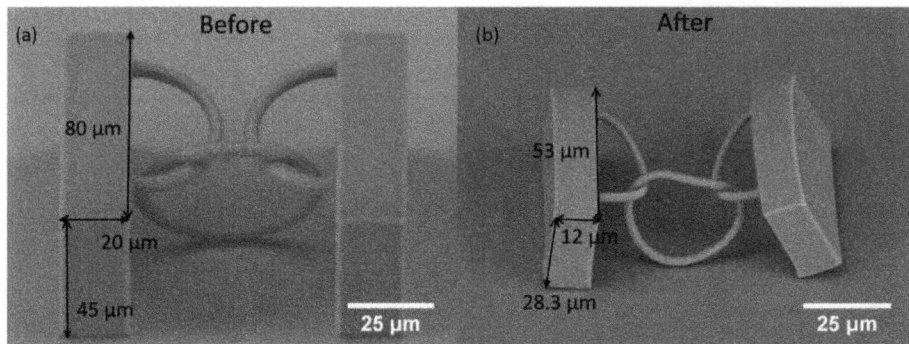

Figure 9. SEM images of the structure consisting of supporting walls and free hanging ring before (a) and after (b) pyrolysis. Magnification is the same. Fabricated objects appear brighter after pyrolysis, which indicates a change in the electrical conductivity of the material.

Next, pyrolysis was employed to tailor photonic crystals with ultra-thin lines. Photonic crystals designed to operate in the visible part of the spectrum require their resolution to be sub-wavelength [31]. 3DLL allows one to achieve the needed feature sizes [32,33], yet is relatively problematic due to the necessity of using ultra-short (<100 fs) laser pulses, tight focusing with oil-immersion lenses, and complicated post-processing techniques, such as critical point drying [34]. Here, we produced woodpile photonic crystals with line widths that can be achieved without any additional post-processing and then reduced in size by the application of pyrolysis. Furthermore, 295 nm wide lines (Figure 10a) were changed to 174 nm (Figure 10a), which is a reduction of ∼40%. The processed structure retained its overall shape. This is demonstrated by the ratio between the photonic crystal period and line width retaining the same ∼1.7 value before and after pyrolysis. Therefore, this technique does not require complicated compensation algorithms to be used during direct laser writing (DLW).

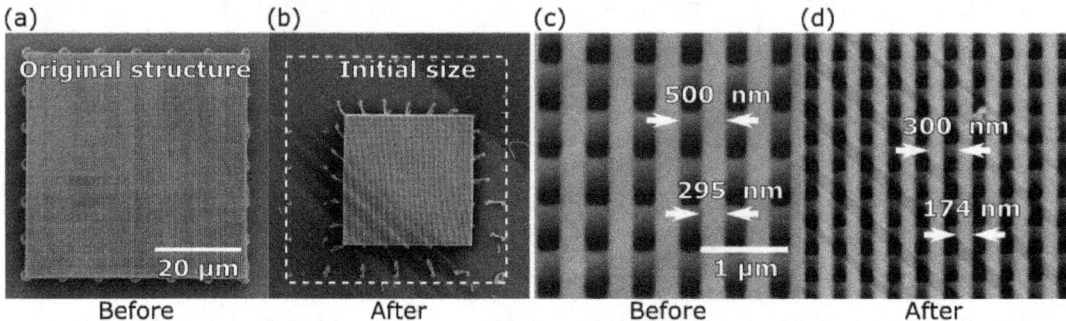

Figure 10. (a) SEM micrograph of a photonic crystal prior to (a) and after (b) pyrolysis; (c) the period and line width initially were 500 nm and 295 nm, respectively, which were shrunk to 300 nm and 174 nm after pyrolysis (d); a change of ∼40%. The *period/width* ratio stayed ∼1.7, indicating a homogeneous reduction in size.

In order to gain further insights into what happens during pyrolysis, a thermal gravimetric analysis (TGA) of a drop of unprocessed SZ2080 was performed (Figure 11). First, the sudden drop in the weight at 100 °C can be attributed to the evaporation of solvent that otherwise would be removed during pre-bake. The weight then remained relatively stable until the temperature reached 350 °C, when it started again to decrease rapidly. The organic component of the hybrid is decomposed at this stage. During this period, two distinct changes in the weight decrease are discerned, hinting at two phases in which the organic component of the material is removed from the hybrid. The overall change in weight was about 28%. This exceeded the volume change of ∼35%–40% and indicates that the lost part of the material is relatively low-density organic compounds. In addition, it is assumed that the remaining material is densified, resulting in a shrunken ceramic structure.

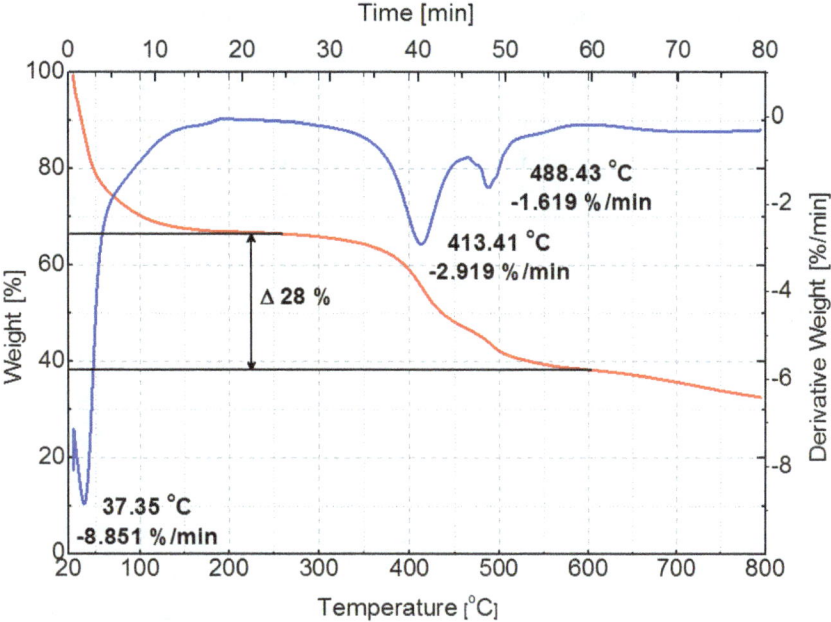

Figure 11. TGA data showing weight loss vs. temperature. The weight loss was 28%; observed shrinkage was ~40%. The organic component was decomposed and removed by heating.

3. Discussion

Here, we discuss the 3DLL of pure materials in detail. First, femtosecond laser photopolymerization of a SZ2080 with IRG PI is recalled. Light breaks the weak single bonds of the PI molecule, which then creates two radicals PI* (Figure 12a (1,2)) [17] . These radicals then react with pre-polymer molecules via double bonds creating a radicalized monomer SZ* (Figure 12a (3,4)). This initiates a chain reaction and growth of an intertwined polymer matrix, which does not dissolve in the organic solvent. In the case of a SZ2080 without PI, such a reaction is induced when nonlinear absorption occurs. The double bond is broken in the pre-polymer (Figure 12b (1,2)), which otherwise would be broken by reactions with PI molecules. Large excitations can build up in the intermediary states because the laser-induced multiphoton excitation rate of pre-polymer species is high in comparison to the thermalization rate, which can be as long as μs [35]. Photochemical (photolytic) processes are dominant compared to photothermal phenomena due to the absence of detectable thermal effects, such as sample vaporization, boiling or thermal expansion, which is ensured by the closure of the reaction volume by the surrounding material. High irradiance exposure is sufficient to directly break chemical bonds of pre-polymerized SZ2080 organic constituents, which initialize formation of an insoluble and rigid organic-inorganic composite.

The difference of ΔI for both pure and PI containing material being only 15.5% is an excellent result, showing that the application of femtosecond pulses and $NA = 1.4$ focusing is sufficient to make experimentation with both materials relatively easy. However, in order to tie this parameter to a more diverse set of materials and setups, we need to normalize it to the I_t, thus acquiring a dynamic fabrication range: $DR = \Delta I / I_t$ [36]. This allows us to describe manufacturing conditions in terms of position and width of ΔI in relation to applied I and I_{th}, which is heavily influenced by all fabrication parameters. In the case of our study, $DR_{IRG} = 2.94$ is more than two times higher than $DR_{Pure} = 1.26$. This result is a consequence of pure material being less photosensitive (higher photon densities are required for avalanche induced direct bond breaking than multi-photon absorption) and having twice as high I_t. Lower DR indicates that, while processing such polymers, special care and skills should be taken when finding the structuring parameters. It is important to note that the two-fold discrepancy in DR is substantially higher than the 15.5% difference in ΔI. Thus, DR is more universal in demonstrating sensitivity to applied fabrication parameters than case-sensitive ΔI. Universal parameters of this kind are advantageous when comparing results achieved in other fabrication setups with different materials and light sources, such as CW lasers [37,38], ps pulses [39] or high repetition rate Ti:Sapphiresystems [36]. On the other hand, this particular parameter still has limitations considering its usage with diverse pulse overlaps caused by variation in v or pulse repetition rate. Further research aimed at better understanding the processes involved in nano-structuring and its numerical evaluation could be one of the directions for further work in the field.

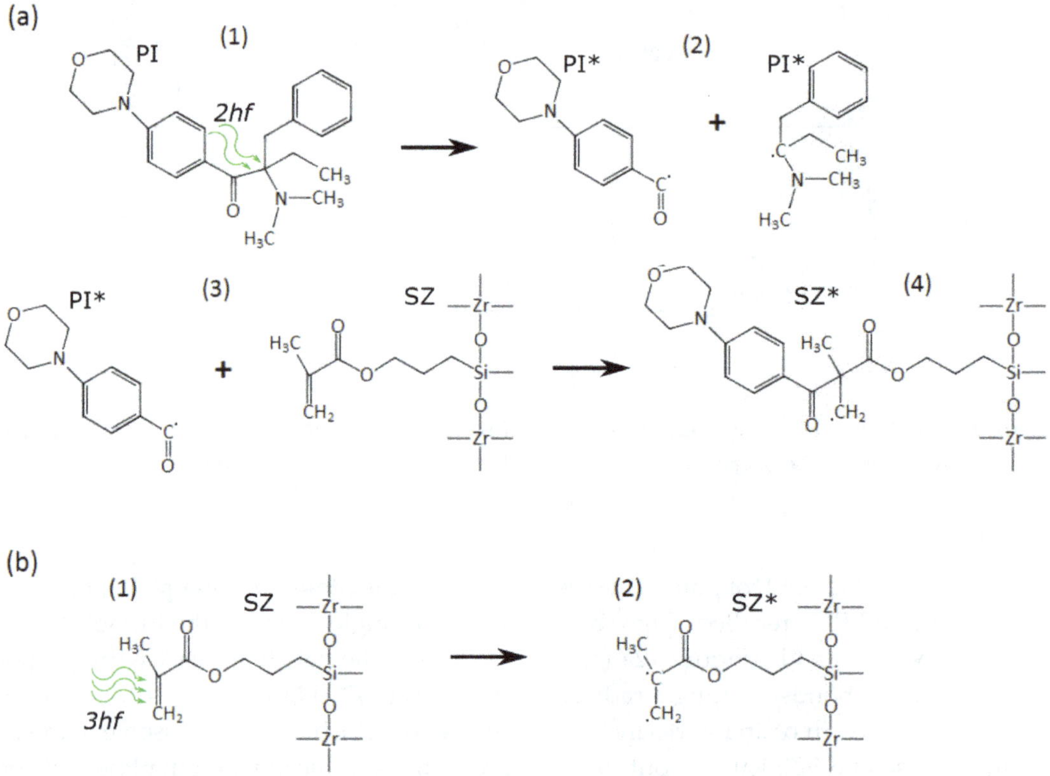

Figure 12.   (a) polymerization reactions initiated by nonlinear absorption of PI molecules and subsequent chemical pathways resulting in a cross linked SZ2080; (b) SZ2080 cross linking without PI. *hf* is the photon energy.

With ever-increasing demand for advanced elements in optical systems, freeform 3D microstructuring techniques become more sophisticated. While advanced glass processing methods show some possibilities for producing relatively small ($\sim$nm–μm) structures [40], they still lack insurface quality and the capability of performing true 3D fabrication. To date, only (DLW) based 3DLL was shown to be capable of true 3D structuring at the microscale. This additive manufacturing technique [41] combines the complete freedom of the architecture of produced objects [42], the possibility of integration on various substrates [43–45] and a large range of materials that can be processed in such a fashion [46]. Here, we have shown that structures produced out of zirconium containing photopolymer SZ2080 can withstand a relatively low $I_A = 8.66$ W/cm^2 short wavelength (405 nm) exposure for prolonged time periods. Additionally, it was demonstrated that micro-optical elements made out of SZ2080 have good resilience to high repetition rate femtosecond laser radiation. This matches well with earlier findings of higher optical damage threshold for SZ2080 [18]. Finally, we have shown that the surface quality of structures produced from non-photosensitized and photosensitized SZ2080 has low surface roughness ($RMS < \lambda/20$ at $\lambda < 400$ nm) and is suitable for applications in micro-optics and opto-fluidics [47,48].

The development of micro-optical elements made from a high-purity glassy hybrid material via a simple low temperature chemical synthesis without use of the PI, which are usually required for light absorption, can help new developments in several fields. For high-intensity laser applications, the use of micro-optical elements has several distinct advantages due to optical damage scaling rules [49]: (1) the damage threshold decreases with increasing spot area $I_{dam} \propto 1/\sqrt{A_{spot}}$; (2) micro-optical elements have low surface roughness, δ_{surf}, which increases the damage threshold $I_{dam} \propto 1/\delta_{surf}^m$ with exponent m between 1 and 1.5. In the field of light filamentation of ultra-short laser pulses in air and gases, the exploration of multiple beamlet generation apertures [50], ring Airy beams [51], and the realization of multifilaments– super-filamentation [52] is an active research area. The peak intensities on the beam-forming optics can reach pre-breakdown I of several TW/cm^2, which demands low absorption materials (no PI in polymerised micro-optics) according to the scaling $I_{dam} = \frac{0.26 \text{ MW/cm}^2}{(\alpha/\text{cm}^{-1})^{0.74}}$, where the absorption coefficient $\alpha = -1/L \ln(T)$ is defined by transmission T and propagation length L [49]. While microlenses have an inherently low L, the result of removing PI also greatly minimizes T. This is the reason why small optical elements using refractive and diffractive beam-forming concepts made from high purity materials are the most promising candidates in the high-power laser field.

The fact that material can be structured without PI with a ΔI comparable to that of photosensitized material is a promising discovery for other science fields as well. Earlier studies dedicated to pure material structuring by femtosecond laser pulses [16,53] have proved complicated and slow. If such pure material can be processed at a relatively high speed ($\sim$mm/s), it would become suitable for biomedical applications, where structure dimensions are in the range of mm–cm [54]. Pure material would guarantee superb biocompatibility, which is a key requirement for tissue engineering, especially taking into account biodegradable implants.

For other diverse applications, SZ2080 can be doped with organic die molecules [55] or noble metal nanoparticles [56], which can be used for polymerization control or increased functionalities. In the recently developed field of astro-photonics [57], where absorption in laser written waveguide lanterns is strong [58], the polymerised waveguides can tackle high loss problems. The photonic wire bonding also improves light confinement, flexibly controls 3D conformation of fiber bundles for phase matched light delivery from the optical image plane to a spectrometer slit, and well defines single-mode operation of the fiber by precise control of the polymerised wire cross section. Unexpectedly, from the perspective of high light intensity applications, large volume optical astro-instrumentation is highly sensitive to detector background counts due to radioactive trace elements in the glass-optics. Potentially, better control of glass forming ingredients can be obtained in the case of photopolymer selection for optical elements prepared via a sol-gel route as SZ2080.

Composition control of the sol-gel resists via different portions of organic and inorganic components provides a tool to tune the refractive index [4] and to create complex micro-optical elements for high-quality optical imaging, as compared with only shape control of composite lenses [44,59–61]. Since the sol-gel route is open to mixing different oxide precursors and concentrations, this opens up the capability of creating the different refractive index PI-free optical elements required for aberration control in multi-component lens optical systems [44,59–61]. With pyrolysis, an even larger range of refractive index tunability is accessible. Current endoscopy and optical imaging applications where micro-optical elements are in contact with live tissue would benefit from the absence of PI due to strong optical absorption and bio-toxicity.

The possibility of using pyrolysis is an additional feature of SZ2080 due to its hybrid organic-inorganic nature that allows it to achieve true 3D glass ceramic structures. This is expected to open up new applications. Shrinkage of the polymer can be used to achieve ultra fine features. The demonstrated homogeneous reduction in size by 40% while keeping a well-defined 3D structure is an improvement compared to the 30% presented in other work [30]. Further studies aimed at better understanding the underlying physical and chemical phenomena during pyrolysis, as well as enhanced control of SZ2080 sintering, will be the basis for our future work.

4. Materials and Methods

The SZ2080 photoresist was acquired from FORTH (Heraklion, Greece) and, as its name implies, contained 20 wt % of inorganic and 80 wt % of organic components. For cross-check experiments, SZ2080 was photosensitized by mixing it with a commercial PI Irgacure 369 (IRG). Samples were prepared by drop casting one droplet of the material on a glass substrate and then pre-backing the sample at 75 °C for 45 min. After fabrication, samples were developed in isobutyl methyl ketone for 45 min and subsequently rinsed in isopropanol for 15 min. The SU8 was processed using a procedure consisting of two pre-bake stages, 30 min at 60 °C and 60 min at 90 °C, and then post-baked at similar temperatures but with half the duration (15 and 30 min, respectively) and developed in propylene glycol monomethyl ether acetate (PGMEA) for 60 min.

Schematics of the 3DLL setup used are shown in Figure 13. The femtosecond laser was Pharos (Light Conversion Ltd., Vilnius, Lithuania) operating at 1030 nm fundamental wavelength, 300 fs pulse duration and 200 kHz repetition rate. Power is controlled by two power control units consisting of a $\lambda/2$ waveplate and Brewster angle polarizer. Such two-stage power attenuation allows for minimizing power fluctuations and provides precise power control during fabrication. A second harmonic at 515 nm wavelength was used for 3D free-form polymerization. The laser beam is expanded by a $2\times$ magnification telescope in order to fill all of the objective aperture. During experiments in which precise feature size control was essential (for instance a photonic crystal), polarization of incident light was kept at a constant 45° degree angle with both horizontal translation axes, this way avoiding any polarization induced anisotropy of fabricated line widths [62]. Structure fabrication is performed with a combination of Aerotech linear stages (ALS130-110-X,Y for positioning in the XY plane, ALS130-60-Z for Z-axis) and a galvanoscanner, operating in sync in an infinite field of view (IFOV) regime, which allows high fabrication speeds ($\sim$mm/s) and superb structure quality. The sample is illuminated by a red LED which enables the fabrication process to be monitored in real time using the CMOS camera.

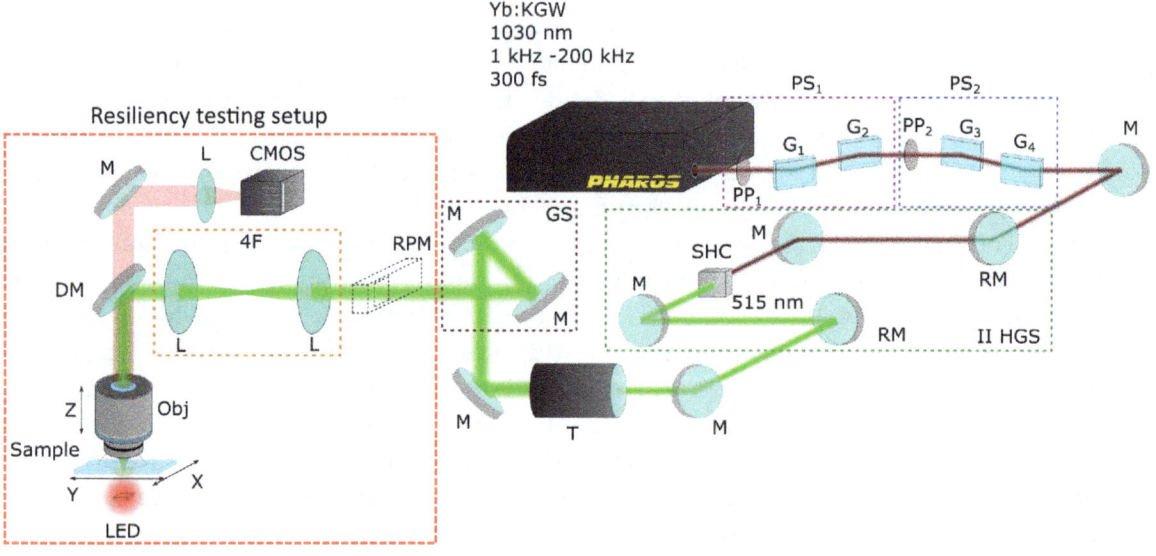

Figure 13. Schematics of setup used for fabrication and microlens degradation experiments: PS—power control stage, PP—phase plate, G—glass plate, M—mirror, RM—removable mirror, SHC—second harmonic crystal, T—telescope, GS—galvanoscanner, RPM—removable power meter, L—lens, 4F—lens system in 4-F configuration, DM—dichroic mirror, CMOS—CMOS camera used to monitor fabrication process, Obj—objective lens, LED—LED used for sample illumination.

The same setup was used to monitor the degradation of micro-optical elements in real time. In this case, the sample was microlenses on a glass slide illuminated by the LED from the bottom. The objective was retraced at some defined distance from them, resulting in a relatively large laser spot on the lenses. Exact values are listed in the text where applicable. This allowed for monitoring lateral intensity distribution projected by the microlens and for illuminating them with femtosecond laser light simultaneously.

The average laser power P was measured before the polymerization and lens degradation experiments and subsequently recalculated at the peak intensity I_p at the center of focal point [15]:

$$I_p = \frac{2PT}{fw^2\pi\tau},$$ (1)

where f is the pulse repetition rate, τ is the pulse duration, and $\omega = 0.61\lambda/NA$ is the waist (radius) of the beam. $T \simeq 0.41$ is the system transparency without glass substrate and pre-polymer for a $63\times$ $NA = 1.4$ objective. Fabrication parameters used for microlenses are $I_{IRG2\%} = 0.48$ TW/cm^2, $I_{IRG1\%} = 0.48$ TW/cm^2, $I_{IRG0.5\%} = 0.61$ TW/cm^2 and $I_{pure} = 0.61$ TW/cm^2 using $v = 50$ μm/s. For AFM analysis, structures were produced at $I_{IRG} = 0.61$ TW/cm^2 and $I_{pure} = 0.86$ TW/cm^2 and $v = 250$ μm/s. The latter parameters were used for fabrication of the sintered ring structure. The photonic crystal for pyrolysis experiments was acquired applying $v = 100$ μm/s and $I = 0.84$ TW/cm^2. The Nd:YAG laser used for IV harmonic ($\lambda = 266$ nm) generation operated at 5 ns pulse duration, 24 mJ pulse energy, 2 Hz repetition rate and was directed onto a 1 cm diameter spot. Both pure SZ2080 and with 1 wt % were exposed to such radiation for 30 min. Pyrolysis was performed in Ar atmosphere at 600 °C temperature for 5 h.

Surface roughness was characterized using SEM TM-1000 (Hitachi, Tokyo, Japan) and AFM Catalyst (Bruker, Billerica, MA, USA) with an Au coated SiN-needle with a $k = 0.06$ N/m stiffness at $F = 18$ kHz and with a tip diameter of 20 nm. For TGA analysis, Pyris 1 TGA (Perkin Elmer, Waltham, MA, USA) equipment was used; the sample was heated in a nitrogen atmosphere from 30 °C to 800 °C with a heating rate of 10 °C/min.

5. Conclusions

A detailed study of laser structuring of non-photosensitized SZ2080 via 3DLL was carried out. The photo-polymer has only a 12.5% lower structure survival rate compared to SZ2080 photosensitized with 1 wt % IRG and comparable $\Delta I \sim 1$ TW/cm^2. Surface roughness of structures produced out of both compositions was in the range of $RMS < 20$ nm, which is sufficient for fabricating micro-optical components. The structuring fidelity of both materials is comparable if parameters from within the ΔI are used. A micro-optical element made out of pure SZ2080 was integrated on the tip of an optical fiber.

Furthermore, investigation of micro-optical element degradation at various laser irradiations was performed. It was shown that SZ2080 in both photosensitized and pure forms was not damaged by tens of hours of exposure to CW 405 nm laser providing $I_A = 8.66$ W/cm^2. On the other hand, microlenses produced out of pure material were shown to survive ~three-fold longer in comparison to those containing 1 wt % IRG, and ~20 times longer than those produced out of SU8 when irradiated by 515 nm 300 fs 200 kHz laser at $I_p = 1.91$ GW/cm^2. If I_p is dropped to 1.27 GW/cm^2 and delivered in an interrupted manner (10 s of exposure followed by 10 s pause), non-photosensitized microlenses showed no signs of degradation even after combined exposure of 15 min, while those containing IRG were damaged substantially. A similar result was achieved with a decrease in the pulse repetition rate, while keeping the $I_p = 1.91$ GW/cm$^2 - 10$ kHz caused no distortions in both pure and photosensitized structures, 100 kHz induced substantial damage only in those containing the IRG, and 200 kHz destroyed both types of lenses. This indicates that, in the case of microlenses, the main deterioration inducing factor is thermo-accumulation and subsequent melting.

Additionally, pyrolysis of hybrid material was performed removing organic constitutes, leaving only densified glass-ceramic structure. Shrinkage during this process was homogeneous and allowed for size reduction by 35%–40%, which is a record high number [30]. Further studies will be focused on the optical and mechanical properties of glass-ceramics and its applications for high irradiance optics and material processing with high I_p of high-repetition rate conditions [63,64]. Potential applications are in the fields of the filamentation of ultra-short laser pulses, the fabrication of fiber-optical elements for sensor applications in the chemically harsh, high temperature, and radioactive environments encountered in nuclear power stations and in optically driven inertial confinement fusion facilities.

Acknowledgments: The authors acknowledge European Commission's Seventh Framework Programme Laserlab-Europe IV JRA support BIOAPP (EC-GA 654148). D.G., S.J. and M.M. acknowledge NATO grant SPS-985048 "Nanostructures for Highly Efficient Infrared Detection".

Author Contributions: L.J. designed and performed all the fabrication and optical element degradation experiments, did most of the measurements and evaluated received data; D.G. carried out pyrolysis of photonic crystals; L.M. contributed AFM measurements; D.S. carried out a high resolution SEM imaging of photonic crystals. S.Š. provided insights into pyrolysis and chemical reactions during photopolymerization processes; S.J. and M.M. proposed the direction and supervised the research. All of the authors contributed to the writing and editing of the manuscript.

Conflicts of Interest: The authors declare no conflict of interest.

References

1. Houbertz, R.; Frohlich, L.; Popall, M.; Streppel, U.; Dannberg, P.; Bräuer, A.; Serbin, J.; Chichkov, B. Inorganic-Organic Hybrid Polymers for Information Technology: From Planar Technology to 3D Nanostructures. *Adv. Eng. Mater.* **2003**, *5*, 551–555.

2. Serbin, J.; Egbert, A.; Ostendorf, A.; Chichkov, B.N.; Houbertz, R.; Domann, G.; Schulz, J.; Cronauer, C.; Fröhlich, L.; Popall, M. Femtosecond laser-induced two-photon polymerization of inorganic-organic hybrid materials for applications in photonics. *Opt. Lett.* **2003**, *28*, 301–303.

3. Lebeau, B.; Innocenzi, P. Hybrid materials for optics and photonics. *Chem. Soc. Rev.* **2011**, *40*, 886–906.

4. Ovsianikov, A.; Viertl, J.; Chichkov, B.; Oubaha, M.; MacCraith, B.; Sakellari, I.; Giakoumaki, A.; Gray, D.; Vamvakaki, M.; Farsari, M.; et al. Ultra-Low Shrinkage Hybrid Photosensitive Material for Two-Photon Polymerization Microfabrication. *ACS Nano* **2008**, *2*, 2257–2262.

5. Schafer, K.J.; Hales, J.M.; Balu, M.; Belfield, K.D.; van Stryland, E.W.; Hagan, D.J. Two-photon absorption cross-sections of common photoinitiators. *J. Photochem. Photobiol. A* **2004**, *162*, 497–502.

6. Rajamanickam, V.P.; Ferrara, L.; Toma, A.; Zaccaria, R.P.; Das, G.; Fabrizio, E.D.; Liberale, C. Suitable photo-resists for two-photon polymerization using femtosecond fiber lasers. *Microelectron. Eng.* **2014**, *121*, 135–138.

7. Harnisch, E.; Russew, M.; Klein, J.; Konig, N.; Crailsheim, H.; Schmitt, R. Optimization of hybrid polymer materials for 2PP and fabrication of individually designed hybrid microoptical elements thereof. *Opt. Mater. Express* **2015**, *5*, 456–461.

8. Lee, K.S.; Yang, D.Y.; Park, S.H.; Kim, R.H. Recent developments in the use of two-photon polymerization in precise 2D and 3D microfabrications. *Polym. Adv. Technol.* **2006**, *17*, 72–82.

9. Sun, H.B.; Kawata, S. Two-Photon Photopolymerization and 3D Lithographic Microfabrication. In *NMR 3D Analysis Photopolymerization*; Springer: Berlin/Heidelberg, Germany, 2006; pp. 169–273.

10. Malinauskas, M.; Žukauskas, A.; Purlys, V.; Gaidukevičiūtė, A.; Balevičius, Z.; Piskarskas, A.; Fotakis, C.; Pissadakis, S.; Gray, D.; Gadonas, R.; et al. 3D microoptical elements formed in a photostructurable germanium silicate by direct laser writing. *Opt. Laser Eng.* **2012**, *50*, 1785–1788.

11. Mačiulaitis, J.; Deveikytė, M.; Rekštytė, S.; Bratchikov, M.; Darinskas, A.; Šimbelytė, A.; Daunoras, G.; Laurinavičienė, A.; Laurinavičius, A.; Gudas, R.; et al. Preclinical study of SZ2080 material 3D microstructured scaffolds for cartilage tissue engineering made by femtosecond direct laser writing lithography. *Biofabrication* **2015**, *7*, 015015.

12. Malinauskas, M.; Gilbergs, H.; Žukauskas, A.; Purlys, V.; Paipulas, D.; Gadonas, R. A femtosecond laser-induced two-photon photopolymerization technique for structuring microlenses. *J. Opt.* **2010**, *12*, 035204.

13. Sun, Q.; Juodkazis, S.; Murazawa, N.; Mizeikis, V.; Misawa, H. Freestanding and movable photonic microstructures fabricated by photopolymerization with femtosecond laser pulses. *J. Micromech. Microeng.* **2010**, *20*, 035004.

14. Farsari, M.; Vamvakaki, M.; Chichkov, B.N. Multiphoton polymerization of hybrid materials. *J. Opt.* **2010**, *12*, 124001.

15. Jonušauskas, L.; Rekštytė, S.; Malinauskas, M. Augmentation of direct laser writing fabrication throughput for three-dimensional structures by varying focusing conditions. *Opt. Eng.* **2014**, *53*, 125102.

16. Buividas, R.; Rekštytė, S.; Malinauskas, M.; Juodkazis, S. Nano-groove and 3D fabrication by controlled avalanche using femtosecond laser pulses. *Opt. Mater. Express* **2013**, *3*, 1674–1686.

17. Malinauskas, M.; Žukauskas, A.; Bičkauskaitė, G.; Gadonas, R.; Juodkazis, S. Mechanisms of three-dimensional structuring of photo-polymers by tightly focussed femtosecond laser pulses. *Opt. Express* **2010**, *18*, 10209–10221.

18. Žukauskas, A.; Batavičiūtė, G.; Ščiuka, M.; Jukna, T.; Melninkaitis, A.; Malinauskas, M. Characterization of photopolymers used in laser 3D micro/nanolithography by means of laser-induced damage threshold (LIDT). *Opt. Mater. Express* **2014**, *4*, 1601–1616.

19. Li, Z.; Torgersen, J.; Ajami, A.; Muhleder, S.; Qin, X.; Husinsky, W.; Holnthoner, W.; Ovsianikov, A.; Stampfl, J.; Liska, R. Initiation efficiency and cytotoxicity of novel water-soluble two-photon photoinitiators for direct 3D microfabrication of hydrogels. *RSC Adv.* **2013**, *3*, 15939.

20. Sakellari, I.; Kabouraki, E.; Gray, D.; Purlys, V.; Fotakis, C.; Pikulin, A.; Bityurin, N.; Vamvakaki, M.; Farsari, M. Diffusion-Assisted High-Resolution Direct Femtosecond Laser Writing. *ACS Nano* **2012**, *6*, 2302–2311.

21. Fischer, J.; Freymann, G.; Wegener, M. The Materials Challenge in Diffraction-Unlimited Direct-Laser-Writing Optical Lithography. *Adv. Mater.* **2010**, *22*, 3578–3582.

22. Jiang, L.; Xiong, W.; Zhou, Y.; Liu, Y.; Huang, X.; Li, D.; Baldacchini, T.; Jiang, L.; Lu, Y. Performance comparison of acrylic and thiol-acrylic resins in two-photon polymerization. *Opt. Express* **2016**, *24*, 13687–13701.

23. Žukauskas, A.; Matulaitienė, I.; Paipulas, D.; Niaura, G.; Malinauskas, M.; Gadonas, R. Tuning the refractive index in 3D direct laser writing lithography: Towards GRIN microoptics. *Laser Photonics Rev.* **2015**, *9*, 706–712.

24. Park, S.H.; Lim, T.W.; Yang, D.Y.; Kim, R.H.; Lee, K.S. Improvement of spatial resolution in nano-stereolithography using radical quencher. *Macromol. Res.* **2006**, *14*, 559–564.

25. Žukauskas, A.; Batavičiūtė, G.; Ščiuka, M.; Balevičius, Z.; Melninkaitis, A.; Malinauskas, M. Effect of the photoinitiator presence and exposure conditions on laser-induced damage threshold of ORMOSIL (SZ2080). *Opt. Mater.* **2015**, *39*, 224–231.

26. Malinauskas, M.; Žukauskas, A.; Purlys, V.; Belazaras, K.; Momot, A.; Paipulas, D.; Gadonas, R.; Piskarskas, A.; Gilbergs, H.; Gaidukevičiūtė, A.; et al. Femtosecond laser polymerization of hybrid/integrated micro-optical elements and their characterization. *J. Opt.* **2010**, *12*, 124010.

27. Schwentenwein, M.; Homa, J. Additive Manufacturing of Dense Alumina Ceramics. *Int. J. Appl. Ceram. Technol.* **2014**, *12*, 1–7.

28. Eckel, Z.C.; Zhou, C.; Martin, J.H.; Jacobsen, A.J.; Carter, W.B.; Schaedler, T.A. Additive manufacturing of polymer-derived ceramics. *Science* **2015**, *351*, 58–62.

29. Tétreault, N.; von Freymann, G.; Deubel, M.; Hermatschweiler, M.; Pérez-Willard, F.; John, S.; Wegener, M.; Ozin, G. New Route to Three-Dimensional Photonic Bandgap Materials: Silicon Double Inversion of Polymer Templates. *Adv. Mater.* **2006**, *18*, 457–460.

30. Li, J.; Jia, B.; Gu, M. Engineering stop gaps of inorganic-organic polymeric 3D woodpile photonic crystals with post-thermal treatment. *Opt. Express* **2008**, *16*, 20073–20080.

31. Rill, M.S.; Plet, C.; Thiel, M.; Staude, I.; Freymann, G.; Linden, S.; Wegener, M. Photonic metamaterials by direct laser writing and silver chemical vapour deposition. *Nat. Mater.* **2008**, *7*, 543–546.

32. Haske, W.; Chen, V.W.; Hales, J.M.; Dong, W.; Barlow, S.; Marder, S.R.; Perry, J.W. 65 nm feature sizes using visible wavelength 3-D multiphoton lithography. *Opt. Express* **2007**, *15*, 3426–3436.

33. Wollhofen, R.; Katzmann, J.; Hrelescu, C.; Jacak, J.; Klar, T.A. 120 nm resolution and 55 nm structure size in STED-lithography. *Opt. Express* **2013**, *21*, 10831–10840.

34. Hengsbach, S.; Lantada, A.D. Direct laser writing of auxetic structures: Present capabilities and challenges. *Smart Mater. Struct.* **2014**, *23*, 085033.

35. Bäuerle, D. *Laser Processing and Chemistry*; Springer: Berlin/Heidelberg, Germany, 2011.

36. Fischer, J.; Mueller, J.B.; Kaschke, J.; Wolf, T.J.A.; Unterreiner, A.N.; Wegener, M. Three-dimensional multi-photon direct laser writing with variable repetition rate. *Opt. Express* **2013**, *21*, 26244–26260.

37. Do, M.T.; Nguyen, T.T.N.; Li, Q.; Benisty, H.; Ledoux-Rak, I.; Lai, N.D. Submicrometer 3D structures fabrication enabled by one-photon absorption direct laser writing. *Opt. Express* **2013**, *21*, 20964–20973.

38. Thiel, M.; Fischer, J.; von Freymann, G.; Wegener, M. Direct laser writing of three-dimensional submicron structures using a continuous-wave laser at 532 nm. *Appl. Phys. Lett.* **2010**, *97*, 221102.

39. Malinauskas, M.; Danilevičius, P.; Juodkazis, S. Three-dimensional micro-/nano-structuring via direct write polymerization with picosecond laser pulses. *Opt. Express* **2011**, *19*, 5602–5610.

40. Wortmann, D.; Gottmann, J.; Brandt, N.; Horn-Solle, H. Micro- and nanostructures inside sapphire by fs-laser irradiation and selective etching. *Opt. Express* **2008**, *16*, 1517–1522.

41. Vaezi, M.; Seitz, H.; Yang, S. A review on 3D micro-additive manufacturing technologies. *Int. J. Adv. Manuf. Technol.* **2012**, *67*, 1721–1754.

42. Maruo, S.; Fourkas, J.T. Recent progress in Multiphoton microfabrication. *Laser Photonics Rev.* **2008**, *2*, 100–111.

43. Lightman, S.; Gvishi, R.; Hurvitz, G.; Arie, A. Shaping of light beams by 3D direct laser writing on facets of nonlinear crystals. *Opt. Lett.* **2015**, *40*, 4460–4463.

44. Gissibl, T.; Thiele, S.; Herkommer, A.; Giessen, H. Sub-micrometre accurate free-form optics by three-dimensional printing on single-mode fibres. *Nat. Commun.* **2016**, *7*, 11763.

45. Rekštytė, S.; Jonavičius, T.; Malinauskas, M. Direct laser writing of microstructures on optically opaque and reflective surfaces. *Opt. Laser Eng.* **2014**, *53*, 90–97.

46. Malinauskas, M.; Žukauskas, A.; Hasegawa, S.; Hayasaki, Y.; Mizeikis, V.; Buividas, R.; Juodkazis, S. Ultrafast laser processing of materials: From science to industry. *Light Sci. Appl.* **2016**, *5*, e16133.

47. Misawa, H.; Juodkazis, S. Photophysics and Photochemistry of a Laser Manipulated Microparticle. *Prog. Polym. Sci.* **1999**, *24*, 665–697.

48. Jonušauskas, L.; Žukauskas, A.; Danilevičius, P.; Malinauskas, M. Fabrication, replication, and characterization of microlenses for optofluidic applications. *Proc. SPIE* **2013**, *8613*, 861318.

49. Menzel, R. *Photonics: Linear and Nonlinear Interactions of Laser Light and Matter*; Springer: Berlin/Heidelberg, Germany, 2001.

50. Kosareva, O.G.; Liu, W.; Panov, N.A.; Bernhardt, J.; Ji, Z.; Sharifi, M.; Li, R.; Xu, Z.; Liu, J.; Wang, Z.; et al. Can we reach very high intensity in air with femtosecond PW laser pulses? *Laser Phys.* **2009**, *19*, 1776–1792.

51. Panagiotopoulos, P.; Papazoglou, D.G.; Couairon, A.; Tzortzakis, S. Sharply autofocused ring-Airy beams transforming into non-linear intense light bullets. *Nat. Commun.* **2013**, *4*, 2622.

52. Point, G.; Brelet, Y.; Houard, A.; Jukna, V.; Milián, C.; Carbonnel, J.; Liu, Y.; Couairon, A.; Mysyrowicz, A. Superfilamentation in Air. *Phys. Rev. Lett.* **2014**, *112*, 223902.

53. Maximova, K.; Wang, X.; Balčytis, A.; Fan, L.; Li, J.; Juodkazis, S. Silk patterns made by direct femtosecond laser writing. *Biomicrofluidics* **2016**, *10*, 054101.

54. Selimis, A.; Mironov, V.; Farsari, M. Direct laser writing: Principles and materials for scaffold 3D printing. *Microelectron. Eng.* **2015**, *132*, 83–89.

55. Žukauskas, A.; Malinauskas, M.; Kontenis, L.; Purlys, V.; Paipulas, D.; Vengris, M.; Gadonas, R. Organic dye doped microstructures for optically active functional devices fabricated via two-photon polymerization technique. *Lith. J. Phys.* **2010**, *50*, 55–61.

56. Jonušauskas, L.; Lau, M.; Gruber, P.; Gokce, B.; Barcikowski, S.; Malinauskas, M.; Ovsianikov, A. Plasmon assisted 3D microstructuring of gold nanoparticle-doped polymers. *Nanotechnology* **2016**, *27*, 154001.

57. Leon-Saval, S.G.; Birks, T.A.; Bland-Hawthorn, J.; Englund, M. Multimode fiber devices with single-mode performance. *Opt. Lett.* **2005**, *30*, 2545–2547.

58. Thomson, R.R.; Birks, T.A.; Leon-Saval, S.G.; Kar, A.K.; Bland-Hawthorn, J. Ultrafast laser inscription of an integrated photonic lantern. *Opt. Express* **2011**, *19*, 5698–5705.

59. Gissibl, T.; Thiele, S.; Herkommer, A.; Giessen, H. Two-photon direct laser writing of ultracompact multi-lens objectives. *Nat. Photonics* **2016**, *10*, 554–560.

60. Juodkazis, S. Manufacturing: 3D printed micro-optics. *Nat. Photonics* **2016**, *10*, 499–501.

61. Žukauskas, A.; Malinauskas, M.; Brasselet, E. Monolithic generators of pseudo-nondiffracting optical vortex beams at the microscale. *Appl. Phys. Lett.* **2013**, *103*, 181122.

62. Rekštytė, S.; Jonavičius, T.; Gailevičius, D.; Malinauskas, M.; Mizeikis, V.; Gamaly, E.G.; Juodkazis, S. Nanoscale Precision of 3D Polymerization via Polarization Control. *Adv. Opt. Mater.* **2016**, *4*, 1209–1214.

63. Efimov, O.; Juodkazis, S.; Misawa, H. Intrinsic single and multiple pulse laser-induced damage in silicate glasses in the femtosecond-to-nanosecond region. *Phys. Rev. A* **2004**, *69*, 042903.

64. Vanagas, E.; Kudryashov, I.; Tuzhilin, D.; Juodkazis, S.; Matsuo, S.; Misawa, H. Surface nanostructuring of borosilicate glass by femtosecond nJ energy pulses. *Appl. Phys. Lett.* **2003**, *82*, 2901–2903.

7

Electrochemical Detection of Hydrogen Peroxide by Inhibiting the *p*-Benzenediboronic Acid-Triggered Assembly of Citrate-Capped Au/Ag Nanoparticles on Electrode Surface

Lin Liu *, Ting Sun and Huizhu Ren

Henan Province of Key Laboratory of New Optoelectronic Functional Materials,
College of Chemistry and Chemical Engineering, Anyang Normal University, Anyang 455000, China;
sunting@aynu.edu.cn (T.S.); ren_huizhu@sohu.com (H.R.)
* Correspondence: liulin@aynu.edu.cn

Academic Editor: Ilaria Fratoddi

Abstract: Metal nanoparticles (NPs) possess unique physicochemical attributes for creating effective recognition and transduction processes in chem/bio-sensing. In this work, we suggested that citrate-capped Au/Ag NPs could be used as the reporters for the design of hydrogen peroxide (H_2O_2) sensors with a simple manipulation principle and an easy detection procedure. Specifically, *p*-benzenediboronic acid (BDBA) induced the aggregation of citrate-capped Au NPs through the cross-linking reaction between citrate and boronic acid of BDBA in solution. By modifying the electrode with a boronic acid derivative, the BDBA-induced assembly of Au NPs was achieved on the electrode surface. This led to a significant decrease in the electron transfer resistance due to the unique conductive ability of Au NPs. However, when the boronate group on the electrode surface was oxidized into its phenol format, the assembly of Au NPs on the electrode surface was not achieved. As a result, a higher electron transfer resistance was observed. The process could be monitored by electrochemical impedance technique. Furthermore, when Ag NPs were used instead of Au NPs in this design, the H_2O_2 concentration could be determined by measuring the linear-sweep voltammetry (LSV) current through the solid-state Ag/AgCl reaction of Ag NPs. The results indicated that NP-based colorimetric assays could be developed into more sensitive electrochemical analysis.

Keywords: boronic acid; hydrogen-peroxide; metal nanoparticles; electrochemical sensors; colorimetric assay

1. Introduction

Simple, cost-effective and highly sensitive chemical and biological sensors feature two functional components. One is the recognition element providing a specific interaction with the target analyte. The other is the transducer component for the sensor signal output. Metal (Au/Ag in particular) nanoparticles (NPs) possess unique physicochemical attributes which have facilitated them in being used for creating effective recognition and transduction processes in chem/bio-sensing in the last decades [1–3]. In contrast to other materials for the design of sensors, Au/Ag NPs offer clear advantages, such as a simple preparation procedure, a size-dependent optical property, facile surface modification, excellent conductivity, and high surface area and/or good catalytic ability [4–6]. The Au/Ag NPs–based sensing techniques usually include colorimetry, fluorescence, electrochemistry, localized surface plasmon resonance, surface enhanced Raman scattering (SERS), quartz crystal microbalance, and bio-barcode assay [7–14]. Among these sensing strategies, colorimetric assays based on target recognition–induced aggregation or redispersion of Au/Ag NPs in particular have been prevalent recently because of their simple manipulation principle and easy detection procedure [8,12]. However, most colorimetric assays show low sensitivity. Thus, the colorimetric examples were expected to re-create existing platforms with improved sensitivity. By simply incorporating the colorimetric principle or technique into another field, new achievements were produced recently [14–18]. For example, based on the high quenching efficiency of dispersed but not aggregated Au NPs on the fluorescence of organic dyes or quantum dots (denoted as the efficient inner filter effect), visual and fluorescent sensors could be developed [14–16]. Furthermore, Wei et al. first suggested that the target-induced aggregation of Ag NPs in solution could be facilely initiated on the solid-liquid surface, which thus converted a liquid-phase Hg^{2+} colorimetric assay into an electrochemical analysis with improved sensitivity [18].

Hydrogen peroxide (H_2O_2) plays an important role in the immune system and functions as a signaling molecule in the regulation of a wide variety of biological processes [19,20]. The concentration change of H_2O_2 has been demonstrated to be closely correlated with many diseases, such as chronic inflammation, diabetes, neurodegenerative disorders and cancers [21,22]. Moreover, H_2O_2 is an enzymatic product used in the laboratory as an indicator to measure the target concentration [23,24]. Thus, both in vivo and in vitro determination of H_2O_2 is of great significance. The most common techniques used for H_2O_2 detection at present are fluorescence and electrochemistry which employ fluorescent probes and enzyme-/nanomaterial-modified electrodes, respectively. We noticed that a few of the fluorescent H_2O_2 sensors are based on the selective and efficient reaction between boronate groups (either boronic acid or boronate ester) and H_2O_2 [25–30]. The boronate group in a fluorescent probe is usually removed by H_2O_2 to yield its corresponding phenol form, thus resulting in an increase/decrease of the fluorescence intensity or a shift of the fluorescence peak. More interestingly, we also noticed that boronic acid can react with citrate to form a boronic acid–citrate complex and the citrate-capped Au/Ag NPs could be simply prepared [31–34]. Inspired by these facts, herein we conceived a simple electrochemical strategy for the sensitive and selective detection of H_2O_2 by converting a liquid-phase colorimetric assay into an electrochemical analysis. In the colorimetric assay, p-benzenediboronic acid (BDBA) induced the assembly of citrate-capped Au NPs through the formation of boronate ester in solution. Once BDBA was oxidized by H_2O_2 into the phenol form, it would lose the ability to trigger the assembly of Au NPs. Since the gold electrode exhibits a superficial microenvironment similar to that of Au/Ag NPs, the BDBA-induced assembly of Au NPs can be facilely initiated on the boronic acid–covered electrode surface through the formation of boronate ester. However, removing the boronate group by H_2O_2 either on the electrode surface or in BDBA would prevent the assembly of Au NPs. Since Au NPs show excellent electrical conductivity, this process could be easily monitored by electrochemical impedance spectroscopy (EIS, a powerful electrochemical technique for studying the surface process and properties) [35–37]. When Ag NPs were used instead of Au NPs, the signal could be measured by linear-sweep voltammetry (LSV) through the solid-state Ag/AgCl reaction of Ag NPs [18]. The electrochemical strategy not only features a simple manipulation principle and an easy detection procedure similar to that of the colorimetric assay but also shows high sensitivity and specificity.

2. Results and Discussion

2.1. Colorimetric Assay of H_2O_2

To demonstrate that BDBA could induce the aggregation of citrate-capped Au NPs and that H_2O_2 could inhibit the BDBA-triggered NP aggregation, we first investigated the color and UV/vis absorption change of the Au NP suspension in the presence of BDBA/H_2O_2. As shown in Figure 1A, compared to the extinction spectrum of citrate-capped Au NPs, the addition of BDBA caused the color change of the Au NP suspension from red to blue, which was accompanied by a decrease in the absorption at 520 nm and the appearance of a new absorption peak at ~660 nm. The red-shift band of the UV/vis spectra and the red-to-blue color change are characteristic of Au NP aggregation.

The aggregation should contribute to the covalent interaction between boronate in BDBA and citrate on the surface of Au NPs (Figure 1B) [31–34]. However, the mixture of $BDBA/H_2O_2$ did not cause significant changes in the color and absorption spectrum of Au NPs. This suggested that the H_2O_2-treated BDBA did not induce Au NP aggregation, which was further confirmed by the TEM observation (Figure 1C): there were aggregated Au NPs in the presence of BDBA and dispersed Au NPs in the presence of the $BDBA/H_2O_2$ mixture. The result can be attributed to the fact that H_2O_2 can react with boronate-derived group to form its phenol form [25–30]. We further investigated the influence of both BDBA and H_2O_2 concentrations on the absorption change of Au NPs. The A_{660}/A_{520} ratio (wherein A_{660} and A_{520} represent the absorption intensity of Au NPs at 660 nm and 520 nm, respectively) increased linearly in the concentration range of 3–300 μM (Figure 2A). In contrast, it decreased with an increased H_2O_2 concentration ranging from 6 to 450 μM (Figure 2B).

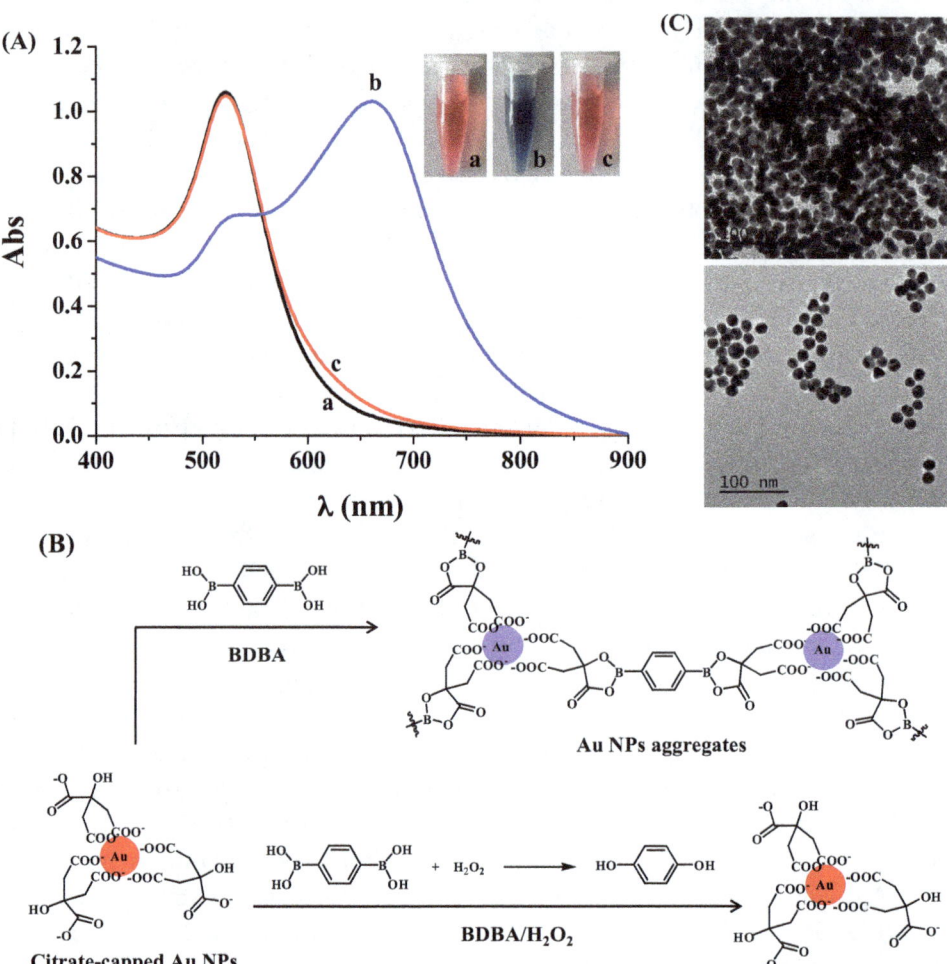

Figure 1. (**A**) UV/vis absorption spectra and optical photographs of 4 nM Au NPs in the absence (curve/tube a) and presence of BDBA (curve/tube b) or H_2O_2-treated BDBA (curve/tube c). The final concentrations of BDBA and H_2O_2 used were 0.6 mM and 1 mM, respectively; (**B**) Schematic illustration of BDBA-induced aggregation of citrate-capped Au NPs and reaction between BDBA and H_2O_2; (**C**) TEM images of Au NP suspension in the presence of BDBA (top) and H_2O_2-treated BDBA (bottom).

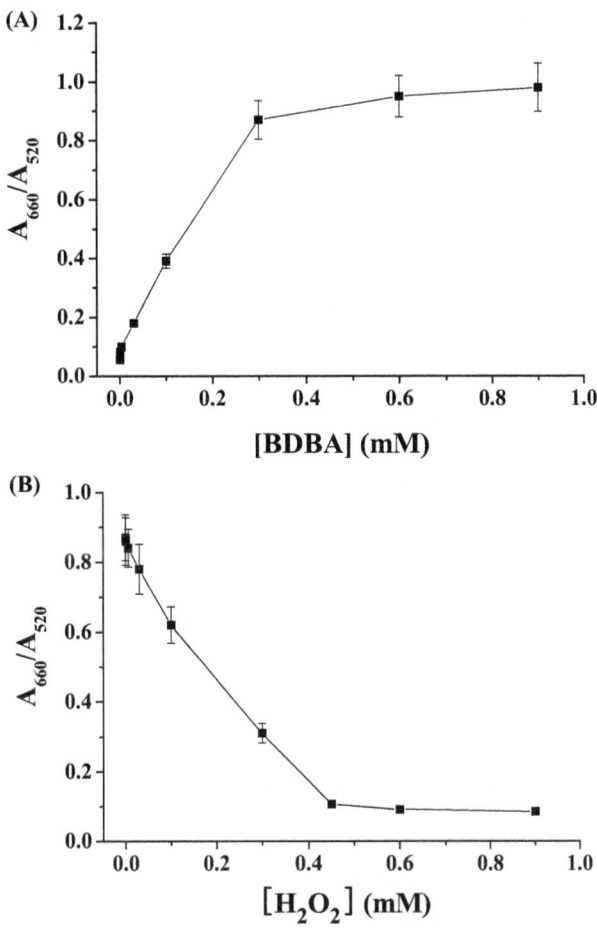

Figure 2. (A) Effect of BDBA concentration on the A_{660}/A_{520} ratio of 4 nM Au NPs; **(B)** Dependence of A_{660}/A_{520} on the concentration of H_2O_2. The final concentration of used BDBA was 0.3 mM.

2.2. Principle of Electrochemical Assay of H_2O_2 by Au NPs

The principle of the Au NP–based electrochemical method is shown in Figure 3. The boronate self-assembled monolayer (SAM) behaves as a barrier for $[Fe(CN)_6]^{3-/4-}$. According to the design, the boronate-covered gold electrode can capture citrate-capped Au NPs and BDBA molecules in solution through the formation of boronate ester bonds. Surface-tethered Au NPs and BDBA can recruit more Au NPs and peptides, thus leading to the formation of a network of NPs-BDBA-NPs-BDBA on the electrode surface. The unique conductive ability of Au NPs may result in a significant change in the charge transfer resistance. However, once the electrode was incubated with H_2O_2, boronate groups on the electrode surface were removed. As a result, the citrate-capped Au NPs could not be captured by the electrode. Note that citrate-capped Au NPs maybe absorb other components in a biological sample and boronate can react with diol-containing biomolecules (e.g., sugar, catechol derivatives and glycoproteins) [38–40]; thus, the detection of H_2O_2 was performed by incubating the electrode with the H_2O_2 sample before introducing Au NPs and BDBA on the electrode surface. Although boronic acid can covalently react with diol-containing biomolecules to form five- or six-membered cyclic ester in an alkaline aqueous solution, the cyclic ester can dissociate when the medium is changed to acidic pH. Thus, the captured diol-containing biomolecules could be detached by rinsing the electrode with 10 mM HCl after the step of treatment by the H_2O_2 sample, thus avoiding the interference of diol-containing biomolecules and facilitating the formation of a network of citrate-capped Au NPs.

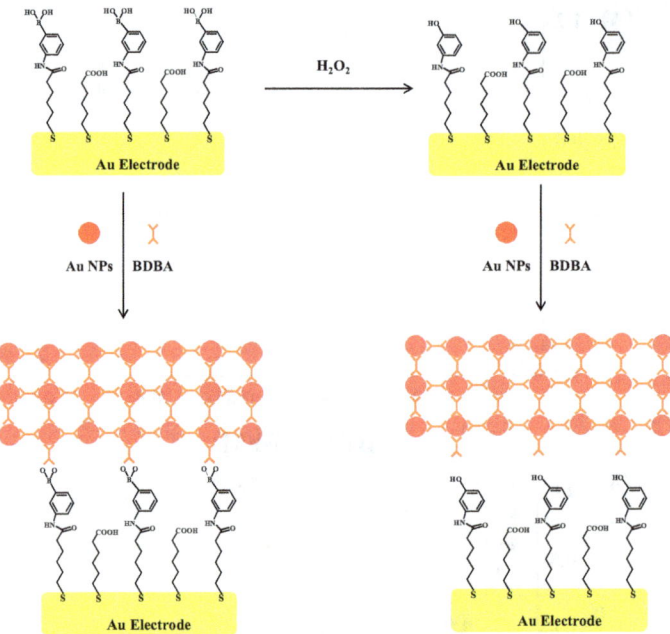

Figure 3. Illustration of the electrochemical strategy for H_2O_2 based on BDBA-induced assembly of citrate-capped Au NPs on boronate-covered gold electrode surface.

2.3. Electrochemical Detection of H_2O_2 Based on the Signal Amplification of Au NPs

EIS has been commonly employed to monitor the molecular assembly on electrode surfaces. Herein, the sensing performances of the modified electrode were characterized by EIS. Figure 4A shows the EIS spectrum of the sensing electrode in different states. Incubating the boronate-covered electrode with the Au NP suspension caused a slight decrease in the diameter of the semicircle of the impedance spectrum (cf. curve a and curve b). The decrease suggested that Au NPs attached onto the electrode surface facilitated the electron transfer. This is attributed to the unique electrical property of Au NPs. Interestingly, incubating the boronate-covered electrode with the mixture of Au NPs and BDBA led to a much smaller semicircle portion in the impedance spectrum (curve d), but no apparent change was observed when incubating the electrode with BDBA only (curve c). The result suggested that more Au NPs were assembled on the electrode surface in the presence of BDBA. The electron transfer resistance (R_{et}) in curve d is much lower than that in curve b, indicating that the signal could be amplified by the network of Au NPs and the BDBA-induced assembly of Au NPs in solution was facilely initiated on the electrode surface. However, incubation of the H_2O_2-treated boronate-covered electrode with the mixture of Au NPs and BDBA did not cause a decrease in the diameter of the semicircle (curve e), indicating that the Au NPs were incapable of assembly on the electrode surface after boronate was converted into its phenol form. Thus, H_2O_2 could be determined based on the change of the charge transfer resistance of the sensing electrode.

Although a high level of BDBA made the aggregation of Au NPs more powerful, a high level of BDBA in the solution could compete with the boronate on the electrode surface to bind citrate-capped Au NPs, thus reducing the sensitivity. Thus, we studied the impact of the concentration ratio of BDBA to Au NPs ([BDBA]/[Au NPs]) on the R_{et}. It was found that R_{et} decreased and then increased with the increasing [BDBA]/[Au NPs] ratio (Figure 4B). The minimum value appeared at 4500:1. In the following detection assay, 4500:1 was chosen as the optimal concentration ratio. The H_2O_2 quantitative detection was performed by monitoring the change of R_{et} (ΔR_{et}). It was observed that ΔR_{et} increased with the increase of the H_2O_2 concentration ([H_2O_2]). The value is proportional to [H_2O_2] in the range of 1 nM~0.6 μM. The regression equation is $\Delta R_{et} = 3731[H_2O_2]$ (μM) + 41, $R = 0.997$. Thus, the Au NPs–based colorimetric assay was converted into an electrochemical analysis with improved sensitivity.

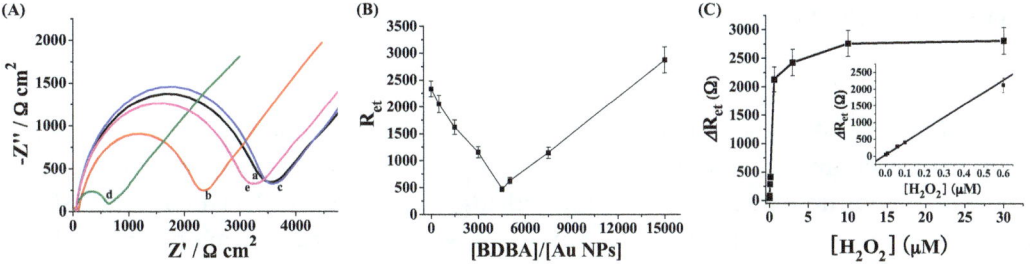

Figure 4. (**A**) EIS of the boronate-covered electrodes before (curve a) and after incubation with Au NPs (curve b), BDBA (curve c) or the mixture of Au NPs and BDBA (curve d). Curve e corresponds to that of the H_2O_2-treated boronate-covered electrode after incubation with the mixture of Au NPs and BDBA. The final concentrations of Au NPs, BDBA and H_2O_2 used were 2 nM, 10 μM and 10 μM, respectively; (**B**) Effect of [peptide]/[Au NPs] ratio on R_{et}; (**C**) Dependence of ΔR_{et} on H_2O_2 concentration (0.001, 0.01, 0.06, 0.1, 0.6, 3, 10 and 30 μM). The inset shows the linear part of the curve. The concentrations of Au NPs and BDBA used were 2 and 9 μM, respectively.

2.4. Electrochemical Detection of H_2O_2 with Ag NPs as the Redox Reporters

Ag NPs have been widely used as the electrochemical elements for signal outputs based on their solid-state Ag/AgCl reaction [41,42]. Particularly, the network of Ag NPs showed amplified LSV signals [13,18]. When Ag NPs were used instead of Au NPs in the proposed method, we found that the electrochemical signal could be measured by LSV based on the solid-state Ag/AgCl reaction from Ag NPs. As shown in Figure 5A, the electrochemical response of the boronate-covered electrode after incubation with Ag NPs and BDBA revealed a reduced peak at around 70 mV (curve a). The peak current was much higher than that after incubation with Ag NPs only (curve b), demonstrating that the signal was amplified by the network of Ag NPs. Expectedly, the H_2O_2-treated sensing electrode showed no significant LSV peak after incubation with the mixture of Ag NPs and BDBA (curve c). The result further confirmed that the assembly of Ag NPs on the electrode surface is dependent upon the formation of boronate ester bonds. We also found that the optimal concentration ratio of BDBA to Ag NPs was 4000:1. Under the optimized ratio, the LSV responses for the determination of various concentrations of H_2O_2 were collected. The quantitative assay was measured based on the LSV current change (ΔI). It was observed that ΔI increased with the increase of the H_2O_2 concentration ($[H_2O_2]$) in a linear range of 1 nM ~0.6 μM (Figure 5B). The regression equation is $\Delta I = 13.3[H_2O_2]$ (μM) + 0.6, $R = 0.998$.

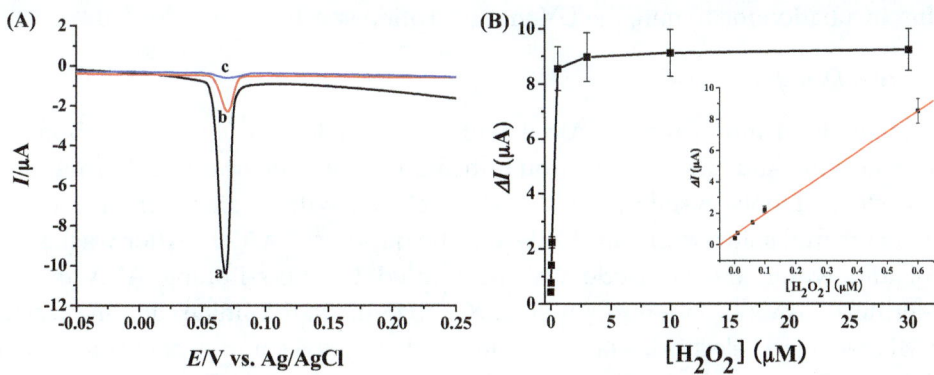

Figure 5. (**A**) LSV responses of the boronate-covered electrodes after incubation with the mixture of Ag NPs and BDBA (curve a) or Ag NPs only (curve b). Curve c corresponds to that of H_2O_2-treated boronate-covered electrode after incubation with the mixture of Ag NPs and BDBA. The final concentrations of Ag NPs, BDBA and H_2O_2 used were 2 nM, 8 μM and 3 μM, respectively; (**B**) Dependence of ΔI on H_2O_2 concentration (0.001, 0.01, 0.06, 0.1, 0.6, 3, 10 and 30 μM). The inset shows the linear part of the curve.

3. Materials and Methods

3.1. Reagents and Materials

The 6-mercaptohexanoic acid (MHA), 3-aminobenzeneboronic acid (ABA), BDBA, *N*-hydroxysulfosuccinimide (NHS), 1-ethyl-3-[3-dimethylaminopropyl]carbodiimide (EDC) hydrochloride, trisodium citrate, sodium borohydride, ethanolamine, KH_2PO_4, and K_2HPO_4 were obtained from Sigma-Aldrich (Shanghai, China). All other chemicals were of analytical grade and obtained from Beijing Chemical Reagent Co. Ltd. (Beijing, China). BDBA and H_2O_2 were diluted with a phosphate-buffered saline solution (PBS buffer, 5 mM, pH 8.2) before use. All solutions were prepared with deionized water purified by using a Millipore system (Simplicity Plus, Millipore Corp., Billerica, MA, USA). The citrate-stabilized Au NPs with a size of 13 nm were synthesized with a trisodium citrate reduction method. The concentration of Au NPs was calculated from the UV-vis absorption spectrum by using an extinction coefficient of 2.7×10^8 $mol^{-1} \cdot cm^{-1}$ at 520 nm. Ag NPs were prepared by the chemical reduction of $AgNO_3$ using sodium borohydride and trisodium citrate as the reducing reagent and the stabilizer, respectively [13]. The Ag NPs concentration was calculated based on the Ag NPs size and the Ag^+ concentration.

3.2. Instruments

The UV-vis spectra were measured by a Cary 60 spectrophotometer using a 1 cm quartz spectrophotometer cell. The electrochemical measurements were conducted on a CHI 660E (CH Instruments, Shanghai, China) electrochemical workstation. The auxiliary and reference electrodes were a platinum wire and an electrode of Ag/AgCl, respectively. The distribution images of Au NPs were recorded by an FEI Tecnai G2 T20 transmission electron microscopy (TEM). The hydrodynamic diameter of the Ag NPs was measured by a Nano ZS laser scattering particles size analyzer (Malvern Instruments Ltd., Malvern, Worcestershire, UK). The photograph images were taken by a Sony Cyber-Shot digital camera (Sony Corp., Tokyo, Japan).

3.3. Colorimetric Detection of H_2O_2

To investigate the BDBA-induced Au NPs aggregation, 400 µL of Au NPs suspensions were added to 600 µL of different concentrations of BDBA solutions. After incubation for 10 min, the UV-vis absorption spectra were collected on the spectrophotometer. For the colorimetric assay of H_2O_2, 300 µL of H_2O_2 solution at a given concentration was first incubated with 300 µL of BDBA solution at ambient temperature for 1 min. Then, 400 µL of citrate-capped Au NPs suspension was added into the mixed solution. After incubation for 10 min, the UV-vis absorption spectra were recorded.

3.4. Electrochemical Detection of H_2O_2

The self-assembled monolayers (SAMs) with medium length (e.g., six carbons) can avoid non-specific absorption and allow for electron transfer. For the preparation of boronic acid-covered electrode, the cleaned polycrystalline gold disk electrode with a diameter of 2 mm) was first immersed in an ethanol solution of 1 mM MHA in the darkness for 12 h. After washed thoroughly with ethanol and water, the electrode was performed by cross-linking ABA molecules onto the carboxy-terminal SAMs surface through the EDC/NHS-mediated amine coupling reaction [43–45]. In brief, the MHA-covered electrode was incubated with a mixed solution comprising of 0.4 M EDC and 0.2 M NHS for 15 min, washed with water and soaked in a ABA solution (0.5 mM) for 4 h. To block the unreacted EDC/NHS-activated carboxy groups, the modified electrode was then soaked in a 1 mM ethanolamine solution for 10 min. After formation of boronic acid-modified SAMs on the electrode surface, the sensing electrode was immersed in a H_2O_2 solution for 1 min. After rinsed with 10 mM hydrochloric acid and water, the sensing electrode was exposed to 40 µL of Au NPs suspension in a homemade plastic cell, followed by addition of 60 µL of BDBA to incubation for 15 min. After washed with water again, the electrode was placed in a mixture of 10 mM $[Fe(CN)_6]^{3-}/^{4-}$ (1:1) and 0.5 M KCl for impedance measurement. The potential was set at 0.245 V and the frequency ranging from 0.01 to 500 kHz. For the LSV experiments, Ag NPs were used instead of Au NPs and the electrochemical signal was collected in a 1 M KCl solution. Other experimental conditions and detection procedures were the same as those of the Au NP–based sensing strategy.

4. Conclusions

In this work, a metal NP–based colorimetric assay was converted into sensitive electrochemical analysis for H_2O_2 detection with a simple manipulation principle and an easy detection procedure. The method is based on two facts: (1) BDBA could induce the assembly of citrate-capped Au/Ag NPs via the boronate-citrate interaction; and (2) the boronate group could be removed by H_2O_2 to yield its corresponding phenol form, thus preventing the assembly of Au/Ag NPs. Since H_2O_2 is an enzymatic product, the sensing electrode could be used to measure H_2O_2 levels in laboratory studies and to design novel chem/bio-sensors. We also envision that the method would be expanded to quantify the levels of exogenous and endogenous H_2O_2 in living cells by using boronate-modified nanometer-sized electrodes.

Acknowledgments: Partial support of this work by the National Natural Science Foundation of China (21205003) and the Program for Science and Technology Innovation Talents at the University of Henan Province (15HASTIT001) is gratefully acknowledged.

Author Contributions: Lin Liu conceived and designed the experiments, analyzed the data and wrote the paper; Ting Sun performed the colorimetric and electrochemical experiments; Huizhu Ren performed parts of the electrochemical experiments.

Conflicts of Interest: The authors declare no conflict of interest.

References

1. Saha, K.; Agasti, S.S.; Kim, C.; Li, X.; Rotello, V.M. Gold nanoparticles in chemical and biological sensing. *Chem. Rev.* **2012**, *112*, 2739–2779. [CrossRef] [PubMed]

2. Ronkainen, N.J.; Okon, S.L. Nanomaterial-based electrochemical immunosensors for clinically significant biomarkers. *Materials* **2014**, *7*, 4669–4709. [CrossRef]

3. Hayat, A.; Catanante, G.; Marty, J.L. Current trends in nanomaterial-based amperometric biosensors. *Sensors* **2014**, *14*, 23439–23461. [CrossRef] [PubMed]

4. Wilson, R. The use of gold nanoparticles in diagnostics and detection. *Chem. Soc. Rev.* **2008**, *37*, 2028–2045. [CrossRef] [PubMed]

5. Sapsford, K.E.; Algar, W.R.; Berti, L.; Gemmill, K.B.; Casey, B.J.; Oh, E.; Stewart, M.H.; Medintz, I.L. Functionalizing nanoparticles with biological molecules: Developing chemistries that facilitate nanotechnology. *Chem. Rev.* **2006**, *113*, 1904–2074. [CrossRef] [PubMed]

6. Khalil, I.; Julkapli, N.M.; Yehye, W.A.; Basirun, W.J.; Bhargava, S.K. Graphene-gold nanoparticles hybrid-synthesis, functionalization, and application in a electrochemical and surface-enhanced Raman scattering biosensor. *Materials* **2016**, *9*, 406. [CrossRef]

7. Bülbül, G.; Hayat, A.; Andreescu, S. Portable nanoparticle-based sensors for food safety assessment. *Sensors* **2015**, *15*, 30736–30758. [CrossRef] [PubMed]

8. Elghanian, R.; Storhoff, J.J.; Mucic, R.C.; Letsinger, R.L.; Mirkin, C.A. Selective colorimetric detection of polynucleotides based on the distance-dependent optical properties of gold nanoparticles. *Science* **1997**, *277*, 1078–1081. [CrossRef] [PubMed]

9. Liu, L.; Zhao, F.; Ma, F.; Zhang, L.; Yang, S.; Xia, N. Electrochemical detection of β-amyloid peptides on electrode covered with N-terminus-specific antibody based on electrocatalytic O_2 reduction by Aβ(1–16)-heme-modified gold nanoparticles. *Biosens. Bioelectron.* **2013**, *49*, 231–235. [CrossRef] [PubMed]

10. Liu, P.D.; Jin, H.Z.; Guo, Z.R.; Ma, J.; Zhao, J.; Li, D.D.; Wu, H.; Gu, N. Silver nanoparticles outperform gold nanoparticles in radiosensitizing U251 cells in vitro and in an intracranial mouse model of glioma. *Int. J. Nanomed.* **2016**, *11*, 5003–5013. [CrossRef] [PubMed]

11. McFarland, A.D.; Van Duyne, R.P. Single silver nanoparticles as real-time optical sensors with zeptomole sensitivity. *Nano Lett.* **2003**, *3*, 1057–1062. [CrossRef]

12. Nam, J.-M.; Thaxton, C.S.; Mirkin, C.A. Nanoparticle-based bio-bar codes for the ultrasensitive detection of proteins. *Science* **2003**, *301*, 1884–1886. [CrossRef] [PubMed]

13. Xia, N.; Wang, X.; Zhou, B.; Wu, Y.; Mao, W.; Liu, L. Electrochemical detection of amyloid-β oligomers based on the signal amplification of a network of silver nanoparticles. *ACS Appl. Mater. Interfaces* **2016**, *8*, 19303–19311. [CrossRef] [PubMed]

14. Xia, N.; Zhou, B.; Huang, N.; Jiang, M.; Zhang, J.; Liu, L. Visual and fluorescent assays for selective detection of beta-amyloid oligomers based on the inner filter effect of gold nanoparticles on the fluorescence of CdTe quantum dots. *Biosens. Bioelectron.* **2016**, *85*, 625–632. [CrossRef] [PubMed]

15. Chang, H.C.; Ho, J.A.A. Gold nanocluster-assisted fluorescent detection for hydrogen peroxide and cholesterol based on the inner filter effect of gold nanoparticles. *Anal. Chem.* **2015**, *87*, 10362–10367. [CrossRef] [PubMed]

16. Xu, J.; Yu, H.; Hu, Y.; Chen, M.Z.; Shao, S.J. A gold nanoparticle-based fluorescence sensor for high sensitive and selective detection of thiols in living cells. *Biosens. Bioelectron.* **2016**, *75*, 1–7. [CrossRef] [PubMed]

17. Xia, N.; Wang, X.; Wang, X.; Zhou, B. Gold nanoparticle-based colorimetric and electrochemical methods for dipeptidyl peptidase-IV activity assay and inhibitor screening. *Materials* **2016**, *9*, 857. [CrossRef]

18. Wei, T.; Dong, T.; Wang, Z.; Bao, J.; Tu, W.; Dai, Z. Aggregation of individual sensing units for signal accumulation: Conversion of liquid-phase colorimetric assay into enhanced surface-tethered electrochemical analysis. *J. Am. Chem. Soc.* **2015**, *137*, 8880–8883. [CrossRef] [PubMed]

19. Sobotta, M.C.; Liou, W.; Stocker, S.; Talwar, D.; Oehler, M.; Ruppert, T.; Scharf, A.N.; Dick, T.P. Peroxiredoxin-2 and STAT3 form a redox relay for H_2O_2 signaling. *Nat. Chem. Biol.* **2015**, *11*, 64–70. [CrossRef] [PubMed]

20. Lippert, A.R.; Van de Bittner, G.C.; Chang, C. Boronate oxidation as a bioorthogonal reaction approach for studying the chemistry of hydrogen peroxide in living systems. *Acc. Chem. Res.* **2011**, *44*, 793–804. [CrossRef] [PubMed]

21. Noh, J.; Kwon, B.; Han, E.; Park, M.; Yang, W.; Cho, W.; Yoo, W.; Khang, G.; Lee, D. Amplification of oxidative stress by a dual stimuli-responsive hybrid drug enhances cancer cell death. *Nat. Commun.* **2015**, *6*, 6907. [CrossRef] [PubMed]

22. Li, R.; Liu, X.; Qiu, W.; Zhang, M. In vivo monitoring of H_2O_2 with polydopamine and prussian blue-coated microelectrode. *Anal. Chem.* **2016**, *88*, 7769–7776. [CrossRef] [PubMed]

23. Gu, X.; Wang, H.; Schultz, Z.D.; Camden, J.P. Sensing glucose in urine and serum and hydrogen peroxide in living cells by use of a novel boronate nanoprobe based on surface-enhanced Raman spectroscopy. *Anal. Chem.* **2016**, *88*, 7191–7197. [CrossRef] [PubMed]

24. Wang, Y.; Hasebe, Y. Carbon felt-based bioelectrocatalytic flow-through detectors: 2,6-Dichlorophenol indophenol and peroxidase coadsorbed carbon-felt for flow-amperometric determination of hydrogen peroxide. *Materials* **2014**, *7*, 1142–1154. [CrossRef]

25. De Gracia Lux, C.; Joshi-Barr, S.; Nguyen, T.; Mahmoud, E.; Schopf, E.; Fomina, N.; Almutairi, A. Biocompatible polymeric nanoparticles degrade and release cargo in response to biologically relevant levels of hydrogen peroxide. *J. Am. Chem. Soc.* **2012**, *134*, 15758–15764. [CrossRef] [PubMed]

26. Michalski, R.; Zielonka, J.; Gapys, E.; Marcinek, A.; Joseph, J.; Kalyanaraman, B. Real-time measurements of amino acid and protein hydroperoxides using coumarin boronic acid. *J. Biol. Chem.* **2014**, *289*, 22536–22553. [CrossRef] [PubMed]

27. Van de Bittner, G.C.; Dubikovskaya, E.A.; Bertozzi, C.R.; Chang, C.J. In vivo imaging of hydrogen peroxide production in a murine tumor model with a chemoselective bioluminescent reporter. *Proc. Natl. Acad. Sci. USA* **2010**, *107*, 21316–21321. [CrossRef] [PubMed]

28. Webb, K.S.; Levy, D. A facile oxidation of boronic acids and boronic esters. *Tetrahedron Lett.* **1995**, *36*, 5117–5118. [CrossRef]

29. Zhang, H.; Ren, T.; Ji, Y.; Han, L.; Wu, Y.; Song, H.; Bai, L.; Ba, X. Selective modification of halloysite nanotubes with 1-pyrenylboronic acid: A novel fluorescence probe with highly selective and sensitive response to hyperoxide. *ACS Appl. Mater. Interfaces* **2015**, *7*, 23805–23811. [CrossRef] [PubMed]

30. Miller, E.W.; Albers, A.E.; Pralle, A.; Isacoff, E.Y.; Chang, C.J. Boronate-based fluorescent probes for imaging cellular hydrogen peroxide. *J. Am. Chem. Soc.* **2005**, *127*, 16652–16659. [CrossRef] [PubMed]

31. Manimala, J.C.; Wiskur, S.L.; Ellington, A.D.; Anslyn, E.V. Tuning the specificity of a synthetic receptor using a selected nucleic acid receptor. *J. Am. Chem. Soc.* **2004**, *126*, 16515–16519. [CrossRef] [PubMed]

32. Bosch, L.I.; Fyles, T.M.; James, T.D. Binary and ternary phenylboronic acid complexes with saccharides and Lewis bases. *Tetrahedron* **2004**, *60*, 11175–11190. [CrossRef]

33. Wiskur, S.L.; Lavigne, J.J.; Metzger, A.; Tobey, S.L.; Lynch, V.; Anslyn, E.V. Thermodynamic analysis of receptors based on guanidinium/boronic acid groups for the complexation of carboxylates, a-hydroxycarboxylates, and diols: Driving force for binding and cooperativity. *Chem. Eur. J.* **2004**, *10*, 3792–3804. [CrossRef] [PubMed]

34. Yang, Y.-C.; Tseng, W.-L. 1,4-Benzenediboronic-acid-induced aggregation of gold nanoparticles: Application to hydrogen peroxide detection and biotin-avidin-mediated immunoassay with naked-eye detection. *Anal. Chem.* **2016**, *88*, 5355–5362. [CrossRef] [PubMed]

35. Miao, P.; Wang, B.; Han, K.; Tang, Y. Electrochemical impedance spectroscopy study of proteolysis using unmodified gold nanoparticles. *Electrochem. Commun.* **2014**, *47*, 21–24. [CrossRef]

36. Yang, Y.; Li, C.; Yin, L.; Liu, M.; Wang, Z.; Shu, Y.; Li, G. Enhanced charge transfer by gold nanoparticle at DNA modified electrode and its application to label-free DNA detection. *ACS Appl. Mater. Interfaces* **2014**, *6*, 7579–7584. [CrossRef] [PubMed]

37. Zhao, J.; Zhu, X.; Li, T.; Li, G. Self-assembled multilayer of gold nanoparticles for amplified electrochemical detection of cytochrome c. *Analyst* **2008**, *133*, 1242–1245. [CrossRef] [PubMed]

38. Anzai, J. Recent progress in electrochemical biosensors based on phenylboronic acid and derivatives. *Mater. Sci. Eng. C-Mater.* **2016**, *67*, 737–746. [CrossRef] [PubMed]

39. Egawa, Y.; Miki, R.; Seki, T. Colorimetric sugar sensing using boronic acid-substituted azobenzenes. *Materials* **2014**, *7*, 1201–1220. [CrossRef]

40. Liu, L.; Xia, N.; Liu, H.P.; Kang, X.J.; Liu, X.S.; Xue, C.; He, X.L. Highly sensitive and label-free electrochemical detection of microRNAs based on triple signal amplification of multifunctional gold nanoparticles, enzymes and redox-cycling reaction. *Biosens. Bioelectron.* **2014**, *53*, 399–405. [CrossRef] [PubMed]

41. Lin, D.; Wu, J.; Wang, M.; Yan, F.; Ju, H. Triple signal amplification of graphene film, polybead carried gold nanoparticles as tracing tag and silver deposition for ultrasensitive electrochemical immunosensing. *Anal. Chem.* **2012**, *84*, 3662–3668. [CrossRef] [PubMed]

42. Singh, P.; Parent, K.L.; Buttry, D.A. Electrochemical solid-state phase transformations of silver nanoparticles. *J. Am. Chem. Soc.* **2012**, *134*, 5610–5617. [CrossRef] [PubMed]

43. Liu, L.; Deng, D.; Xing, Y.; Li, S.; Yuan, B.; Chen, J.; Xia, N. Activity analysis of the carbodiimide-mediated amine coupling reaction on self-assembled monolayers by cyclic voltammetry. *Electrochim. Acta* **2013**, *89*, 616–622. [CrossRef]

44. Liu, S.; Wollenberger, U.; Halámek, J.; Leupold, E.; Stçcklein, W.; Warsinke, A.; Scheller, F.W. Affinity Interactions between phenylboronic acid-carrying self-assembled monolayers and flavin adenine dinucleotide or horseradish peroxidase. *Chem. Eur. J.* **2005**, *11*, 4239–4246. [CrossRef] [PubMed]

45. Yildiz, H.B.; Freeman, R.; Gill, R.; Willner, I. Electrochemical, photoelectrochemical, and piezoelectric analysis of tyrosinase activity by functionalized nanoparticles. *Anal. Chem.* **2008**, *80*, 2811–2816. [CrossRef] [PubMed]

8

Microwave-Assisted Polyol Synthesis of Water Dispersible Red-Emitting Eu^{3+}-Modified Carbon Dots

Hailong Dong [1], Ana Kuzmanoski [1], Tobias Wehner [2], Klaus Müller-Buschbaum [2] and Claus Feldmann [1,*]

[1] Karlsruhe Institute of Technology (KIT), Institut für Anorganische Chemie, Engesserstrasse 15, 76131 Karlsruhe, Germany; hailong.dong@kit.edu (H.D.); anan19nana85@hotmail.com (A.K.)

[2] Institute of Inorganic Chemistry, University of Würzburg, Am Hubland, D-97074 Würzburg, Germany; tobias.wehner@uni-wuerzburg.de (T.W.); k.mueller-buschbaum@uni-wuerzburg.de (K.M.-B.)

* Correspondence: claus.feldmann@kit.edu

Academic Editor: Javier Narciso

Abstract: Eu^{3+}-modified carbon dots (C-dots), 3–5 nm in diameter, were prepared, functionalized, and stabilized via a one-pot polyol synthesis. The role of Eu^{2+}/Eu^{3+}, the influence of O$_2$ (oxidation) and H$_2$O (hydrolysis), as well as the impact of the heating procedure (conventional resistance heating and microwave (MW) heating) were explored. With the reducing conditions of the polyol at the elevated temperature of synthesis (200–230 °C), first of all, Eu^{2+} was obtained resulting in the blue emission of the C-dots. Subsequent to O$_2$-driven oxidation, Eu^{3+}-modified, red-emitting C-dots were realized. However, the Eu^{3+} emission is rapidly quenched by water for C-dots prepared via conventional resistance heating. In contrast to the hydroxyl functionalization of conventionally-heated C-dots, MW-heating results in a carboxylate functionalization of the C-dots. Carboxylate-coordinated Eu^{3+}, however, turned out as highly stable even in water. Based on this fundamental understanding of synthesis and material, in sum, a one-pot polyol approach is established that results in H$_2$O-dispersable C-dots with intense red Eu^{3+}-line-type emission.

Keywords: carbon dot; europium; microwave; polyol; surface conditioning

---

1. Introduction

Carbon dots (C-dots) have recently attracted considerable attention due to their unique properties (e.g., inexpensive nature, chemical stability, adaptable surface functionalization, high biocompatibility, intense photoluminescence (PL)) and a wide range of potential applications (e.g., bioimaging/biosensing, optoelectronics, catalysis) [1–6]. Typically, C-dots show intense, broad PL in the blue to green spectral range [7–9]. Intense and stable red emission, in contrast, was only reported in few papers [10,11]. Efficient red emission, on the other hand, is most essential for full-colour emission and additive colour mixing to white light, as well as for biomedical application since long-wavelength excitation and emission are less harmful for tissue and show deep-tissue penetration. Finally, green autofluorescence of tissue is avoided [12].

Recently, we could present lanthanide-modified C-dots and their PL for the first time [13,14]. Thus, Eu^{3+}- and Tb^{3+}-modified C-dots were prepared via the polyol method [15] by in situ thermal decomposition of the solvent (e.g., polyethylene glycol 400/PEG400) and showed excellent quantum yields for line-type red (75%) and green (85%) emission [13]. Meanwhile PEG-modified C-dots are considered as most promising for biomedical applications [16]. Although line-type red f-f emission of Eu^{3+}-modified C-dots is highly promising, several restrictions limit their use by now: (i) the PL is of limited reproducibility; (ii) the emission is rapidly quenched by humidity, which is not acceptable for biomedical application; and (iii) the achievable yield of C-dots is very low.

To protect Eu^{3+} centers against H_2O-driven quenching, advanced synthesis techniques and sophisticated coordination chemistry were suggested [17–19]. Thus, Zhou et al. have modified C-dots by attaching Eu^{3+}-coordination complexes with diethylenetriamine pentaacetic acid as a ligand [17]. Song et al. have modified the C-dot surface by self-assembled Eu^{3+} and 5'-guanosine monophosphate [18]. Ye et al. pre-synthesized the coordination complex 4,4'-bis (1'',1'',1'',2'',2'',3'',3''-heptafluoro-4'',6''-hexanedion-6''-yl)chloro-sulfo-o-terphenyl-Eu^{3+}, which was, thereafter, deposited on the C-dot surface [19]. An alternative approach, moreover, suggested nanocomposites composed of Eu^{3+}-doped LaF_3 and C-dots and results in a quantum yield of 11% [20].

Taken together, fundamental understanding of the PL of Eu^{3+}-modified C-dots and their stabilization is still lacking but essential for rational material optimization. In the following, we illustrate the role of Eu^{2+}/Eu^{3+}, the influence of O_2 and H_2O as well as the remarkable difference between conventional resistance heating and MW-heating on the surface functionalization and the resulting PL (Figure 1). With this knowhow, a one-pot, MW-mediated polyol synthesis of Eu^{3+}-modified C-dots showing stable red emission in water is realized.

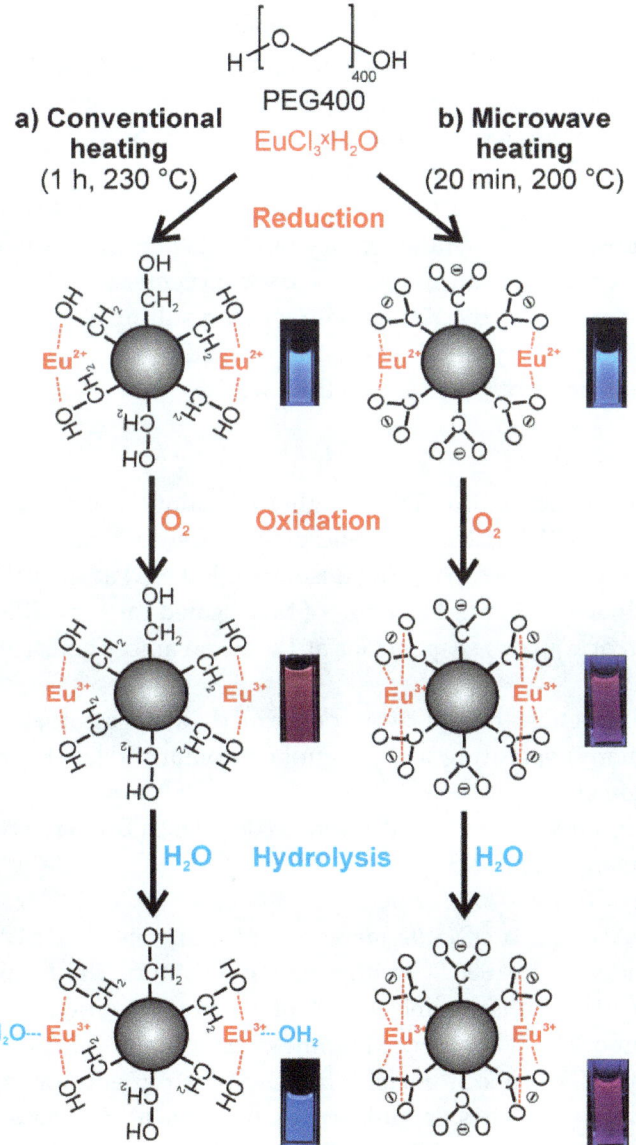

Figure 1. One-pot polyol synthesis of Eu-modified C-dots with the role of Eu^{2+}, Eu^{3+}, O_2, H_2O, and the influence of (**a**) conventional resistance heating and (**b**) MW-heating on the surface functionalization and PL (excitation via UV-LED, λ_{max} = 365 nm, SI: Figure S2).

2. Experimental

2.1. Synthesis

Synthesis of Eu^{3+}-modified C-dots via conventional resistance heating: In a standard recipe for preparing Eu^{3+}-modified C-dots, 0.5 mmol of $EuCl_3 \times 6H_2O$ were dissolved in 10 mL of PEG400. This solution was heated via a heating mantle in a round-bottomed flask to 230 °C in an argon atmosphere [13]. This temperature was maintained for 1 h resulting in a transparent and colloidally stable suspension of Eu-modified C-dots in excess PEG400. Due to the reducing properties of the polyol at high temperature, Eu^{3+} was reduced to Eu^{2+}. To study the re-oxidation of Eu^{2+} to Eu^{3+}, dry air was bubbled through the as-prepared suspension (30–35 air bubbles per minute). This re-oxidation is slow and proceeds on a timescale of several days. To evaluate the influence of humidity and water, a low amount of water (0.2 mL) was added.

Synthesis of Eu^{3+}-modified C-dots via microwave (MW) heating: In a standard recipe for preparing Eu^{3+}-modified C-dots, 0.2 mmol of $EuCl_3 \times 6H_2O$ were dissolved in 50 mL of PEG400. This solution was heated via a microwave oven (1200 W, MLS Rotaprep, MLS, Leutkirch, Germany) in a round-bottomed flask to 200 °C in an argon atmosphere. This temperature was reached in 3 min to 200 °C (at 1200 W) and, thereafter, kept for 20 min (at 800 W). MW heating resulted in an opaque suspension of Eu-modified C-dots in excess PEG400. It is to be noted that a shorter MW treatment (e.g., 10 min) results in very limited amounts of C-dots, whereas a longer period of MW treatment (e.g., 30 min) leads to significantly larger C-dots that are known for weak luminescence. In contrast conventional heating, the Eu^{3+}-modified C-dots made via MW-heating do not show any sensitivity to humidity and water. Subsequent to synthesis, they can be dispersed in water and—in contrast to the conventional heating procedure—still show Eu^{3+}-based red emission. Accordingly, the MW-heated Eu^{3+}-modified C-dots can be directly diluted with water resulting in slightly yellowish, colloidally stable suspensions. Alternatively, yellowish powders were collected after centrifugation and washing (i.e., three times redispersion/centrifugation in/from water).

2.2. Analytical Tools

Transmission electron microscopy (TEM), high-angle annular dark-field scanning transmission electron microscopy (HAADF-STEM), and energy dispersive X-ray spectroscopy (EDXS) were conducted with a FEI Osiris microscope (FEI, Hillsboro, OR, USA) at 200 kV, equipped with a Bruker Quantax system (XFlash detector). TEM samples of MW-heated Eu^{3+}-modified C-dots were prepared by vacuum evaporation of aqueous suspensions at 120 °C on amorphous carbon (lacey-) film-coated copper grids.

X-ray powder diffraction (XRD) was performed with a Stoe STADI-P diffractometer (Stoe, Darmstadt, Germany) operating with Ge-monochromatized Cu-Kα-radiation (λ = 1.54178 Å) and Debye-Scherrer geometry.

Fourier-transform infrared spectra (FT-IR) were recorded on a Bruker Vertex 70 FT-IR spectrometer (Bruker, Ettlingen, Germany) using KBr pellets.

Thermogravimetry (TG) was performed with a Netzsch STA 409C instrument (Netzsch, Selb, Germany) applying α-Al_2O_3 as a crucible material and reference. The MW-heated Eu^{3+}-modified C-dots were heated under air to 1000 °C with a rate of 5 K/min. The resulting data were baseline corrected by subtracting the results of a measurement of an empty crucible.

Fluorescence lifetime: The fluorescence lifetimes were obtained as process decay times with an Edinburgh Instruments (FLS920) spectrometer (Edinburgh Instruments, Livingston, UK). The samples were prepared in quartz glass cuvettes under inert-gas atmosphere. The decay times were recorded by time-correlated single-photon counting (TCSPC) with a 375 nm pulsed laser diode or a microsecond flash lamp) with an excitation wavelength of 375 nm. The fluorescence emission was collected at right angles to the excitation source, and the emission wavelength was selected with a monochromator and detected by a single-photon avalanche diode (SPAD). The resulting intensity decays were calculated through tail fit analysis (Edinburgh F900 analysis software). The quality of the fits was evidenced by low χ^2 values ($\chi^2 < 1.4$).

Fluorescence spectroscopy (FL) and determination of quantum yield: Excitation and emission spectra were recorded using a photoluminescence spectrometer Horiba Jobin Yvon Spex Fluorolog 3 (Horiba Jobin Yvon, Bensheim, Germany), equipped with a 450 W Xenon lamp, double monochromators, Ulbricht sphere, and photomultiplier as the detector (90° angle between excitation source and detector). Determination of the absolute quantum yield was performed as suggested by Friend [21]. First, the diffuse reflection of the sample was determined under excitation. Second, the emission was measured for the respective excitation wavelength. Integration over the reflected and emitted photons in wavelength range of 390–720 nm by use of an Ulbricht sphere allows calculating the absolute quantum yield. Standard corrections were used for the spectral power of the excitation source, the reflection behaviour of the Ulbricht sphere and the sensitivity of the detector. The QY was obtained from dispersion of the Eu^{3+}-modified C-dots in H_2O that were adjusted to an absorbance of 0.1. The sample holder for determining the absolute quantum yield of suspensions in an Ulbricht sphere was constructed according to Friend and is shown in SI: Figure S1 [21].

UV and blue light emitting diodes (UV- and blue-LED): UV-LED and blue-LED light sources were purchased from Zweibrüder Optoelectronics (Zweibrüder Optoelectronics, Solingen, Germany). The UV-LED operates at a wavelength range of 350–380 nm with $\lambda_{max} = 365$ nm (SI: Figure S2a). The blue-LED operates at a wavelength range of 440–500 nm with $\lambda_{max} = 465$ nm (SI: Figure S2b).

3. Results and Discussion

3.1. Eu-Modified C-Dots via Polyol Synthesis and Conventional Resistance Heating

Following our previous work [13], we started with Eu^{3+}-modified C-dots that were prepared by conventional resistance heating of solutions of $EuCl_3 \times 6H_2O$ in PEG400 (1 h, 230 °C). The resulting C-dots show variable blue and/or red emission depending on the conditions of heating (e.g., duration, temperature, atmosphere, amount of Eu^{3+}). Although some samples show excellent quantum yields (75%) in the polyol, they also show rapid PL quenching in water. Moreover, the yield is limited to about 1 mg of C-dots per 20 mL of PEG400. Considering the reducing properties of the polyols at high temperatures (>180 °C) [15] and the comparably-low electrochemical potential of the $Eu^{3+} \rightarrow Eu^{2+}$ reduction (–0.36 V) [22], the variable emission may result from different amounts of Eu^{2+} and Eu^{3+}.

To elucidate the correlation of synthesis conditions and PL, Eu-modified C-dots were prepared under strict inert conditions (Ar) and, thereafter, stored under Ar (Figures 1a and 2a). Even after two months the suspensions only show the typical broad blue emission of the C-dots upon UV-LED excitation (Figure 2a, SI: Figure S3a). Just a weak peak at 614 nm indicates $f \rightarrow f$ emission of Eu^{3+}. Lifetime measurements ($\lambda_{exc} = 375$ nm, $\lambda_{em} = 440$ nm) could be fitted by a multi-exponential equation with decays of $\tau_1 = 0.6$, $\tau_2 = 2.6$ and $\tau_3 = 11.3$ ns (Table 1; Figure 3) that are in good agreement with previously reported lifetime data of C-dots [23–25]. It is to be noted that an emission of Eu^{2+} is not to be expected since it would also occur in the blue to green spectral range. Moreover, the C-dot emission is significantly faster ($\tau_{C-dot} \sim 1$–10 ns) and, therefore, much more efficient than Eu^{2+} emission ($\tau_{Eu2+} \sim 0.5$–0.9 µs) [26].

In contrast to inert conditions, the Eu-modified C-dots in dry air show a slow, continuous increase of the Eu^{3+} f-f lines and an obvious PL shift from blue to red (Figures 1a and 2a; SI: Figure S3b). In fact, the red Eu^{3+} emission becomes about 25-times more intense than the blue C-dot emission (Figure 2b,c; SI: Figure S4). Whereas the Eu^{3+} emission increases exponentially until saturation with all europium oxidized, the C-dot emission remains at constant level (Figure 2b; SI: Figure S5). Lifetime measurements ($\lambda_{exc} = 375$ nm, $\lambda_{em} = 440$ and 615 nm) again show the typical decay of the C-dots ($\tau_1 = 0.5$, $\tau_2 = 2.5$, $\tau_3 = 8.4$ ns) as well as significantly longer lifetimes ($\tau_1 = 474.0$, $\tau_2 = 874.6$ µs, Table 1; Figure 3) that are indicative of Eu^{3+} emission ($\tau_{Eu3+} \sim 500$–1000 µs) [27]. Taken together, the observed PL can be rationalized by reduction to Eu^{2+} during the polyol synthesis and subsequent re-oxidation to Eu^{3+} in the presence of oxygen. Only in the presence of Eu^{3+}, however, an efficient energy transfer from the C-dots is possible resulting in an intense line-type $f{\rightarrow}f$ emission.

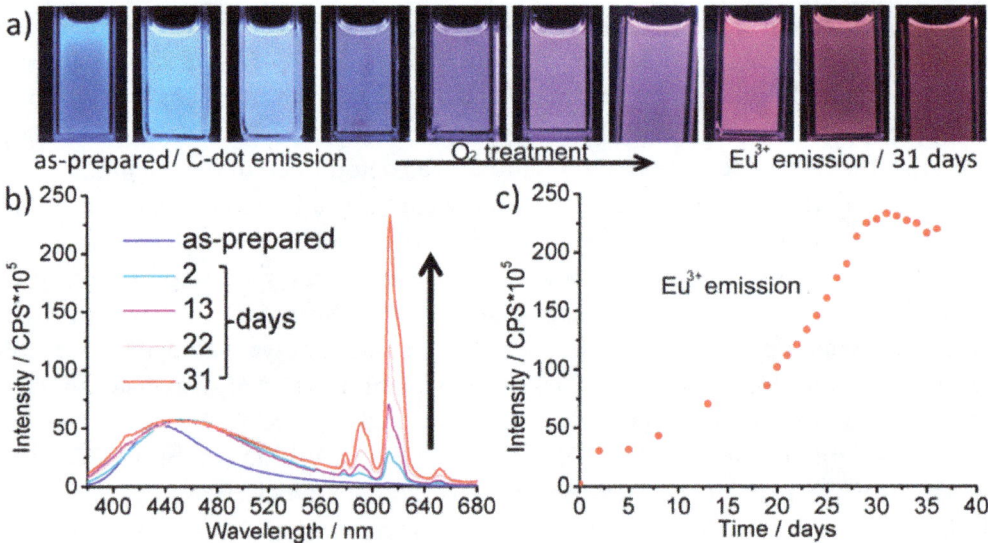

Figure 2. Dry air treatment of conventionally heated Eu-modified C-dots: (**a**) photographs with time-depending emission (UV-LED excitation); (**b**) PL spectra (normalized on C-dot emission at 440 nm); and (**c**) emission intensity of Eu^{3+} at 614 nm (all spectra with λ_{exc} = 366 nm).

Table 1. Photoluminescence lifetimes of Eu-modified C-dots (suspensions in PEG400, λ_{exc} = 375 nm).

Sample	λ_{em}/nm	B_1/%	τ_1/ns	B_2/%	τ_2/ns	B_3/%	τ_3/ns	χ^2
As-prepared (Eu^{2+}-modified)	440	14.9	0.57 ± 0.01	45.3	2.56 (±0.02)	39.7	11.31 ± 0.10	1.23
Dry-air treatment (Eu^{3+}-modified)	440	11.9	0.53 ± 0.01	37.0	2.52 (±0.03)	51.1	8.43 ± 0.04	1.23
Dry-air treatment (Eu^{3+}-modified)	615	93.7	474×10^3 (±1.0)	6.3	875×10^3 (±15)	/	/	1.36

λ_{em}: Emission wavelength at which the decay was monitored; B: Percentage contribution of different decay processes; χ^2: Wellness of the exponential fit in comparison to raw data. Values in parentheses indicate the standard deviations of the respective decay time.

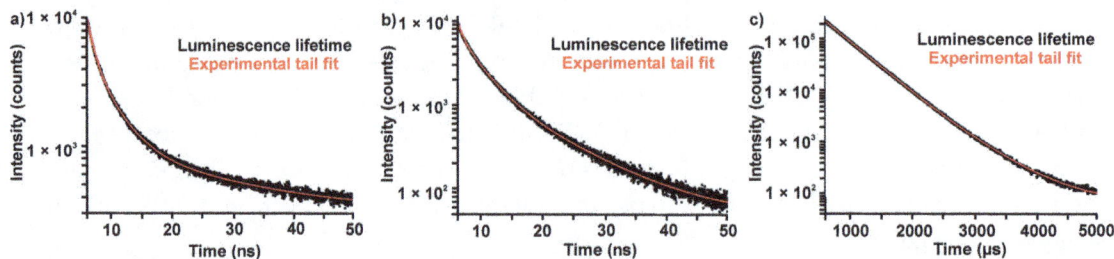

Figure 3. Photoluminescence decay curves Eu-modified C-dots (in PEG400, λ_{exc} = 375 nm): (**a**) as-prepared (Eu^{2+}-modified, λ_{em} = 440 nm); (**b**) treatment with dry air (2 months, λ_{em} = 440 nm); and (**c**) treatment with dry air (two months, λ_{em} = 615 nm).

It is well-known that the parity-forbidden $f{\rightarrow}f$ transitions on Eu^{3+} are efficiently quenched by O–H vibrational relaxation, especially, if H_2O is directly coordinated to the Eu^{3+} center [27]. This effect is here also observed if small portions of water were added to the Eu^{3+}-modified C-dots (Figure 4a) or if the as-prepared C-dots were treated with humid air instead of dry air (Figure 4b). In both cases the Eu^{3+} emission at 614 nm is rapidly quenched and only the blue C-dot emission remains (SI: Figure S5).

Successive treatment of the Eu-modified C-dots, first, with dry air, and second, with humid air also reproducibly shows an increase (i.e., oxidation to Eu^{3+}) followed by a rapid decrease (i.e., H_2O-driven quenching) of the red emission (Figure 1a; SI: Figure S6).

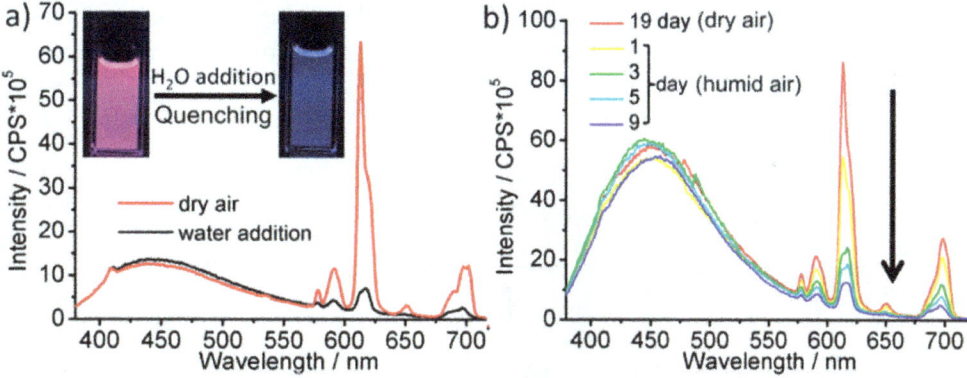

Figure 4. H_2O-driven quenching of conventionally-heated Eu-modified C-dots: (**a**) photographs and PL spectra with time-depending emission upon addition of water (UV-LED excitation); and (**b**) PL spectra during humid air treatment (λ_{exc} = 366 nm; normalized on C-dot emission at 440 nm).

3.2. Eu^{3+}-Modified C-Dots via Polyol Synthesis and MW-Heating

After elucidating the role of Eu^{2+}/Eu^{3+} and the influence of O_2/H_2O on the PL of Eu-modified C-dots, we intended to increase the yield of polyol-made C-dots and to stabilize the Eu^{3+}-based red emission, preferentially, in the presence of water. Our previous studies have shown that an increased duration of heating (>2 h) and/or an elevated temperature of heating (>230 °C) indeed support the thermal decomposition of PEG400. Both measures, however, mainly result in a formation of larger C-dots (>10 nm) that only show weak PL [1–6,13]. As an alternative to conventional resistance heating we have, therefore, applied MW-heating, which is well known for extremely fast heating, and which is optimal for controlled nucleation of nanoparticles [28]. Hence, a solution of $EuCl_3 \times 6H_2O$ in PEG400 was MW-heated for 20 min at 200 °C (Figure 1b). With these conditions opaque suspensions were obtained that result in yellowish powders subsequent to centrifugation and washing with 40-times higher yield (40 mg C-dots per 20 mL, Figure 5a). Most importantly, MW-heated, polyol-made Eu^{3+}-modified C-dots instantaneously show intense red emission as powder samples (Figure 5a,), and even more interestingly, also after redispersion in water (Figure 5a). This is a remarkable difference of MW-heated and conventionally heated Eu^{3+}-modified C-dots.

According to high-angle annular dark-field scanning transmission electron microscopy (HAADF-STEM), C-dots made via MW-heating exhibit diameters of 2–4 nm at narrow size distribution (Figure 6a). High resolution (HR)TEM images validate the particle size and the crystallinity of the C-dots (Figure 6c). In fact, this diameter is identical to the size of conventionally heated, polyol-made C-dots [13]. Element mappings, furthermore, indicate a uniform Eu distribution (Figure 6b; SI: Figure S7). Total combustion analysis (thermogravimetry, 1000 °C, air) proves a total carbon content of 68 wt % (SI: Figure S8a). The solid remnant (32 wt %)—according to X-ray diffraction—was identified as Eu_2O_3 (SI: Figure S8b). As a result, the Eu^{3+}-modified C-dots can be concluded to contain 12 mol % Eu^{3+}.

Whereas the particle size of the as-prepared C-dots, the total amount of Eu^{3+} and the C-dot→Eu^{3+} energy transfer with Eu^{3+}-based emission were already confirmed, the difference between conventionally heated and MW-heated C-dots and the stable PL of the latter in water remain surprising. To this respect, it is well known that coordination complexes of Eu^{3+} can show intense PL in water if all coordination sites are blocked by stronger ligands preventing energy transfer into vibronic states of water, for instance, by chelating Eu^{3+} with carboxylate ligands [17,18,29,30]. Fourier-transform infrared spectroscopy (FT-IR) indeed indicates a significant difference between the surface conditioning of the polyol-made C-dots (Figures 1 and 6d). Thus, ν(O–H) (3600–3000 cm^{-1}) represents the dominating vibration for conventional resistance heating, whereas ν(C=O) (1700–1400 cm^{-1}) is dominating for MW-heated C-dots (Figure 6d). Obviously, the C-dot surface is either hydroxyl-terminated due to polyol functionalization or carboxylate-terminated by oxidized polyols (Figure 6d). Carboxylate-termination of the C-dot surface guarantees for efficient coordination of Eu^{3+} near to the C-dot surface with both fast Eu$^{2+}\rightarrow$Eu^{3+} oxidation after synthesis and shielding of Eu^{3+} against H$_2$O.

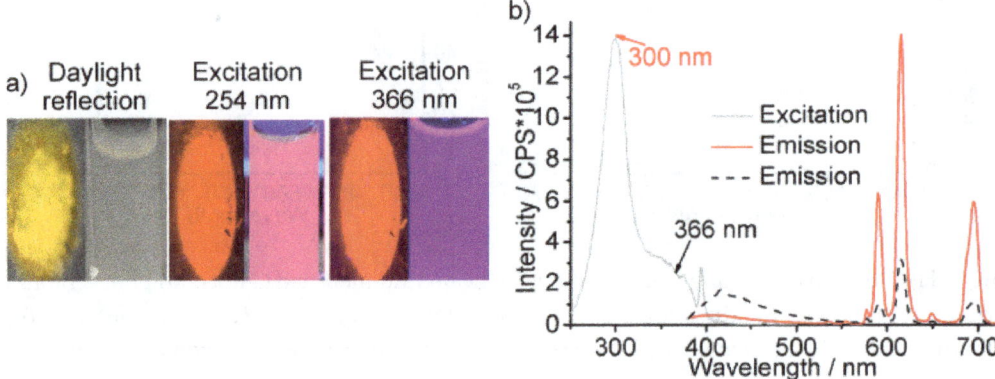

Figure 5. Fluorescence of MW-heated Eu^{3+}-modified C-dots: (**a**) photos of powder samples and aqueous suspension (under daylight and under excitation); and (**b**) excitation (λ_{em} = 615 nm) and emission (red line: λ_{exc} = 300 nm; black dash: λ_{exc} = 366 nm) spectra of aqueous suspensions.

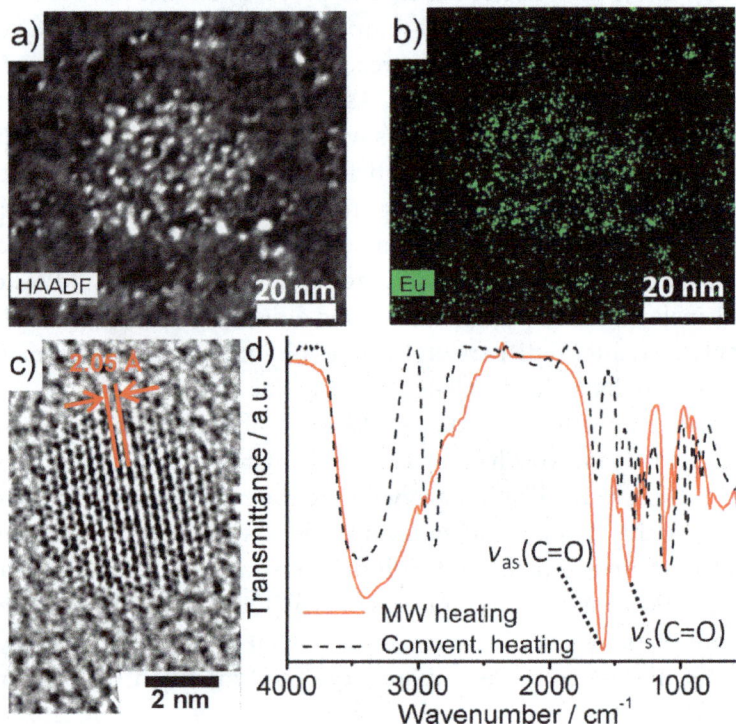

Figure 6. Composition and surface functionalization of MW-heated Eu^{3+}-modified C-dots: (**a**) HAADF-STEM image; (**b**) Eu elemental mapping; (**c**) HRTEM image with lattice distance; and (**d**) FT-IR spectra of MW-heated C-dots (**red**) in comparison to conventionally heated C-dots (**black**).

Excitation and emission spectra confirm the red emission of MW-heated Eu^{3+}-modified C-dots in aqueous suspensions (Figure 5b). For excitation at 366 nm, certain C-dot emission occurs (400–500 nm).

Excitation at maximum absorption of the C-dots (300 nm), moreover, shows intense line-type $f \rightarrow f$ emission of Eu^{3+} only, indicating efficient Förster resonance energy transfer (FRET) from the C-dot to Eu^{3+}. The absolute quantum yield (determined according to *Friend*, SI: Figures 1 and 2) [21] is 18% for aqueous suspensions (λ_{exc} = 366 nm). Notably, the intense red emission of Eu^{3+}-modified, polyol-made C-dots is observed in aqueous suspensions without the need of any additional stabilizing agent.

4. Conclusions

A one-pot polyol synthesis of Eu^{3+}-modified C-dots is presented. Microwave (MW) heating turned out as a key-factor to increase the yield and to obtain carboxylate-functionalized C-dots that can be directly dispersed in water without complete quenching of the red emission of Eu^{3+}. Moreover, the role of Eu^{2+}/Eu^{3+}, the influence of O_2/H_2O and the importance of the surface functionalization on the photoluminescence are correlated and result in a fundamental understanding of synthesis and material properties. Based on this study a knowledge-based synthesis is possible.

Due to their intense line-type red emission, polyol-made Eu^{3+}-modified C-dots are highly interesting for optoelectronics (full-colour emission and colour mixing to white light), as well as for optical imaging (sufficient tissue penetration and low background emission for line-type red emission). Finally, the MW-mediated polyol synthesis can be transferred to C-dots showing other rare-earth metal modification and emission.

Supplementary Materials
Figure S1: Sample holder for determining the absolute quantum yield of suspensions in an Ulbricht sphere according to Friend. Figure S2: Emission spectra of: (a) UV-LED and (b) blue-LED. Figure S3: Conventionally heated Eu-modified C-dots after storing for 2 month in argon (a), and treatment with dry air thereafter (b) (emission spectra and photographs with λ_{exc} = 366 nm). Figure S4: Excitation spectra of Eu-modified C-dots (suspensions in PEG400) made via conventional heating: (a) Storing in argon for 2 months (recorded at λ_{em} = 439 nm); (b) Storing in argon for 2 months (recorded at λ_{em} = 614 nm); (c) Treatment with dry air for 2 months (recorded at λ_{em} = 439 nm); (c) Treatment with dry air for 2 months (recorded at λ_{em} = 614 nm). Figure S5: Excitation spectra of Eu-modified C-dots during treatment with dry air: (a) normalized excitation spectra (λ_{em} = 445–460 nm); (b) Normalized emission of C-dots at 445–460 nm (λ_{exc} = 366 nm). Figure S6: The influence of humidity on Eu-modified C-dots (PEG400 suspensions): (a) Normalized excitation spectra during bubbling of dry and humid air (λ_{exc} = 440–460 nm); (b) Emission intensity of Eu^{3+} at 614 nm; (c) Photographs during bubbling of dry (days 0–19) and humid (days 20–28) air (λ_{exc} = 366 nm). Figure S7: EDXS images with Eu (a) and C (b) element mappings. Figure S8: Chemical composition of Eu-modified C-dots: (a) Thermogravimetry (TG); (b) XRD pattern of TG-remnant (ICDD 00-034-0392/Eu_2O_3 as a reference).

Acknowledgments: H.D. and C.F. are grateful to the Deutsche Forschungsgemeinschaft (DFG) for funding in the project Carbon Nanoparticles (C-Dots) (FE911/8).

Author Contributions: H.D. and C.F. conceived and designed the experiments; H.D. performed the synthesis, and A.K. performed fluorescence spectroscopy and analyzed the data; T.W. and K.M.B. performed and analyzed decay measurements; K.M.B. and C.F. wrote the paper.

Conflicts of Interest: The authors declare no conflict of interest.

References

1. Lim, S.Y.; Shen, W.; Gao, Z. Carbon quantum dots and their applications. *Chem. Soc. Rev.* **2015**, *44*, 362–382. [CrossRef] [PubMed]
2. Hong, G.; Diao, S.; Antaris, A.L.; Dai, H. Carbon nanomaterials for biological imaging and nanomedicinal therapy. *Chem. Rev.* **2015**, *115*, 10816–10906. [CrossRef] [PubMed]
3. Hola, K.; Zhang, Y.; Wang, Y.; Giannelis, E.P.; Zboril, R.; Rogach, A.L. Carbon dots-emerging light emitters for bioimaging, cancer therapy and optoelectronics. *Nano Today* **2014**, *9*, 590–603. [CrossRef]
4. Song, Y.; Zhu, S.; Yang, B. Bioimaging based on fluorescent carbon dots. *RSC Adv.* **2014**, *4*, 27184–27200. [CrossRef]
5. Wang, Q.H.; Bellisario, D.O.; Drahushuk, L.W.; Jain, R.M.; Kruss, S.; Landry, M.P.; Mahajan, S.G.; Shimizu, S.F.E.; Ulissi, Z.W.; Strano, M.S. Low dimensional carbon materials for applications in mass and energy transport. *Chem. Mater.* **2014**, *26*, 172–183. [CrossRef]

6.   Sun, Z.; Li, Q.; Zhu, Y.; Tan, B.; Xu, Z.P.; Dou, S.X. Ultra-small fluorescent inorganic nanoparticles for bioimaging. *J. Mater. Chem. B* **2014**, *2*, 2793–2918.

7.   Jiang, K.; Sun, S.; Zhang, L.; Lu, Y.; Wu, A.; Cai, C.; Lin, H. Red, green, and blue luminescence by carbon dots: Full-color emission tuning and multicolor cellular imaging. *Angew. Chem. Int. Ed.* **2015**, *54*, 5360–5363. [CrossRef] [PubMed]

8.   Ge, J.; Jia, Q.; Liu, W.; Guo, L.; Liu, Q.; Lan, M.; Zhang, H.; Meng, X.; Wang, P. Red-emissive carbon dots for fluorescent, photoacoustic, and thermal theranostics in living mice. *Adv. Mater.* **2015**, *27*, 4169–4177. [CrossRef] [PubMed]

9.   Nie, H.; Li, M.; Li, Q.; Liang, S.; Tan, Y.; Sheng, L.; Shi, W.; Zhang, S.X. Carbon Dots with continuously tunable full-color emission and their application in Ratiometric pH sensing. *Chem. Mater.* **2014**, *26*, 3104–3112. [CrossRef]

10. Zhang, X.; Zhang, Y.; Wang, Y.; Kalytchuk, S.; Kershaw, S.V.; Wang, Y.; Wang, P.; Zhang, T.; Zhao, Y.; Zhang, H.; et al. Color-switchable electroluminescence of carbon dot light-emitting diodes. *ACS Nano* **2013**, *7*, 11234–11241. [CrossRef] [PubMed]

11. Tao, H.Q.; Yang, K.; Ma, Z.; Wan, J.M.; Zhang, Y.J.; Kang, Z.H.; Liu, Z. In vivo NIR fluorescence imaging, biodistribution, and toxicology of photoluminescent carbon dots produced from carbon nanotubes and graphite. *Small* **2012**, *8*, 281–290. [CrossRef] [PubMed]

12. Fujimoto, J.G.; Farkas, D. *Biomedical Optical Imaging*; Oxford University Press: Oxford, UK, 2009.

13. Dong, H.; Kuzmanoski, A.; Gößl, D.M.; Popescu, R.; Gerthsen, D.; Feldmann, C. Polyol-mediated C-dot formation showing efficient Tb^{3+}/Eu^{3+} emission. *Chem. Commun.* **2014**, *50*, 7503–7506. [CrossRef] [PubMed]

14. Dong, H.; Roming, M.; Feldmann, C. Unexpected fluorescence of polyols and PEGylated nanoparticles derived from carbon dot formation. *Part. Part. Syst. Charact.* **2015**, *32*, 467–475. [CrossRef]

15. Dong, H.; Chen, Y.C.; Feldmann, C. Polyol synthesis of nanoparticles: Status and options regarding metals, oxides, chalcogenides, and non-metal elements. *Green Chem.* **2015**, *17*, 4107–4132. [CrossRef]

16. Havrdova, M.; Hola, K.; Skopalik, J.; Tomankova, K.; Petr, M.; Cepe, K.; Polakova, K.; Tucek, J.; Bourlinos, A.; Zboril, R. Toxicity of carbon dots—Effect of surface functionalization on the cell viability, reactive oxygen species generation and cell cycle. *Carbon* **2016**, *99*, 238–248. [CrossRef]

17. Zhou, Z.; Wang, Q.; Wang, J.; Zhang, C. Imaging two targets in live cells based on rational design of lanthanide organic structure appended carbon dots. *Carbon* **2015**, *93*, 671–680. [CrossRef]

18. Song, Y.; Chen, J.; Hu, D.; Liu, F.; Li, P.; Li, H.; Chen, S. Ratiometric fluorescent detection of biomakers for biological warfare agents with carbon dots chelated europium-based nanoscale coordination polymers. *Sens. Actuators B* **2015**, *221*, 586–592. [CrossRef]

19. Ye, Z.; Tang, R.; Wu, H.; Wang, B.; Tan, M.; Yuan, J. Preparation of europium complex-conjugated carbon dots for ratiometric fluorescence detection of copper(II) ions. *New J. Chem.* **2014**, *38*, 5721–5726. [CrossRef]

20. Samanta, T.; Hazra, C.; Mahalingam, V. C-dot sensitized Eu^{3+} luminescence from Eu^{3+}-doped LaF_3-C dot nanocomposites. *New J. Chem.* **2016**, *39*, 106–109. [CrossRef]

21. De Mello, J.C.; Wittmann, H.F.; Friend, R.H. An improved experimental determination of external photoluminescence quantum efficiency. *Adv. Mater.* **1997**, *9*, 230–232. [CrossRef]

22. Huheey, J.E.; Keiter, E.A.; Keiter, R.L. *Inorganic Chemistry: Principles of Structure, Reactivity*; Pearson: New York, NY, USA, 2008.

23. Sun, Y.P.; Zhou, B.; Lin, Y.; Wang, W.; Shiral, K.A.; Fern, O.; Pathak, P.; Meziani, M.J.; Harruff, B.A.; Wang, X.; et al. Quantum-Sized Carbon Dots for Bright and Colorful Photoluminescence. *J. Am. Chem. Soc.* **2006**, *128*, 7756–7757. [CrossRef] [PubMed]

24. Mondal, S.; Chatti, M.; Mallick, A.; Purkayastha, P. pH triggered reversible photoinduced electron transfer to and from carbon nanoparticles. *Chem. Commun.* **2014**, *50*, 6890–6893. [CrossRef] [PubMed]

25. Gonçalves, H.; Esteves da Silva, J.C. Fluorescent carbon dots capped with PEG200 and mercaptosuccinic acid. *J. Fluoresc.* **2010**, *20*, 1023–1028. [CrossRef] [PubMed]

26. Mao, X.; Zheng, H.; Long, Y.; Du, J.; Hao, J.; Wang, L.; Zhou, D. Study on the fluorescence characteristics of carbon dots. *Spectrochim. Acta A* **2010**, *75*, 553–557. [CrossRef] [PubMed]

27. Blasse, G.; Grabmaier, C. *Luminescent Materials*; Springer: Berlin, Germany, 1994.

28. Baghbanzadeh, M.; Carbone, L.; Cozzoli, P.D.; Kappe, C.O. Microwave-assisted synthesis of colloidal inorganic nanocrystals. *Angew. Chem. Int. Ed.* **2011**, *50*, 11312–11359. [CrossRef] [PubMed]

29. Zhou, J.M.; Shi, W.; Xu, N.; Cheng, P. Highly selective luminescent sensing of fluoride and organic small-molecule pollutants based on novel lanthanide metal-organic frameworks. *Inorg. Chem.* **2013**, *52*, 8082–8090. [CrossRef] [PubMed]

30. Thielemann, D.T.; Wagner, A.T.; Birtalan, E.; Kölmel, D.; Heck, J.; Rudat, B.; Neumaier, M.; Feldmann, C.; Schepers, U.; Bräse, S.; et al. A luminescent cell-penetrating pentadecanuclear lanthanide cluster. *J. Am. Chem. Soc.* **2013**, *135*, 7454–7457. [CrossRef] [PubMed]

9

Ultrasonic Monitoring of the Interaction between Cement Matrix and Alkaline Silicate Solution in Self-Healing Systems

Mohand Ait Ouarabi [1,2], **Paola Antonaci** [3,*], **Fouad Boubenider** [2], **Antonio S. Gliozzi** [1] and **Marco Scalerandi** [1]

1   Department of Applied Science and Technology, Condensed Matter and Complex Systems Physics Institute, Politecnico di Torino, 10129 Torino, Italy; aitouarabi@gmail.com (M.A.O.); antonio.gliozzi@polito.it (A.S.G.); marco.scalerandi@infm.polito.it (M.S.)
2   Laboratoire de Physique des Matériaux, Université des Sciences et de la Technologie Houari Boumediene, BP 32 El Alia, Bab Ezzouar 16111, Algeria; fboubenider@yahoo.fr
3   Department of Structural, Geotechnical and Building Engineering, Politecnico di Torino, 10129 Torino, Italy
*   Correspondence: paola.antonaci@polito.it

Academic Editor: Nele De Belie

Abstract: Alkaline solutions, such as sodium, potassium or lithium silicates, appear to be very promising as healing agents for the development of encapsulated self-healing concretes. However, the evolution of their mechanical and acoustic properties in time has not yet been completely clarified, especially regarding their behavior and related kinetics when they are used in the form of a thin layer in contact with a hardened cement matrix. This study aims to monitor, using linear and nonlinear ultrasonic methods, the evolution of a sodium silicate solution interacting with a cement matrix in the presence of localized cracks. The ultrasonic inspection via linear methods revealed that an almost complete recovery of the elastic and acoustic properties occurred within a few days of healing. The nonlinear ultrasonic measurements contributed to provide further insight into the kinetics of the recovery due to the presence of the healing agent. A good regain of mechanical performance was ascertained through flexural tests at the end of the healing process, confirming the suitability of sodium silicate as a healing agent for self-healing cementitious systems.

Keywords: self-healing concrete; sodium silicate; cracks; NDT; nonlinear ultrasonic inspection; scaling subtraction method

1. Introduction

Due to their attractive potential and practical value, self-healing materials have been extensively investigated over the last ten years, and significant advances have been achieved [1]. In particular, self-healing in cementitious materials has been targeted through different strategies, from enhancing intrinsic healing [2,3] to exploiting bacterial metabolic reactions [4], up to developing novel autonomic vascular [5] or capsule-based systems [6–9].

Among all of them, the capsule-based approach appears to be one of the most promising due to its versatility. It consists of adding to the concrete mix a variable amount of micro- or macro-capsules that sequester one or more healing agents inside them, in such a way that when the capsules are ruptured by damage, the self-healing mechanism is triggered through the release and reaction of the healing agent(s) in the region of damage. The type of healing agent to be stored in the capsules can be changed according to the desired healing mechanism to be generated, ranging from mono-component systems (in which a single healing agent is present and the healing process is expected to take place upon its reaction with air, or with ambient humidity, or with the cement matrix itself) to multi-component systems (in which two or more healing agents are present, reacting upon contact with each other).

This paper takes its starting point from the observation of mono-component systems that display a lower level of complexity with respect to the multi-component counterpart, also minimizing the risks of incomplete reaction due to insufficient mixing of the healing agents. In particular, attention is focused on mono-component encapsulated systems using an alkaline silicate solution as a healing agent. Indeed, such alkaline solutions as sodium, potassium or lithium silicates are considered to be very promising for the purpose of developing successful self-healing cementitious systems because of their low viscosity, which enables them to effectively diffuse through the cracks once released at the damage site, and also owing to their chemical affinity with the cement matrix. Their good compatibility with cementitious composites is well known by the construction industry, where soluble silicates, in particular sodium silicate (also referred to as "water-glass"), already have several applications. For example, they are used as moisture reducers in the wet kiln process of clinker production [10], as binders for cold consolidation of silica-based aggregates [11], as active ingredients in the production of dense or lightened geopolymers [12] and as soil stabilizers [13,14]. Furthermore, one of their foremost uses is as concrete sealers. Unlike other sealants, soluble sodium silicate penetrates the concrete surface to form new solid species. As a result, the surface has enhanced properties, such as decreased permeability, increased hardness and overall increased durability [10].

Despite their common use in the cement industry, the exact mechanisms by which the silicates act to improve the performance of concretes is still controversial. Excluding the gelation/polymerization reactions that occur rapidly when the pH of liquid silicate falls below 10.7 (which is not the case of the concrete environment, which is known to be highly alkaline), one argument is that sodium silicates behave as efficient sealers because SiO_2 precipitates in the pores upon evaporation or absorption of water, thus contributing to densify the concrete microstructure [15]. A second standpoint is that the silicates are able to form an expansive gel similar to that generated during alkali-silicate reactions to fill the concrete voids by swelling [10,16]. A third theory is that the active silicates can react with the excess portlandite or calcium hydroxide (CH) available in the concrete to yield relatively insoluble calcium silicate hydrates (C-S-H gels) [10,16–18]. Similar reactions are theorized for potassium and lithium silicates, as well. Various authors agree that the latter is the main mechanism of self-healing by alkaline silicate solutions [8,19,20], because the delayed creation of C-S-H gel, which is a binding material very similar to the main product of the hydration of Portland cement, is primarily responsible for the recovery of strength and stiffness.

Anyway, some factors still need to be explored before alkaline silicate solutions can be confirmed as suitable healing agents for self-healing concrete applications. In particular, it has to be assessed if the expected reactions can actually be manifested in the specific conditions of encapsulated systems and how long would be needed for them to take place. The role of contact with air regarding the manifestation of the healing reactions needs to be clarified, since this interaction would be negligible in the case of encapsulated systems, where the soluble silicates are released directly into the cracks from the capsules dispersed in the material bulk. Furthermore, the age of concrete at the time of damage can play a role, since the amount of free calcium hydroxide available for the healing reactions tends to decrease in time with concrete aging.

The progression of the healing reactions in the conditions above described is quite difficult to characterize. Indeed, a destructive test would be necessary to analyze the chemical species and quantify the extent of the mechanical recovery, but on the other hand, it would compromise the integrity of the observed system, inhibiting the possibility to further monitor its evolution. Ultrasonic tests could be a possible alternative to estimate indirectly the occurrence of specific interactions between healing agents and air or healing agents and concrete matrix. Ultrasound might help to shed light on the duration of processes, such as gelification of the alkaline solution, and link such a duration to external factors, like the age of concrete, etc. This could for instance help to consider what factors are to be controlled or optimized in order to improve the effectiveness of the healing process.

Indeed, several parameters that can be measured in ultrasonic experiments are known to change when the microstructure of the investigated sample is evolving, as during healing, thus providing information on the evolving process itself. Two main classes of acoustic parameters could be identified, essentially as a result of linear and nonlinear acoustic experiments, respectively. On one side, linear acoustics has been extensively applied to investigate the presence and evolution of damage in concrete [21–23] and to monitor phase transition processes [24,25]. Among the various (linear) indicators, it is worth recalling the measurement of the transmission coefficient across a distributed crack region [26], the velocity variation [27], the softening-induced shift of the resonance frequency due to prestresses [28] or crack distributions [23]. On the other side, nonlinear ultrasonic methods have been recently proven to be even more sensitive than linear methods in order to monitor damage evolution (or even recovery due to autogenous healing mechanisms) in concrete [29–33]. Here, as well, several different indicators have been defined: second and third harmonics amplitude [22,34–36], the shift of the resonance frequency [37–40], coda wave properties [41,42] or the nonlinear indicator based on the Scaling Subtraction Method (SSM) [43–46]. They all have been widely used in the literature to monitor the evolution of damage and could be used to monitor structural transitions in the behavior of the healing agent in contact with a cementitious matrix.

In this paper, a combined linear-nonlinear ultrasonic approach was adopted to analyze the behavior of a cracked cementitious mortar in the presence of a thin layer of sodium silicate solution. The outline of the whole experimental setup and the characteristics of the specimens are described in Section 2, while the details of the single experiments are discussed in Section 3, together with the main results achieved. Verification of the mechanical recovery is discussed in Section 4, and concluding remarks are reported in Section 5.

2. Materials and Methods

2.1. Specimens

The specimens used for this study were in the shape of standard ($4 \times 4 \times 16$ cm^3) mortar prisms. The mortar matrix was produced using a common CEM II/A-LL 42.5 R cement, a water-to-cement ratio of 0.5 and a cement-to-sand ratio of 1:3 by weight, in accordance with Standard UNI EN 196-1 Methods of testing cement, Part 1: Determination of strength. The age of the mortar prisms at the time of testing was about two years, thus guaranteeing that the cement hydration process was virtually completed and the amount of free calcium hydroxide stabilized.

The samples were preliminarily tested using the ultrasonic protocol described below in their "intact state". Afterwards, a pass-through crack was generated at mid-span by means of three-point-bending tests up to failure, in such a way to create a complete disconnection between the two halves of each prism. A thin layer of sodium silicate solution was then applied manually to the crack surface, with the aim to reassemble the two residual fragments, thanks to the activation of the healing process, which would induce the progressive creation of a stable bond between the opposite edges of the fractured zone, bridging them together and eventually restoring the material integrity. The residual crack width after reassembling was evaluated by means of a $20\times$ optical microscope, resulting in around 400 μm, on average. Such a residual crack width was deemed adequately representative of common damage states in real-sized concrete structures since, based on the indications reported in the European Standard EN 1992-1-1: Eurocode 2: Design of concrete structures—Part 1-1: General rules and rules for buildings, crack sizes of 200–400 μm (depending on the structure type and exposure class) have to be regarded as threshold values discriminating between acceptable damage in the serviceability limit state (cracks smaller than this do not need a specific control) and unacceptable crack widths (that require crack control).

In continuity with previous studies [8,47], the sodium silicate used was provided by Sigma Aldrich and was characterized by a 10.6 wt % proportion of Na_2O, a 26.5 wt % proportion of SiO_2 and a 62.9 wt % of water.

Intentionally, the geometric configuration of the specimens described above was rather simplified with respect to the real case of self-healing encapsulated cementitious systems. Indeed, the latter differs due to the presence of the ruptured capsules at the site of the crack and also because the healing agent released from the capsules does not necessarily flow up to covering the fracture surfaces entirely. However, in order to monitor the changes in the acoustic behavior of the interface between concrete and healing agent, this simplified system was easier to analyze because of the reduced number of testing variables which are not controllable. Furthermore, the chosen configuration was still realistic, in that it corresponded to the optimal diffusion of the silicate and the contact with air was limited to the borders of the silicate layer, as in real conditions.

2.2. Ultrasonic Testing Configurations

A set of three experiments was conducted one after another to analyze the acoustic and elastic properties of each sample. The cycle of three experiments was repeated at different times to monitor the effects of the evolution of self-healing during recovery. In particular, the system was monitored in the following states:

- in the initial state, referred to as "0" (i.e., when the prisms were still intact, before the creation of the transversal crack at mid-span via the initial three-point-bending test);
- in the damaged state, referred to as "1" (i.e., immediately after the generation of the mechanical disconnection at mid-span and before the application of the healing agent on the fracture surface);
- at regular time intervals denoted as "2" to "n", during the healing process. The first set of measures during healing was taken a few minutes after the healing process has started, and the monitoring procedure continued up to three weeks; though major changes in the acoustic behavior turned out to be established in the first few days, as will be detailed in the following.

Mechanical testing was also performed after three weeks.

At each time, as mentioned, three experiments were performed to characterize both the linear and nonlinear elastic properties of the system. The three experiments, differing for the type of excitation source used, will be described in detail in Section 3. Here, it is sufficient to mention that the characterization consisted of:

- Resonant modes analysis under frequency sweep excitation;
- Linear and nonlinear analysis (transmission coefficients and harmonics generation) under continuous wave excitation;
- Nonlinear analysis according to the Scaling Subtraction Method (SSM) under pulse excitation.

2.3. Ultrasonic Testing Experimental Setup

In all cases, the experimental configuration for the ultrasonic measurements was the same (see Figure 1): the specimens were equipped with two identical narrowband piezoelectric transducers (Matest C370-02), characterized by a central frequency of 55 kHz, a bandwidth of 5 kHz and a sufficiently flat (though necessarily attenuated) response in the frequency range out of the baseband bandwidth. They worked one as an emitting source and the other as a receiver. They were glued to the 4 by 4 cm^2 opposite sides of the prisms by means of phenyl salicylate (a coupling agent with excellent linearity) and were not removed for the entire duration of the testing campaign.

The emitting transducer was connected to an arbitrary waveform generator (Agilent Technologies 33500B, Santa Clara, CA, USA), in series with a 20× fixed gain high voltage linear amplifier (FLC Electronics A400DI, Partille, Sweden), so that it was forced to vibrate according to a prescribed excitation function (different for the three characterization protocols). In parallel, it was connected also to a 4-channel digitizing oscilloscope (Agilent Technologies DSO9024H Infiniium), so that the input signals could be recorded and used to trigger the data acquisition. The receiving transducer was directly connected to the oscilloscope for data acquisition.

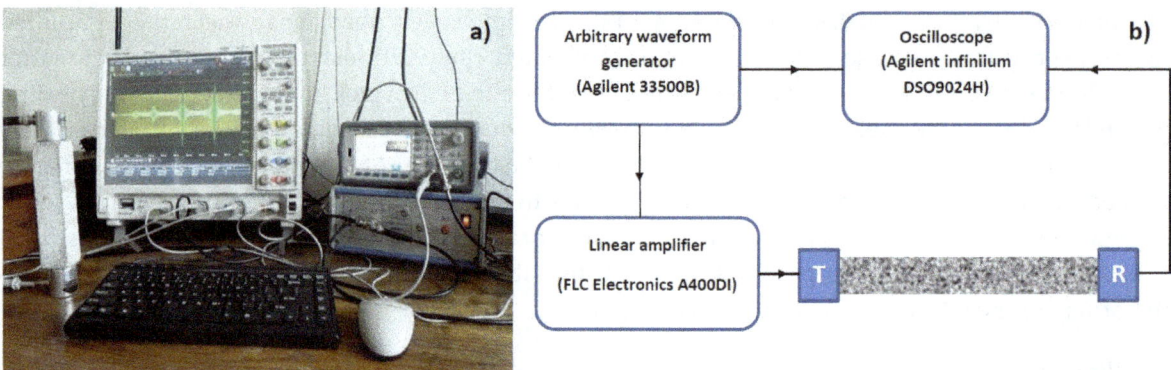

Figure 1. Experimental setup: (**a**) Photograph of the implemented configuration; (**b**) Sketch of the electronic configuration.

The linearity of the acquisition system was tested and verified in [33]. The same study reports the robustness of many of the ultrasonic methods used here with respect to small temperature and humidity fluctuations, as occurring in indoor conditions. Based on this conclusion, it was decided here to perform the ultrasonic testing without imposing any control of temperature and humidity, whose values ranged approximately between 19 and 24 °C and 35% and 45%, respectively.

2.4. Mechanical Testing

All of the specimens were subjected to mechanical tests by means of three-point-bending immediately after the initial ultrasonic tests and at the end of the healing process, in such a way so as to determine the flexural strength in the original state and after repair. This testing procedure was used to confirm the information about the settling of the healing reactions and to formulate a final judgment on the quality of the repair process. The tests were performed using a 25-kN servo-controlled universal testing machine working in deflection control, with a test velocity of 0.015 mm/s.

3. Ultrasonic Tests: Results and Discussion

3.1. Resonant Modes Analysis under Frequency Sweep Excitation

A first set of ultrasonic experiments was performed in order to analyze the resonant modes of the specimens under a frequency sweep excitation. For the resonance experiments, the excitation function was set to a sine sweep ranging from 1–30 kHz in 1 s. A linear sweep was chosen. In this way, it was possible to excite the first three longitudinal vibration modes of the intact specimen (centered around 8, 16 and 24 kHz, respectively) and observe their evolution in time, from the intact state to a repaired configuration. The response of the specimens to such a sweep excitation was recorded using a time window of 1 s with a sampling rate of 400 kSa/s. Measurements were performed using a fixed amplitude of excitation $A_{sweep} = 12$ V.

For each measurement, the spectral response was calculated based on the recorded data by means of a fast Fourier transform algorithm implemented through the MATLAB® FFT function (with Blackman windowing and 4×10^5 points, corresponding to the entire duration of the signal). Results are shown in Figure 2, where the evolution of the spectral magnitude with the progression of the healing process is shown for Specimen No. 1. A similar behavior was manifested by the other tested specimens also, but it was not reported here for the sake of brevity.

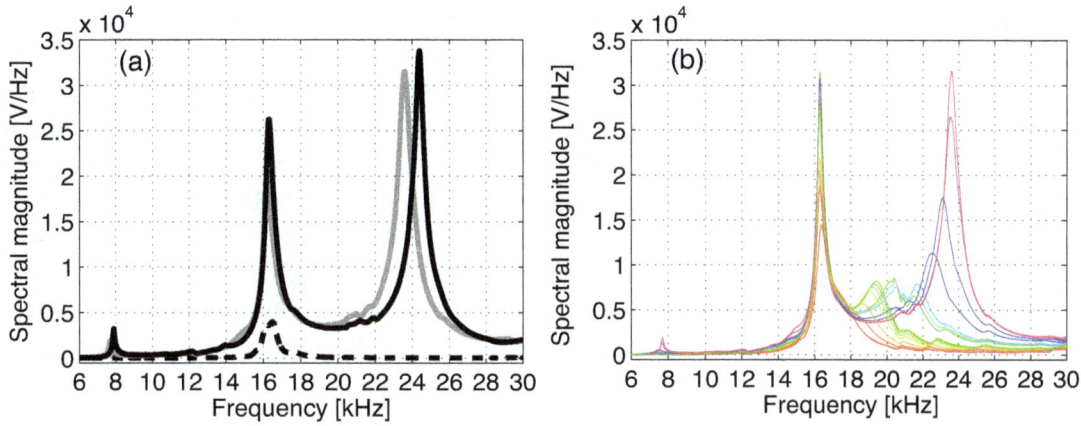

Figure 2. Evolution of the output signal spectra as a function of time. Data refer to Specimen No. 1. (**a**) Spectra at time "0" (intact state, solid black line), time "1" (damaged state, dashed black line) and time "n" (repaired state, solid gray line); (**b**) Spectra during the healing process, from time "2" to time "n".

The spectra for the intact and damaged samples are shown in Figure 2a, together with the spectrum at the end of the healing process. It is easy to distinguish a sharp difference in the spectral response of the specimen between time "0" (intact state, solid black line) and time "1" (fully-damaged state, dashed black line). At time "1", the sample presents a crack crossing the entire transversal section, and as a consequence, transmission is strongly affected with a much smaller vibration energy transmitted at time "1" with respect to time "0". As a result, most of the modes disappear within the noise level, and only the peak corresponding to the mode at 16 kHz could be appreciated. Note that, since cracking occurs roughly in the middle of the sample, this mode is amplified with respect to the others, because it corresponds to the first resonance mode of the two halves of the sample: indeed, the crack reduces the free vibration length to half of the initial one, the first resonance frequency being redoubled in turn. After approximately five days (solid gray line), the spectrum shows again all peaks as in the intact state, with almost the same amplitude and slightly different resonance frequencies.

Data recorded at successive time intervals from "2" to "n" (Figure 2b) showed an intermediate behavior, with a progressive transition from a state similar to that of the cracked specimen up to a state very close to the intact one. It is worth pointing out that the measurement taken at time "2" (red line, corresponding to the lowest amplitude of the peak at 16 kHz) was performed a few minutes after the application of the sodium silicate solution to the fracture surface, and yet, evidence of a much improved signal transmission can be noticed (as will also be discussed later). Even though the healing agent is still in a completely fluid configuration, the magnitudes of the spectral peaks were significantly higher than at time "1" (though lower than at time "0"), and resonance peaks begin to appear close to 8, 16 and 24 kHz (corresponding to the peaks for a full-length specimen) as in the case of the intact specimen.

Both peaks at 8 kHz (slightly visible because of their low amplitude) and 24 kHz show the same behavior: immediately after application of the healing agent, some energy begins to be detected at the two frequencies, but without evidence of a clear peak. Later, when solidification of the healing agent is partially completed, the spectrum assumes a clearer shape (blue line) with peaks occurring at a frequency that increases (hardening) with the solidification process up to getting close to that of the intact state.

Slightly different is the behavior of the peak at 16 kHz (which corresponds to the second mode of the specimen and the first mode of each of the two halves). The frequency of the peak is rather insensitive to the evolution of the healing agent, while its amplitude seems to be diminishing with the solidification of the interfacial layer between the two halves of the sample, a consequence of the fact that the resonances within the two parts of the sample become less and less important.

A more accurate observation of the spectral response during the process of sodium silicate hardening was performed for the peak at 8 kHz and Specimen No. 1 (see Figure 3). A similar behavior can be appreciated for the other specimens and for the peak at 24 kHz, as well, though not reported here for the sake of brevity. In Figure 3, the frequency at which the first mode is detected (i.e., the frequency of the maximum of the spectrum around 8 kHz) is shown vs. time. Apart from a drop towards lower frequencies occurring in the first few hours after the application of the sodium silicate solution (that could be ascribed to the presence of humidity supplied by the aqueous solution itself), the resonance peak moved towards higher values, eventually tending to the resonance frequency detected for the intact state. This resonance shift process seemed to show an asymptotic evolution, which can be considered substantially completed in about seven days (about 10^4 min) after the beginning of the healing process.

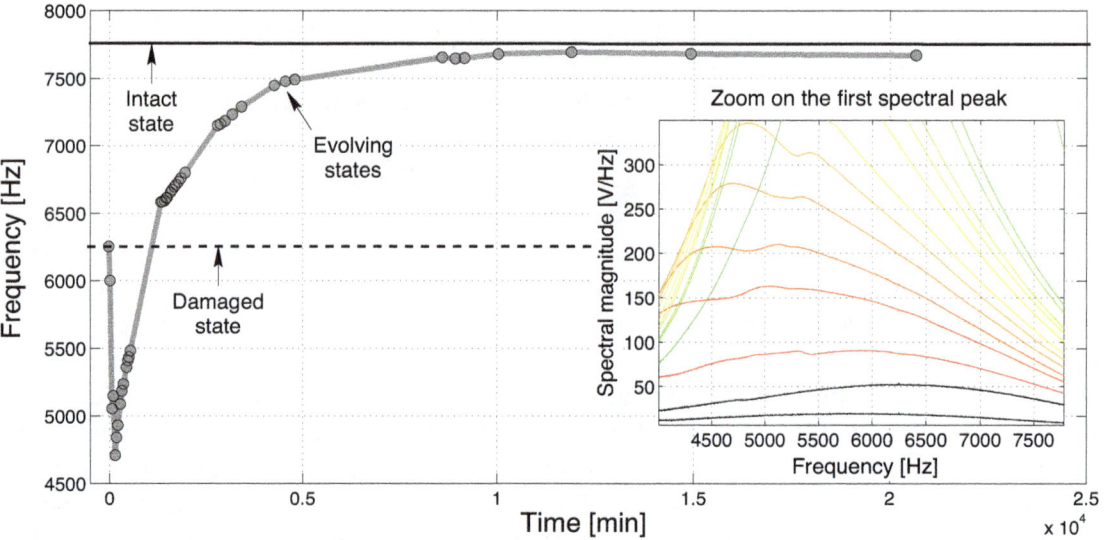

Figure 3. Resonance frequency of the first longitudinal mode as a function of time detected with a sweep source. Data refer to Specimen No. 1.

In the inset, a zoom of the spectrum around 8 kHz is shown for the first instances of the evolution to highlight the decrease in the frequency at short times. Note that as long as the sodium silicate is still in a liquid configuration, the peak is far from being sharp.

Based on the discussion reported above, a first conclusion can be drawn: the acoustic response of a cracked cementitious mortar element, and in particular, its resonance behavior, changes in time due to the presence of a layer of sodium silicate in adherence to the fracture surfaces. Its evolution seems to converge towards the response originally manifested in the intact state, i.e., prior to crack formation, thus indicating a successful healing of the sample as long as its elastic/acoustic properties are concerned.

3.2. Linear and Nonlinear Analysis under Continuous Wave Excitation

A second testing procedure was implemented changing the type of excitation, which was set to a continuous sine wave with frequency of 8 kHz and an amplitude of 3 V. The sampling rate was set to 5 MSa/s. The response signal $u(t)$ was detected once standing wave conditions were reached, and a fast Fourier transform of the signal was performed. A typical signal at a given healing time is shown in Figure 4a, and its spectrum is shown in Figure 4b. Note that the experiments under such a continuous wave excitation have to be regarded more precisely as a combination of transmission and resonance conditions, rather than pure transmission.

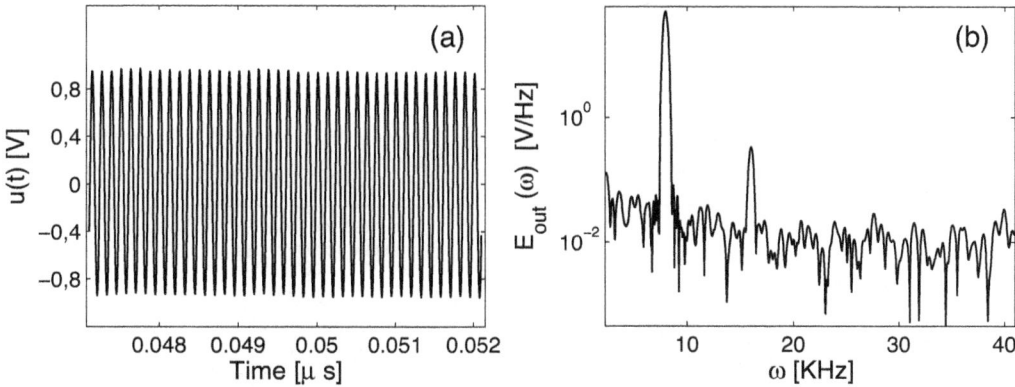

Figure 4. (**a**) Typical signal recorded with a sinusoidal source function and (**b**) relative spectrum.

Two different indicators could be extracted from the spectrum. First, a sort of linear transmission indicator could be defined, related to the energy transmitted across the interface and detected at the receiver position. To estimate the transmission properties, the spectral magnitude of the output signal was integrated in a frequency window of $\delta = 1$ kHz centered around the frequency of the excitation wave ($\omega_0 = 8$ kHz):

$$E_\omega = \int_{\omega_0 - \delta/2}^{\omega_0 + \delta/2} E_{out}(\omega)d\omega \tag{1}$$

The transmission coefficient T is defined by dividing E_ω by the spectral amplitude of the input signal integrated in the same time window:

$$T = \frac{E_\omega}{E_\omega^{inp}} = \frac{\int_{\omega_0 - \delta/2}^{\omega_0 + \delta/2} E_{out}(\omega)d\omega}{\int_{\omega_0 - \delta/2}^{\omega_0 + \delta/2} E_{inp}(\omega)d\omega} \tag{2}$$

It has to be pointed out that the frequency of the excitation and hence the central frequency value at which the output contribution was evaluated were intentionally set to 8 kHz in order to enhance the perceptibility of the output signals by improving the signal-to-noise ratio, since this frequency corresponded roughly to the first resonance mode of the intact specimens, as discussed in Section 3.1.

The evolution in time of T is illustrated in Figure 5 for two of the analyzed samples. The values of the transmission coefficient for the intact and damaged samples (states "0" and "1") are reported as solid lines. It is easy to observe an asymptotic behavior during healing: indeed, T increased in time from a very low value immediately after the application of the sodium silicate solution to the fracture surface up to a steady value that was reached again in about seven days (10^4 min). The final value of the transmission coefficient is very close to that of the intact sample. The same evolution with the same time scale was captured through the resonance experiments (see Section 3.1). Therefore, apparently the sodium silicate was able to interact with the cracked cement matrix from a physical-chemical point of view, and as a consequence, the acoustic behavior of the cracked specimens changed in time as long as the healing reactions due to the presence of the alkaline solution were taking place, with a progressive recovery of the acoustic properties initially shown by the intact specimen. The increase of the output spectral magnitude in the transmission experiments, as well as the spectral changes in the resonance experiments were proofs of such a healing phenomenon.

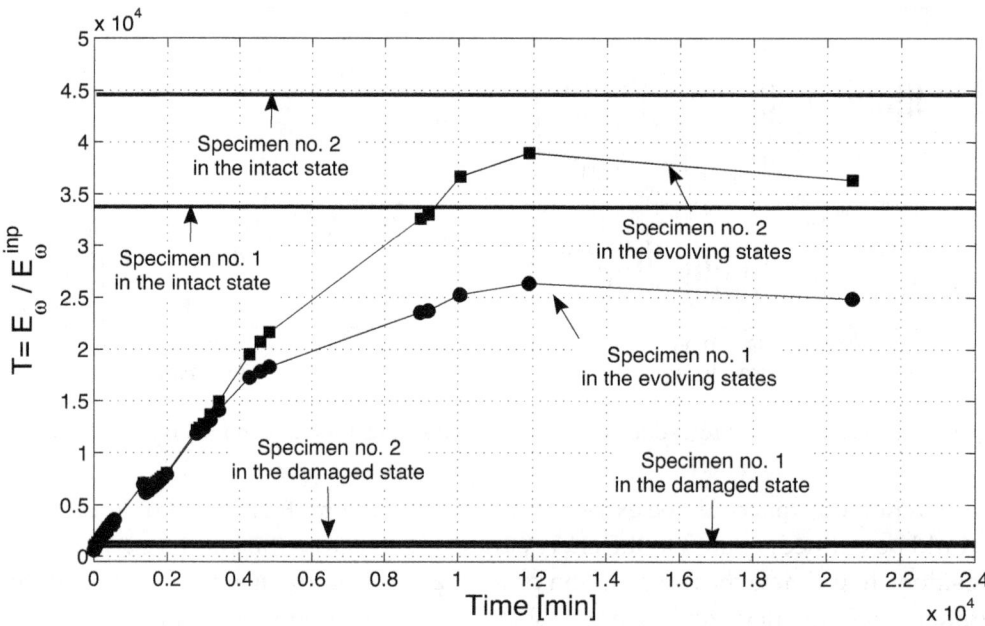

Figure 5. Evolution in time of the transmission coefficient T at the fundamental frequency for Specimen No. 1 and Specimen No. 2.

The spectrum of the signal shown in Figure 4b shows also the emergence of a nonlinear effect, that is the generation of higher order harmonics. In particular, attention was focused here on the analysis of the second order harmonic, i.e., the peak at a frequency value doubled with respect to the fundamental one (16 kHz). As usually observed, the higher the excitation amplitude, the more evident was the nonlinear harmonic generation phenomenon (as will be discussed later). To quantify the effect, it is possible to define a parameter $E_{2\omega}$ representing the amplitude of the second harmonic, obtained integrating the spectrum around the second harmonic peak in a similar way to Equation (1). This quantity, divided by the fundamental harmonic amplitude to avoid spurious effects from the increase of the transmission coefficient discussed above, allows one to define a nonlinear parameter:

$$E_{NL} = \frac{E_{2\omega}}{E_\omega} = \frac{\int_{2\omega_0-\delta/2}^{2\omega_0+\delta/2} E_{out}(\omega)d\omega}{\int_{\omega_0-\delta/2}^{\omega_0+\delta/2} E_{out}(\omega)d\omega} \tag{3}$$

The parameter E_{NL} is shown vs. time in Figure 6 for two samples. The values for the intact and damaged states are reported as solid lines for reference. It is remarkable that the nonlinear effect tended to disappear in time with the progression of sodium silicate hardening: the amount of harmonics generated at time "0" corresponds roughly to the maximum of harmonics generation, as observed on the cracked sample, while it reduces to substantially the same value as the intact sample at the end of the process. The observation is in counter trend with respect to the spectral magnitude corresponding to the fundamental frequency of 8 kHz. In principle, one would expect second harmonic generation to be enhanced by the improved transmission, since in this way, a higher exciting signal is able to reach the fractured region. The result obtained goes in the opposite direction. Therefore, since the appearance of nonlinear effects is known to be related to the presence and the extent of damage, the reduction of nonlinear effects has to be interpreted as a sign of the partial recovery of the material integrity at the fractured interface. Both the solidification of the healing agents (solids in general exhibit less nonlinear features than fluids) and the reduction of clapping between the concrete surfaces are indeed contributing to the reduction in harmonics generation.

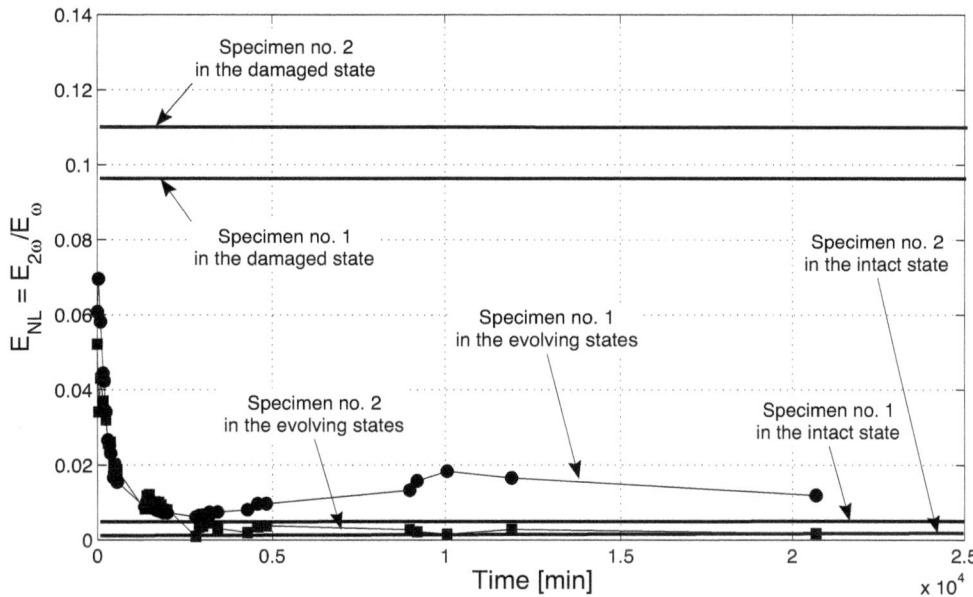

Figure 6. Nonlinear effect of second harmonic generation under continuous sine excitation as a function of healing time for Specimen No. 1 and Specimen No. 2.

A non-negligible difference in time scales with respect to the results of the linear measurement discussed above is however noticeable. The stabilization of the specimens response was reached in approximately four days (about 5000 min), as visible in Figure 6. This could indicate that the healing process is actually a multifaceted phenomenon that involves changes in viscosity of the silicate solution in parallel to chemical reactions with the cement matrix, with the possible formation of different chemical species at different times (see also Section 3.4). The nonlinear analysis is able to capture the variations in the material microstructure that occur in the earlier stages of such a complex healing process, while the linear analysis is probably more suitable to describe the macroscopic changes that finally lead to a mechanical recovery.

The nonlinear analysis discussed above was further developed with an additional test, in which the amplitude of the exciting sinusoidal wave was progressively increased in the range 50 mV–10 V, and the dependence of the harmonics generation as a function of the strain in the sample was analyzed. This corresponds to a standard Nonlinear Elastic Wave Spectroscopy (NEWS) approach [48]. At each time during the healing process, the quantity $E_{2\omega}/E_\omega$ is plotted versus E_ω (see Figure 7). The results found are in agreement with previous studies, which have reported that the output amplitude E_{NL} corresponding to the second harmonic component normalized to the output amplitude at the fundamental frequency has a power-law dependence on the output amplitude E_ω at the fundamental frequency [48,49]:

$$E_{NL} = \frac{E_{2\omega}}{E_\omega} = aE_\omega^b \tag{4}$$

The power-law behavior can be easily appreciated using a dB scale as in Figure 7, where the power-law correlations between the two quantities are represented by straight lines, the slope of each corresponding to the respective exponents in the power-law plot. The generation of the second harmonic (at each given time) increases with increasing strain energy in the sample (each value of E_ω corresponds to a given excitation amplitude). Once again, it is possible to notice that the extent of the nonlinear phenomena tended to decrease in time (curves flattening to the right-bottom side of the plot). In addition, the slope of the curves showed a variation in time, changing from a slope close to one for the initial measurements to a slope approximately equal to 2/3 at the end of the healing process, in analogy with some cases reported in the literature [50].

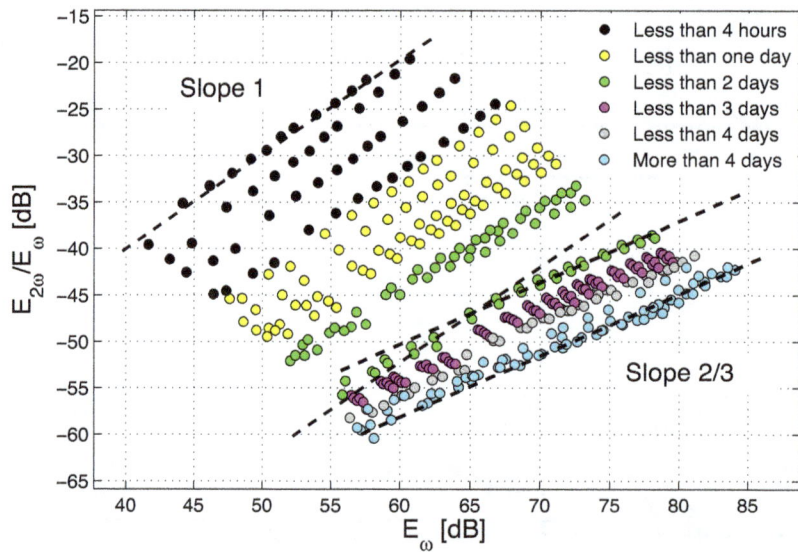

Figure 7. $E_{NL} = E_{2\omega}/E_\omega$ versus E_ω for variable-amplitude sine excitations as a function of healing time for Specimen No. 1.

Quite surprisingly, the curves corresponding to the measurements performed in a time interval between one day and two days after the application of the sodium silicate to the fracture surface displayed a double slope, meaning that two different slope values seemed to be suitable to fit the data for low excitation amplitudes and high excitation amplitudes, respectively. This phenomenon has already been observed in the literature for non-consolidated granular media [51,52].

3.3. Nonlinear Analysis According to the Scaling Subtraction Method under Pulse Excitation

A final nonlinear analysis using pulse excitations at various amplitudes was implemented to complete the ultrasonic investigation on the effect of a sodium silicate solution on a cracked cementitious matrix. It was conducted in accordance with the Scaling Subtraction Method (SSM) procedure and therefore allowed taking into account the contributions to nonlinearity due not only to higher order harmonics generation (see Section 3.2), but also to the nonlinear attenuation and phase shift phenomena, as pointed out in [43]. As fully detailed in previous works [30,44,45,53], the SSM protocol consisted of:

- exciting the specimens using a sequence of pulses at various amplitudes A_i in the range 50 mV–15 V (by means of the same equipment described in Section 2.3). More specifically, a rectangular pulse of width $t = 10$ μs was used;
- at each time from "2" to "n", recording the output response $v_i(t)$ of the specimen to such a variable-amplitude excitation (in a time window of 10 ms, with a sampling rate of 10 MSa/s);
- calculating the so-called "reference signals" at injection amplitude A_i, defined as $v_{ref}(t) = A_i/A_1 v_1(t)$ where $v_1(t)$ is the signal detected at the lowest excitation amplitude A_1;
- computing the "scaled-subtracted signals" obtained as the difference in time between the actual output signals at the various excitation amplitudes and the reference signals at the same amplitudes: $w_i(t) = v_i(t) - v_{ref}(t)$; examples are reported in Figure 8;
- summarizing the information contained in the whole temporal signals using compact indicators: in this case, in continuity with [33], the root mean square (RMS) of the signals ($v(t)$ and $w(t)$) over a prescribed time interval including only the first arrivals was used. It represents the average power of the signals over the assigned time interval and was denoted as x (RMS of the output signal $v(t)$) or η (RMS of the scaled-subtracted signal $w(t)$), while the ratio η/x was referred to as y;

- analyzing the relation between y and x and its evolution in time, thus providing information on the type and extent of nonlinearity in the system as a function of the progression of the healing process.

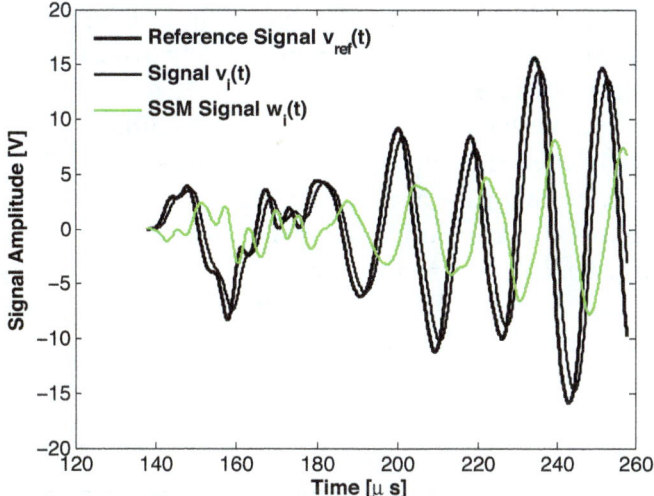

Figure 8. Example of reference signal $v_{ref}(t)$ (black line), detected signal $v_i(t)$ (red line) and scaled-subtracted signal $w_i(t)$ (green line) in a short time window close to the first arrivals time. Data refer to a sample at an intermediate level of healing and intermediate excitation amplitude $A_i = 12$ V.

The signals $v_{ref}(t)$ (black line), $v_i(t)$ (red line) and $w_i(t)$ (green line) are shown in Figure 8 for $A_i = 12$ V. It is possible to observe the nonlinear phase shift and attenuation that make the signal $v_i(t)$ different from $v_{ref}(t)$. The SSM approach allows one to define a nonlinear signal $w_i(t)$, which presents a good signal to noise ratio (much higher than harmonics).

The results of the SSM analysis confirmed the conclusions achieved through the previous nonlinear experiments (see Section 3.2). Plotting y vs. x (Figure 9) revealed that nonlinear phenomena, globally including secondary harmonic generation, nonlinear attenuation and phase shift, were present in the system because the nonlinear indicator y extracted from the scaled-subtracted signals increases with amplitude; thus, the difference between reference and recorded signals is not due to noise effects alone [54].

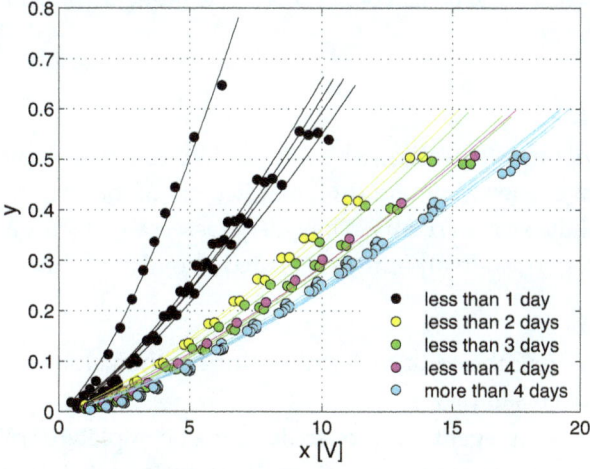

Figure 9. y versus x plot for variable-amplitude pulse excitations in Scaling Subtraction Method (SSM) experiments at different healing times.

Nonetheless, such a nonlinear behavior decreases significantly in time, as proven by the shifting of the y versus x curves to the right-bottom side of the plot, meaning that the material was actually experiencing a consolidation in time due to the occurrence of the healing process, so that the effect of the crack, which was primarily responsible for the nonlinear manifestations observed during the experiments, had to be considered as negligible at the end of the healing process. A stabilization in the nonlinearity decrement seemed to be reached in approximately five days (about 7000 min), in good accordance with the results reported in Section 3.2.

This behavior is similarly visible when considering the evolution in time of the exponent of the power-law fitting that can be used to interpolate the y versus x data [48,50,51,55]. As highlighted in Figure 10, the exponent decreases in time of about 15%–20% and reaches a plateau in approximately four days (same as inferred from nonlinear harmonics measurements). This behavior is qualitatively similar for all specimens analyzed and is reported in Figure 10. The exponent found at the end of the healing process is about 1.2, very close to the exponent found from measurements in the intact case, reported as a reference as a solid line. The exponent at the beginning of the healing process is only slightly smaller than the exponent measured on the cracked sample (again reported for reference as a solid line).

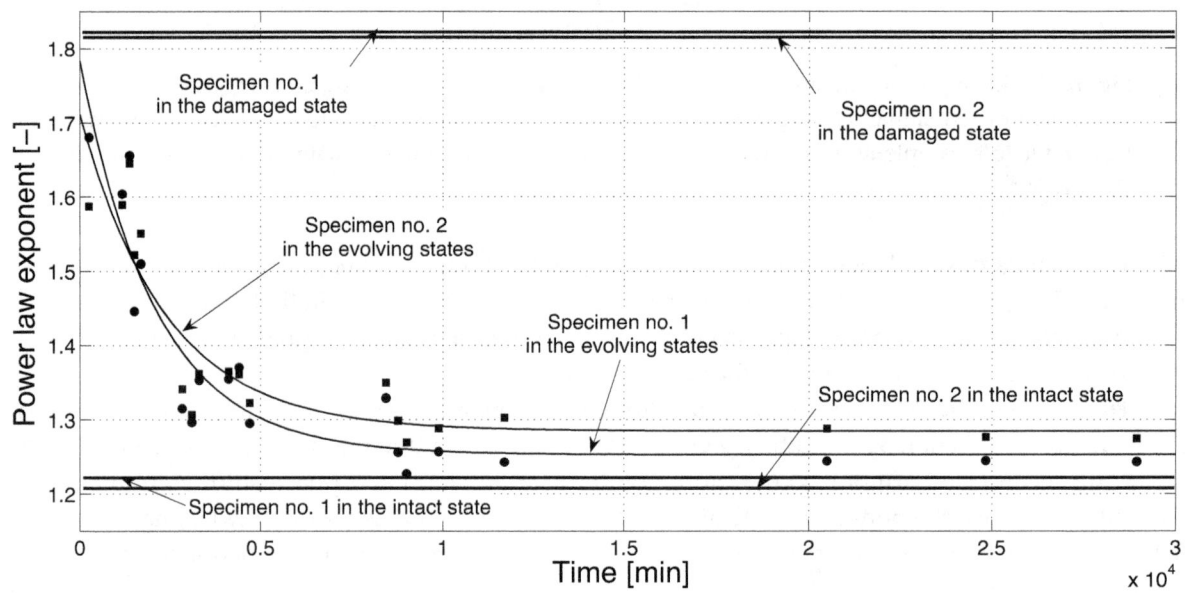

Figure 10. Time evolution of the exponent of the power-law fitting used to interpolate the y versus x data from SSM experiments.

3.4. Discussion

On the whole, the experiments performed point out two relevant results, concerning (a) the time scale over which significant events occur and (b) the indicators of the amount of damage in the sample. In all cases, the recovery of the system seems to be clearly assessed, at least as concerns the recovery of the elastic and acoustic properties. In all figures reported above, both linear and nonlinear indicators show that:

- Immediately after applying the sodium silicate solution, the value of all of the indicators is very close to that of the broken sample;
- After a few days, the healing agent is supposedly almost completely solidified, and the ultrasonic parameters are very close to those measured on the sample in its initial intact state.

As concerns the time scales, acoustic recovery seems to be completed within a few days, i.e., in a much shorter time than expected for full mechanical recovery (20–30 days, according to literature data). The time scale for the recovery of the nonlinear properties seems slightly shorter than that for the recovery of the linear elastic properties: 4–5 days vs. seven days. At this stage, the meaningfulness of this difference has still to be confirmed and analyzed. Anyway, a possible interpretation could be found in the essentially different mesoscopic elements mostly influencing nonlinear vs. linear processes in similar systems. It has been shown for instance in [52] that the weakest contacts in a granular assembly (thus, the smallest and softest elastic elements in the system) are mostly responsible for the nonlinear elastic properties, while the strongest contacts mostly define the structure strength and linear properties. More recently, in [56], the relaxation effect of slow dynamics has been analyzed, which can be regarded, to some extent, as a self-healing process of elastic property recovery: here, the essentially different roles of weak contacts and strong contacts are highlighted in the different time scales involved in the multiple relaxation processes. Therefore, in the present healing process, it can be speculated that the two time scales for nonlinear and linear elastic parameters' recovery correspond to the onset and evolution of different chemical-physical processes, e.g., gelification and penetration into the cracks (soft contacts) and the formation of solid reaction products (hard contacts).

The evolution of the exponent of the power-law expressing the dependence of the nonlinear indicator on the excitation amplitude (Figures 7 and 10) also could give information about the physical processes occurring in the sample. In all of the cases here examined, the power-law exponent decreases with healing time. Elsewhere, it has been shown that an increase of the power-law exponent was possibly ascribed to the appearance of open cracks in a sample [50]. Thus, the results shown here seem to indicate that the sodium silicate here used as a healing agent contributes significantly to crack closure. This effect, taking place in a quite short time, could be an indication of the good quality of the healing agent chosen.

4. Mechanical Tests: Results and Discussion

As anticipated in Section 2.4, at the end of the ultrasonic testing, the specimens were subjected to a final three-point-bending test to determine their flexural strength after repair in comparison to the initial one. Such a mechanical test being destructive, it could not be performed at different times during the healing process as the ultrasonic measurements, but it had to be executed just once, when the acoustic response was proven to be unquestionably stabilized. As a matter of fact, it was performed three weeks after the generation of the crack at mid-span and the subsequent beginning of the healing process. Such a decision was made because from the literature data, it was inferred that at least three to four weeks is the curing time normally required when using sodium silicate in concrete applications, for example as a sealant for concrete pavements. For the same reason, the ultrasonic observation was globally prolonged for three weeks, as well, though stabilization of the acoustical properties turned out to be ensured in a much shorter time.

An example of the mechanical results achieved is reported in Figure 11. Here, the load versus deflection curve for one of the repaired prisms at the end of the healing process is shown in comparison with the one initially obtained for the same prism in the intact state, as measured during the loading process applied to break the prism into two halves. All of the repaired specimens showed a very satisfactory mechanical performance, with a load recovery in the range 66%–76% of the original strength.

The recovery was even more impressive when observing the aspect of the repaired specimens at the end of the flexural tests: as visible in Figure 12, a new crack path was created during the second three-point-bending test, which only partially coincided with the one previously generated at the beginning of the testing campaign. The recovery was limited to a fraction of the original strength probably because the manual application of the sodium silicate to the fracture surface was imperfect at the edges, thus possibly reducing the active cross-section in the repaired sample with respect to the intact one. However, on the whole, the mechanical tests provided clear evidence of the effectiveness of sodium silicate as a healing agent, since the fractured/repaired region perfectly resisted a subsequent flexural test up to failure, both from a qualitative and a quantitative point of view.

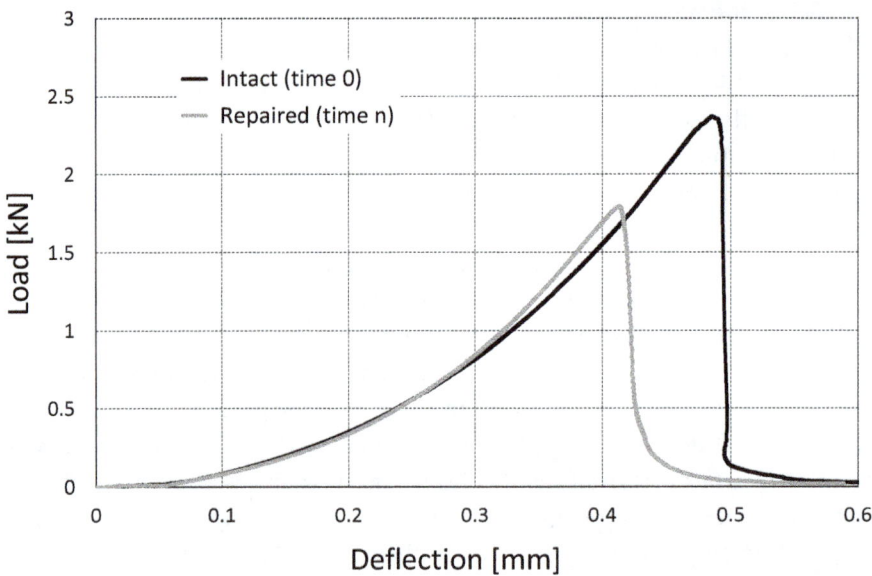

Figure 11. Load versus deflection curves from three-point-bending tests in the intact state (solid black line) and in the damaged state (solid gray line).

Figure 12. Detail of one of the specimens after the final three-point-bending test, with the creation of a new crack path partially separated from the one generated via the first flexural test.

5. Conclusions

The interaction between a cracked cementitious matrix and a solution of sodium silicate was analyzed in terms of acoustic behavior by means of linear and nonlinear ultrasonic techniques. The alkaline solution was used as a repairing agent to consolidate a pass-through crack deliberately generated at mid-span of some mortar prisms by means of three-point-bending tests.

The linear ultrasonic analysis investigated the resonance modes and the acoustic transmission properties of the prisms in their intact, cracked and progressively repaired states, revealing that some changes were induced in time by the presence of the sodium silicate and that the healing mechanism thus created was able to virtually turn the acoustic behavior of the prisms back to their initial state, prior to crack formation. The nonlinear analysis by means of NEWS and SSM confirmed such effects, also allowing one to speculate about the kinetics of the healing reactions induced by the alkaline solution, though the complexity of the healing mechanism itself and the impossibility to perform destructive characterizations during the healing process make it difficult to provide an exhaustive explanation, and further investigation is still to be done.

The flexural tests performed at the end of the healing process revealed a satisfactory recovery of mechanical properties, up to 76% of the initial strength in just three weeks. This finding not only confirms the suitability of sodium silicate to be used as a repairing agent for self-healing concrete applications, but also validates the ability of the ultrasonic techniques here used to detect the significant microstructural changes that eventually affect the mechanical behavior.

The relation between the recovery time for the acoustic properties (here observed to be restored in 5–7 days) and the recovery time for the structural properties of the sample (here tested at 21 days) still needs to be further explored. Additional investigations of the ultrasonic behavior in a long time range (more than three weeks), together with an early mechanical testing of the structural recovery (at 5–7 days) are planned in order to shed light on this issue. Linking the recovery behavior of the acoustic properties to the final mechanical recovery will eventually allow one to:

- define a measure of the expected final recovery without the need for a destructive testing;
- speed-up the optimization of the healing agents by sensibly reducing the time needed to assess the recovery in different conditions and for different healing agents;
- define a protocol to ultrasonically monitor the healing process when the healing agent is released from broken capsules, as in real conditions.

Acknowledgments: This study was conducted in the framework of the research project *SHEcrete—Development of advanced Self-HEaling concrete systems with improved reliability and durability*. The financial support offered by Politecnico di Torino and Fondazione CRT to cover the research activities and the open access publishing costs is gratefully acknowledged. One of the authors M.A.O. gratefully acknowledges the financial support from Le Ministère Algérien de l'Enseignement Supérieur et de la Recherche Scientifique (Bourse résidentielle à l'étranger dans le cadre du P.N.E. 2015/2016).

Author Contributions: P.A. defined the general setup for materials and specimens and designed and performed the mechanical experiments. M.S. and A.S.G. conceived of and designed the ultrasonic experiments. M.A.O. performed the ultrasonic experiments. M.S., A.S.G and M.A.O. analyzed the ultrasonic data. All authors contributed equally in discussing the results, writing and revising the paper.

Conflicts of Interest: The authors declare no conflict of interest.

References

1. Van Tittelboom, K.; De Belie, N. Self-healing in cementitious materials—A review. *Materials* **2013**, *6*, 2182–2217.
2. Huang, H.; Ye, G. Simulation of self healing by further hydration in cementitious materials. *Cem. Concr. Compos.* **2012**, *34*, 460–467.
3. Sisomphon, K.; Copuroglu, O.; Koenders, E.A.B. Self-healing of surface cracks in mortars with expansive additive and crystalline additive. *Cem. Concr. Res.* **2012**, *34*, 566–574.
4. Jonkers, H.M.; Thijssen, A.; Muyzer, G.; Copuroglu, O.; Schlangen, E. Application of bacteria as self-healing agent for the development of sustainable concrete. *Ecol. Eng.* **2010**, *36*, 230–235.
5. Mihashi, H.; Kaneko, Y.; Nishiwaki, T.; Otsuka, K. Fundamental study on development of intelligent concrete characterized by self-healing capability for strength. *Trans. Jpn. Concr. Inst.* **2000**, *22*, 441–450.
6. Hillouin, B.; Van Tittelboom, K.; Gruyaert, E.; De Belie, N.; Loukili, A. Design of polymeric capsules for self-healing concrete. *Cem. Concr. Compos.* **2015**, *55*, 298–307.

7. Maes, M.; Van Tittelboom, K.; De Belie, N. The efficiency of self healing cementitious materials by means of encapsulated polyurethane in chloride containing environments. *Constr. Build. Mater.* **2014**, *71*, 528–537.

8. Formia, A.; Terranova, S.; Antonaci, P.; Pugno, N.M.; Tulliani, J.M. Setup of extruded cementitious hollow tubes as containing/releasing devices in self-healing systems. *Materials* **2015**, *8*, 1897–1923.

9. Antonaci, P.; Pugno, N.M.; Tulliani, J.M. Production and mechanical behavior of micro-encapsulated self-healing cement: A case study. In Proceedings of the 5th World Tribology Congress, Torino, Italy, 8–10 September 2013; pp. 1399–1402.

10. LaRosa Thompson, J.; Silsbee, M.R.; Gill, P.M.; Scheetz, B.E. Characterization of silicate sealers on concrete. *Cem. Concr. Res.* **1997**, *7*, 1561–1567.

11. Kouassi, S.S.; Tognonvi, M.T.; Soro, J.; Rossignol, S. Consolidation mechanism of materials obtained from sodium silicate solution and silica-based aggregates. *J. Non-Cryst. Solids* **2011**, *357*, 3013–3021.

12. Palmero, P.; Formia, A.; Antonaci, P.; Brini, S.; Tulliani, J.M. Geopolymer technology for application-oriented dense and lightened materials. Elaboration and characterization. *Ceram. Int.* **2015**, *41*, 12967–12979.

13. Latifi, N.; Eisazadeh, A.; Marto, A. Strength behavior and microstructural characteristics of tropical laterite soil treated with sodium silicate-based liquid stabilizer. *Environ. Earth Sci.* **2014**, *72*, 91–98.

14. Pakir, F.; Marto, A.; Yunus, N.Z.M.; Tajudin, S.A.A.; Tan, C.S. Effect of sodium silicate as liquid based stabilizer on shear strenght of marine clay. *J. Teknol.* **2015**, *76*, 45–50.

15. McGettigan, E. Silicon-based weatherproofing materials. *Concr. Int.* **1992**, *14*, 52–56.

16. Song, Z.; Xue, X.; Li, Y.; Yang, J.; He, Z.; Shen, S.; Jiang, L.; Zhang, W.; Xu, L.; Zhang, H.; et al. Experimental exploration of the waterproofing mechanism of inorganic sodium silicate-based concrete sealers. *Constr. Build. Mater.* **2016**, *104*, 276–283.

17. Jiang, L.; Xue, X.; Zhang, W.; Yang, J.; Zhang, H.; Li, Y.; Zhang, R.; Zhang, Z.; Xu, L.; Qu, J.; et al. The investigation of factors affecting the water impermeability of inorganic sodium silicate-based concrete sealers. *Constr. Build. Mater.* **2015**, *93*, 729–736.

18. Prasetia, I. The ineffectiveness of sodium based surface treatment on the mitigation of alkali silica reaction. *Int. J. Adv. Eng. Technol.* **2015**, *8*, 466–474.

19. Huang, H.; Ye, G. Application of sodium silicate solution as self-healing agent in cementitious materials. In Proceedings of the International RILEM Conference on Advances in Construction Materials Through Science and Engineering, Hong Kong, China, 5–7 September 2011; pp. 530–536.

20. Pelletier, M.M.; Brown, R.; Shukla, A.; Bose, A. Self Healing Concrete with a Microencapsulated Healing Agent. *Cem. Concr. Res.* **2011**.

21. Aggelis, D.G.; Barkoula, N.M.; Matikas, T.E.; Paipetis, A.S. Acoustic structural health monitoring of composite materials: Damage identification and evaluation in cross ply laminates using acoustic emission and ultrasonics. *Compos. Sci. Technol.* **2012**, *72*, 1127–1133.

22. Van den Abeele, K.; Desadeleer, W.; De Schutter, G.; Wevers, M. Active and passive monitoring of the early hydration process in concrete using linear and nonlinear acoustics. *Cem. Concr. Res.* **2009**, *739*, 426–432.

23. Li, Z.J. Microcrack characterization in concrete under uniaxial tension. *Mag. Concr. Res.* **1996**, *48*, 219–228.

24. Musyoka, N.M.; Petrik, L.F.; Hums, E.; Baser, H.; Schwieger, W. In situ ultrasonic monitoring of zeolite a crystallization from coal fly ash. *Catal. Today* **2012**, *190*, 38–46.

25. Vlad, M.; González, L.; Gómez, S.; López, J.; Carlson, J.E.; Fernández, E. Ultrasound monitoring of the setting of calcium-based bone cements. *J. Mater. Sci. Mater. Med.* **2012**, *23*, 1563–1568.

26. Bui, D.; Kodjo, S.A.; Rivard, P.; Fournier, B. Evaluation of concrete distributed cracks by ultrasonic travel time shift under an external mechanical perturbation: Study of indirect and semi-direct transmission configurations. *J. Nondestruct. Eval.* **2013**, *32*, 25–26.

27. Zhang, Y.; Abraham, O.; Grondin, F.; Loukili, A.; Tournat, V.; Duff, A.L.; Lascoup, B.; Durand, O. Study of stress-induced velocity variation in concrete under direct tensile force and monitoring of the damage level by using thermally-compensated coda wave interferometry. *Ultrasonics* **2012**, *52*, 1038–1045.

28. Lesage, J.C; Sinclair, A.N. Characterization of prestressed concrete cylinder pipe by resonance acoustic spectroscopy. *J. Pipeline Syst. Eng. Pract.* **2015**, *6*, 04014011.

29. Payan, C.; Ulrich, T.J.; Le Bas, P.Y.; Saleh, T.; Guimaraes, M. Quantitative linear and nonlinear resonance inspection techniques and analysis for material characterization: Application to concrete thermal damage. *J. Acoust. Soc. Am.* **2014**, *136*, 537–546.

30. Antonaci, P.; Bruno, C.L.E.; Scalerandi, M.; Tondolo, F. Effects of corrosion on linear and nonlinear elastic properties of reinforced concrete. *Cem. Concr. Res.* **2013**, *51*, 96–103.

31. Hilloulin, B.; Legland, J.B.; Lys, E.; Abraham, O.; Loukili, A.; Grondin, F.; Durand, O.; Tournat, V. Monitoring of autogenous crack healing in cementitious materials by the nonlinear modulation of ultrasonic coda waves, 3D microscopy and X-ray microtomography. *Constr. Build. Mater.* **2016**, *123*, 143–152.

32. Baccouche, Y.; Bentahar, M.; Mechri, C.; El Guerjouma, R.; Hédi Ben Ghozlen, M. Hysteretic nonlinearity analysis in damaged composite plates using guided waves. *J. Acoust. Soc. Am.* **2013**, *133*, EL256–EL261.

33. Antonaci, P.; Bruno, C.L.E.; Gliozzi, A.S.; Scalerandi, M. Monitoring evolution of compressive damage in concrete with linear and nonlinear ultrasonic methods. *Cem. Concr. Res.* **2010**, *40*, 1106–1113.

34. Bouchaala, F.; Payan, C.; Garnier, V.; Balayssac, J.P. Carbonation assessment in concrete by nonlinear ultrasound. *Ceme. Concr. Res.* **2011**, *41*, 557–559.

35. Shah, A.A.; Ribakov, Y. Non-destructive evaluation of concrete in damaged and undamaged states. *Mater. Des.* **2009**, *30*, 3504–3511.

36. Kim, G.; Kim, J.Y.; Kurtis, K.E.; Jacobs, L.J.; Le Pape, Y.; Guimaraes, M. Quantitative evaluation of carbonation in concrete using nonlinear ultrasound. *Mater. Struct.* **2016**, *49*, 399–409.

37. Bentahar, M.; El Aqra, H.; El Guerjouma, R.; Griffa, M.; Scalerandi, M. Hysteretic elasticity in damaged concrete: Quantitative analysis of slow and fast dynamics. *Phys. Rev. B* **2006**, *73*, 014116.

38. Chen, J.; Kim, J.Y.; Kurtis, K.E.; Jacobs, L.J. Theoretical and experimental study of the nonlinear resonance vibration of cementitious materials with an application to damage characterization. *J. Acoust. Soc. Am.* **2011**, *130*, 2728–2734.

39. Chen, J.; Zhang, L. Experimental study of effects of water-cement ratio and curing time on nonlinear resonance of concrete. *Mater. Struct.* **2015**, *148*, 423–433.

40. Scalerandi, M.; Gliozzi, A.S.; Ait Ouarabi, M.; Boubenider, F. Continuous waves probing in dynamic acoustoelastic testing. *Appl. Phys. Lett.* **2016**, *108*, 214103.

41. Hilloulin, B.; Zhang, Y.; Abraham, O.; Loukili, A.; Grondin, F.; Durand, O.; Tournat, V. Small crack detection in cementitious materials using nonlinear coda wave modulation. *NDT E Int.* **2014**, *68*, 98–104.

42. Zhang, Y.; Abraham, O.; Tournat, V.; Le Duff, A.; Lascoup, B.; Loukili, A.; Grondin, F.; Durand, O. Validation of a thermal bias control technique for Coda Wave Interferometry (CWI). *Ultrasonics* **2013**, *53*, 658–664.

43. Bruno, C.L.E.; Gliozzi, A.S.; Scalerandi, M.; Antonaci, P. Analysis of elastic nonlinearity using the scaling subtraction method. *Phys. Rev. B* **2009**, *79*, 064108.

44. Antonaci, P.; Bruno, C.L.E.; Gliozzi, A.S.; Scalerandi, M. Evolution of damage-induced nonlinearity in proximity of discontinuities in concrete. *Int. J. Solids Struct.* **2010**, *47*, 1603–1610.

45. Scalerandi, M.; Griffa, M.; Antonaci, P.; Wyrzykowski, M.; Lura, P. Nonlinear elastic response of thermally damaged consolidated granular media. *J. Appl. Phys.* **2013**, *113*, 154902.

46. Antonaci, P.; Formia, A.; Gliozzi, A.S.; Scalerandi, M.; Tulliani, J.M. Diagnostic application of nonlinear ultrasonics to characterize degradation by expansive salts in masonry systems. *NDT E Int.* **2013**, *55*, 57–63.

47. Formia, A.; Irico, S.; Bertola, F.; Canonico, F.; Antonaci, P.; Pugno, N.M.; Tulliani, J.M. Experimental analysis of self-healing cement-based materials incorporating extruded cementitious hollow tubes. *J. Int. Mater. Syst. Struct.* **2016**, *27*, 2633–2652.

48. Van den Abeele, K.; Sutin, A.; Carmeliet, J.; Johnson, P.A. Micro-damage diagnostics using nonlinear elastic wave spectroscopy (NEWS). *NDT E Int.* **2001**, *34*, 239–248.

49. Van den Abeele, K.; De Visscher, J. Damage assessment in reinforced concrete using spectral and temporal nonlinear vibration techniques. *Cem. Concr. Res.* **2000**, *30*, 1453–1464.

50. Scalerandi, M.; Idjimarene, S.; Bentahar, M.; El Guerjouma, R. Evidence of evolution of microstructure in solid elastic media based on a power law analysis. *Commun. Nonlinear Sci. Numer. Simul.* **2015**, *22*, 334–347.

51. Tournat, V.; Gusev, V.E.; Zaitsev, V.Yu.; Castagnède, B. Acoustic second-harmonic generation with shear to longitudinal mode conversion in granular media. *Europhys. Lett.* **2004**, *66*, 798–804.

52. Tournat, V.; Zaitsev, V.; Gusev, V.; Nazarov, V.; Béquin, P.; Castagnède, B. Probing weak forces in granular media through nonlinear dynamic dilatancy: Clapping contacts and polarization anisotropy. *Phys. Rev. Lett.* **2004**, *92*, 085502.

53. Ait Ouarabi, M.; Boubenider, F.; Gliozzi, A.S.; Scalerandi, M. Nonlinear coda wave analysis of hysteretic elastic behavior in strongly scattering elastic media. *Phys. Rev. B* **2016**, *94*, 134103.

54. Bentahar, M.; El Guerjouma, R.; Idijmarene, S.; Scalerandi, M. Influence of noise on the threshold for detection of elastic nonlinearity. *J. Appl. Phys.* **2013**, *113*, 043516.

55. Scalerandi, M.; Gliozzi, A.S.; Olivero, D. Discrimination between cracks and recrystallization in steel using nonlinear techniques. *J. Nondestruct. Eval.* **2014**, *33*, 269–278.

56. Zaitsev, V.Y.; Gusev, V.E.; Tournat, V.; Richard, P. Slow relaxation and aging phenomena at the nanoscale in granular materials. *Phys. Rev. Lett.* **2014**, *112*, 108302.

10

Experimental Study on Mechanical Properties and Porosity of Organic Microcapsules Based Self-Healing Cementitious Composite

Xianfeng Wang, Peipei Sun, Ningxu Han and Feng Xing *

Guangdong Provincial Key Laboratory of Durability for Marine Civil Engineering,
College of Civil Engineering, Shenzhen University, Shenzhen 518060, Guangdong, China;
xfw@szu.edu.cn (X.W.); sunpeipei@agile.com.cn (P.S.); nxhan@szu.edu.cn (N.H.)
* Correspondence: xingf@szu.edu.cn

Academic Editor: Nele De Belie

Abstract: Encapsulation of healing agents embedded in a material matrix has become one of the major approaches for achieving self-healing function in cementitious materials in recent years. A novel type of microcapsules based self-healing cementitious composite was developed in Guangdong Provincial Key Laboratory of Durability for Marine Civil Engineering, Shenzhen University. In this study, both macro performance and the microstructure of the composite are investigated. The macro performance was evaluated by employing the compressive strength and the dynamic modulus, whereas the microstructure was represented by the pore structure parameters such as porosity, cumulative-pore volume, and average-pore diameter, which are significantly correlated to the pore-size distribution and the compressive strength. The results showed that both the compressive strength and the dynamic modulus, as well as the pore structure parameters such as porosity, cumulative-pore volume, and average-pore diameter of the specimen decrease to some extent with the amount of microcapsules. However, the self-healing rate and the recovery rate of the specimen performance and the pore-structure parameters increase with the amount of microcapsules. The results should confirm the self-healing function of microcapsules in the cementitious composite from macroscopic and microscopic viewpoints.

Keywords: microcapsules; healing rate; recovery rate; pore size; dynamic modulus

---

1. Introduction

Concrete has been one of the most widely used building materials in the world owing to its low energy consumption, low cost, and relatively high durability [1]. However, in the natural environment, there is a risk of erosion. Materials' age and environmental effects result in concrete microcracks, local damage, and fracture. In particular, in actual concrete structures, micro-cracks are difficult to detect accurately because of the limitations of the detection technology; moreover, the conventional method cannot effectively repair the internal structure of these invisible microcracks. If these microcracks are not effectively repaired, it will affect the normal performance and service life of the structure, and may lead to macroscopic cracking and cause structural brittle fracture, and even lead to a catastrophic accident [2]. Hence, it is necessary to develop new repairing techniques and materials that are able to perceive material damage, and passively and automatically repair the damaged site, thereby restoring the mechanical properties and durability of concrete.

There are several self-healing approaches, such as microbiology, shape memory alloy or polymer, extending hydration of cementitious admixtures, and microencapsulation technology, which are possible because the material has a core-shell tiny container structure. Xing et al. [3] developed a self-healing technique using organic microcapsules for cement paste. In their study, the integrity of organic microcapsules was maintained while preparing the cement paste, and the microcapsules ruptured when the cracks passed through them. An element analysis provided definite proof of the healing phenomenon on the crack faces.

Microcapsules, as temporary vessels, hold the healing agent until damage induced trigger occurs. The repair principle of microcapsules based on the self-healing cementitious composites is similar to the principle of bionics, wherein a crack is initiated and then propagated. The embedded microcapsules ruptured under stresses, and subsequently, the healing agent was released into the crack plane through the capillary action to achieve the healing function, thereby inhibiting crack propagation and repairing the crack, even restoring or improving the effect of the material strength [4–6]. It is efficient and has unique advantages from the viewpoint of durability. This method can heal cracks induced at places difficult to access; moreover, the method is relatively low cost [7]. Although the method needs to be further developed and investigated in the future, the encapsulation scheme seems to be a promising approach for self-healing [8,9].

In recent years, a considerable number of studies have been conducted following this route. Su et al. [10–12] used a type of rejuvenator as the core materials to investigate the mechanical healing behaviors of bitumen using a modified beam on elastic foundation method, wherein three types of microcapsules with different mean sizes and shell thicknesses were considered. Dong et al. [13,14] investigated the self-healing capacity of a cementitious composite containing organic microcapsules by evaluating the crack healing effect, mechanical property, as well as chloride permeability. Their experimental results revealed that the crack-healing ratios were 20%–45%, and the healing ratios of the compressive strength and impermeability were approximately 13% and 19.8%, respectively. They [15] also developed a chemical self-healing system, for which experiments were conducted in a stimulated concrete pore solution. The smart release behavior of the healing agent in the microcapsule, characterized by the ethylene diamine tetra-acetic acid titration method, was a function of time, and controlled by the wall thickness of the microcapsule. Lv et al. [16] developed a type of polymeric microcapsule with phenol–formaldehyde resin for the shell and dicyclopentadiene as the healing agent for the self-healing of microcracks in cementitious materials. The chemical stability of the microcapsules and the trigger performance were verified in a simulated concrete pore solution and hardened cement paste specimens.

De Belie's group [17,18] studied the self-healing concrete by employing the microencapsulated bacterial spores, wherein the breakage of the microcapsules upon cracking was verified using scanning electron microscope (SEM), and the self-healing capacity was evaluated via the crack healing ratio and the water permeability. Their results showed that the healing ratios in the specimens were 48%–80%. They also studied the microstructure of the capsules containing self-healing materials by using micro-computed tomography. The three-dimensional distribution and de-bonding of the microcapsules in their native state in a polymer system with self-healing properties were indicated.

In addition, there are a number of review papers [19–25] in the field of self-healing materials, and five international conferences were conducted on self-healing materials [26,27]. Souradeep and Kua [24] suggested eight factors that affect the effectiveness of self-healing by encapsulation, which included the following: (1) robustness during mixing and (2) probability of cracks encountering the capsules. They indicated that there is a lack of research on the efficacy of self-healing in an actual application environment. As some fundamental issues, such as the control of fabrication, have not been clarified, most research is still in the laboratory level. Muhammad et al. [25] reviewed the self-healing measurement methods, particularly concerning the healing effect of the width, depth, and length of cracks. They indicated that few studies on the healing efficiency were conducted at the microstructure or nanostructure level. It is also found that, for microcapsules based self-healing materials, there are still fewer systematic studies on the healing behavior from the viewpoints of porosity and the correlation between the macro behavior and microstructure.

Moreover, with the hardening of the concrete structure, the evaporation of free water inside the concrete may generate pores, and the presence of pores of different sizes is an important component of hardened concrete structure, which is an important factor influencing its performance. Mercury intrusion porosimetry (MIP) [28–30] is commonly used to evaluate the pore structure of cementitious materials. By using MIP, the pore-size distribution may be determined mainly based on the relationship between the amount of mercury flowing into the porous system of the concrete materials and the applied pressure. It is capable of measuring a vast range of pore entry radii varying from 6 nm to 400 μm.

In this study, the organic microcapsules were prepared and used to make the self-healing cementitious composite specimens. This type of microencapsulation approach is based on the physical trigger. The organic microcapsules with epoxy core can produce ductility for the cementitious composites and produce relatively high healing rate with the amount of microcapsules. This is the advantage. However, embedding the microcapsules may weaken the initial strength of the specimen; hence, an optimum dosage should be determined before industrial application. The healing efficiency of the specimens was investigated based on the strength, dynamic modulus, as well as the pore-size distribution via MIP test. The damage was inflicted to the specimens by applying uniform compression. The pore-size distributions at intact, damaged, and healed states were measured, and the healing ratios, as well as the recovery ratios were determined. Then, the pore parameters, such as porosity, cumulative-pore volume, and average-pore diameter, were obtained, and further research on the pore-size distribution model of mortars was conducted, from which the healing effect of the microcapsules present in the cementitious materials was validated.

2. Experimental Scheme

2.1. Materials and Specimens

2.1.1. Preparation of Microcapsules

The organic microcapsules were synthesized at Guangdong Provincial Key Laboratory of Durability for Marine Civil Engineering, Shenzhen University. The shell material is urea formoldehyde (UF), and the core-healing agent is epoxy. The materials used for the synthesis are urea (analytical reagent: AR) obtained from Tianjin Jinfeng Chemical Ltd. Co. (Tianjin, China), formaldehyde solution (AR) obtained from Tianjin Baishi Chemical Ltd. Co. (Tianjin, China), triethanolamine (AR) obtained from Tianjin Fuyu Chemical Ltd. Co. (Tianjin, China), butyl glycidyl ether (BGE) obtained from Shanghai Bangcheng Chemical Ltd. Co. (Tianjin, China), a type of epoxy resin E-51 obained from Shenzhen Yoshida Chemical Ltd. Co. (Shenzhen, China), and sulfuric acid obtained from Tianjin Guangfu Institute of Fine Chemicals. The details of the synthesis can be found in a previous study [6].

Table 1 lists the detailed parameters of the microcapsules. Figure 1a shows an image of the microcapsules analyzed using a SEM. Figure 1b shows the particle-size distribution of the microcapsules, wherein the mean diameter is 121.66 μm. The diameter distribution was determined by using an optical microscope and SEM for accounting the samples of 300 microcapsules. The shell thickness of the organic microcapsules was tested using SEM images when the microcapsules were ruptured by grinding. The capsule core content was obtained by extraction [31]. The procedure was as follows: weighing a certain amount of dried microcapsules, full grinding out the core agent, and then placing the ground material in acetone and placed 3d; during the period, acetone was changed every 24 h so that the core content fully flowed out. Next, the ground material was dried in a drying oven, and the remaining material formed the wall of the microcapsule. The core content w_{cc} can be calculated as

$$w_{cc} = \frac{\left(m_{capsule} - m_{shell}\right)}{m_{capsule}} \times 100\% \tag{1}$$

where $m_{capsule}$ is the mass of the measured microcapsules, and m_{shell} is the mass of the microcapsule wall.

Table 1. Microcapsules parameters.

Rotation Velocity in Synthesis (r/min)	Mean Diameter (μm)	Wall Thickness (μm)	Capsule Core Content (%)
600	121.66	5.46	67.8

(a)

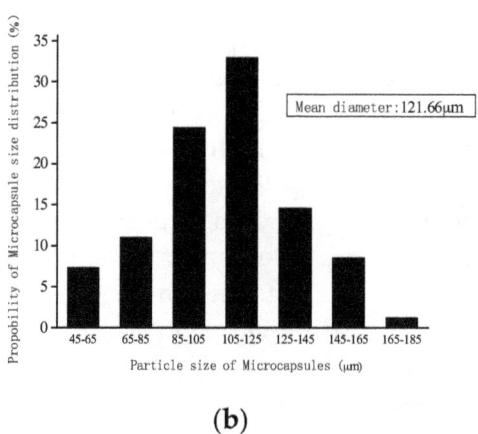

(b)

Figure 1. Organic microcapsules: (**a**) SEM images of the microcapsules; (**b**) Particle-size distribution of the microcapsules.

2.1.2. Preparation of the Microcapsule-Based Cementitious Materials and Specimens

To build the microcapsule-based self-healing cementitious composite, the following materials were used: China portland Cement GB-175-2007 [32] PII42.5R type from Guangzhou Zhujiang Cement Ltd. Company (Guangzhou, China); drinkable tap water; GB/T17671-1999 ISO standard sand [33] from Xiamen Isiou Ltd. Company (Xiamen, China); a curing agent MC120D from Guangzhou Kawai Electronic Materials Ltd. Company (Guangzhou, China).

The mortar specimens were prepared using the mix proportion, given in Table 2, wherein the water/cement and the binder/sand ratios were 0.5 and 1:3, respectively. The microcapsule size (diameter of 121.66 μm) and the content (0%, 3%, 6%, and 9% to cement mass) were considered. The amount of catalyst MC120D used was half the amount of the organic microcapsules.

Table 2. Mix proportions of specimens.

Sample No.	1	2	3	4
Particle diameter (μm)	N/A	121.66	121.66	121.66
Microcapsule content (to cement mass)	0%	3%	6%	9%

The prismatic specimens of dimensions 40 mm × 40 mm × 160 mm were prepared by mixing the microcapsules as well as the catalyst MC120D with water, cement, and sands, as shown in Figure 2. The specimens were demolded after 24 h and cured for 28 days under the same conditions as that of the standard mortar tests (temperature 20 °C, humidity >90%). Figure 3 shows the microcapsules dispersed in the cement mortar observed using SEM.

Figure 2. Specimens for test.

Figure 3. Microcapsules dispersed in the cementitious composite.

2.2. Experimental Methods

2.2.1. Compressive Strength Test

The compressive strength test was conducted using the testing machine RGM-4010 (REGEL Corp., Shenzhen, China) based on the standard given in a previous study [33]. The loading speed was 2.4 kN/s. The compression strength is the ratio of the maximum compression load to the loading area (40 × 40 mm). Three specimens were grouped as one sample in the test, wherein the average value was used for the representative one.

2.2.2. Preparation of Specimens for Self-Healing Test

Three groups of specimens under the same mix proportion were prepared. The first group was for the compressive strength and pore structure tests at the intact state. The second group was for the test at the damaged state, which was obtained by applying a pre-load of 60% σ_{max} (maximum compressive strength) to the specimens. The third group was for the test at the healed state, which was obtained by curing the wrapped damaged specimens in a curing box below a temperature of 50 °C for 7 days.

The sizes of the specimens for the dynamic mechanical analysis (DMA) test and pore-structure test were much smaller than the standard cement mortar; moreover, they were different for each test. The corresponding specimens were obtained using fine cutting, and were immersed in ethanol for 7 days to terminate hydration. Thereafter, they were placed in a drying box at 60 °C, which were ready for the test. Table 3 gives the test number and the types of the tests types, wherein the sample No. is referred from Table 2.

Table 3. Test No. and types.

No. \ Sample No.	1	2	3	4	Test Type
1	1-1-1	2-1-1	3-1-1	4-1-1	Compression test,
2	1-1-2	2-1-2	3-1-2	4-1-2	DMA, MIP test
3	1-2-1	2-2-1	3-2-1	4-2-1	DMA test
4	1-2-2	2-2-2	3-2-2	4-2-2	MIP test to damaged sample
5	1-3-1	2-3-1	3-3-1	4-3-1	Compression test to healed sample
6	1-3-2	2-3-2	3-3-2	4-3-2	DMA, MIP test

2.2.3. Dynamic Mechanical Analysis (DMA) Test

The specimens for the DMA test were prepared to dimensions 30 mm × 30 mm × 30 mm. The testing machine was DMA + 1000 from ACOEM Metravib Ltd. Co. (Limonest, France). The frequency was 1 Hz, and the static and the dynamic force were 120 N and 100 N, respectively. The DMA tests were conducted for samples No. 1–4 with a microcapsule size of 121.66 µm. The dynamic moduli were then obtained.

2.2.4. MIP Test

The MIP test has been widely used to test the pore-structure parameters of cement-based materials, such as porosity, critical pore size, threshold pore size, and mean diameter. In this study, Auto Pore 9500 type testing machine from Micromeritics Ltd. Co. (Norcross, GA, USA). was used. The dimensions of the samples were below 10 mm × 10 mm × 10 mm. Both the low and high-pressure analyses were conducted. First, a vacuum condition of less than 40 µm Hg air pressure was obtained. Then, the mercury was pressurized at a pressure range of 0.54 to 29.98 psia. The corresponding pore size ranges were from 6 to 350 µm. After the low-pressure test, the high-pressure test was implemented in a similar manner. The high-pressure range of 36.4–29,906.6 psia corresponded to the measured pore size range of 6.05–4964 nm.

The basic principle of MIP test is to convert the mercury pressure to the pore size by applying the Washburn, as expressed in equation [30].

$$d = \frac{4\gamma \cos \theta}{P} \tag{2}$$

where d represents the pore size (µm); γ denotes the mercury surface tension, 0.484 N/m; θ signifies the contact angle of the mercury with the pore wall, 117°; P is the mercury pressure of the injected sample (Pa). With the Washburn equation, the pore-size distribution as well as the related pore structure data can be calculated.

3. Results and Discussion

3.1. Compression Test Results and Discussions

The compressive strength was obtained by using $\sigma = \frac{F}{b^2}$, where F is the maximum compressive load and b is the lateral length of the specimen (40 mm). Figure 4 shows the variation in the compressive strength with the amount of microcapsules. Herein, for convenience of expression, sample No. 1 was treated as a reference. The compressive strength was 46.1 MPa. The error bars in the figure indicated the standard deviation, wherein the maximum value was 3.4 MPa in the case of 9% microcapsules.

The compressive strengths are observed to decrease with the amount of microcapsules. The Young's modulus of UF is less than 10 GPa, and the Young's modulus of the cement mortar is in the range of 10–30 GPa. Hence, the lower values of the Young's modulus and the interface strength make the microcapsules perform as weak phases. The larger the microcapsules, the greater the effect [34]. The compressive strength can be decreased up to 14.5% in the case of 9% microcapsules. However, in the case of small particle sizes, a small amount of microcapsules (3%) may lead to a slight increase (1%) in the strength or not affect the variation. This may be the cause of the filler effect of the microcapsules, which coincide with the results of a previous study [6].

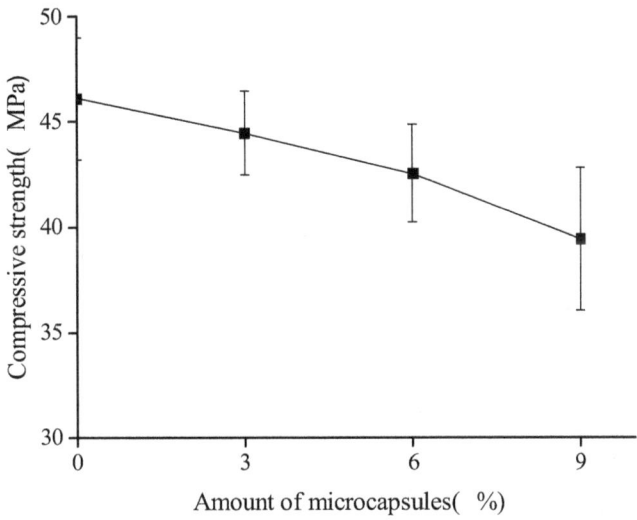

Figure 4. Variation in the compressive strength with the amount of microcapsules.

As mentioned previously, the mechanism of the self-healing function achieved by the microcapsules involves the cracking induced rupture of the microcapsules. The core-healing agent flows out and reacts with the catalyst; thereafter, it solidifies and glues to the cracks. To investigate the self-healing efficiency of the mechanical behavior, the recovery rate η_{S-ROC} and the healing rate η_{S-HEA} are defined, respectively, as

$$\eta_{S-ROC} = \frac{f_{healed}}{f_{original}} \times 100\% \,, \tag{3}$$

$$\eta_{S-HEA} = \frac{f_{healed} - f_{damaged}}{f_{damaged}} \times 100\% \,, \tag{4}$$

where f_{healed} denotes the specimen strength after the healing process (MPa); $f_{original}$ is the original strength of the specimen at 28 days (MPa); $f_{damaged}$ is the strength of the specimen at the damaged state because of the pre-loading (MPa).

Figures 5 and 6 show the healing and recovery rates of the compressive strength, respectively. The figures show that, in the absence of microcapsules (reference specimen), the healing and recovery rates of the compressive strength are −3.1% and 96.90%, respectively, which means that, from an overall point of view, the strength is difficult to recovery because of the effect of relatively large cracks, though the negative value arises from the measurement dispersion. The healing and recovery rates are greater than 0% and 100%, respectively, for the specimens containing the microcapsules, which represents a good healing efficiency of the microcapsules. The adhesive reaction glues the cracks and improves the compressive strength. For the cement mortar with the same size, but different content of the microcapsules, which are in the range of 3% to 9%, the healing and recovery rates increased with the increase in the content of microcapsules. The content of microcapsules increases the content of the healing agent, thereby increasing the healing rate. The healing rate of the cement mortars with the microcapsules of content 9% reached 5.42%. This should strongly demonstrate the self-healing effect on the compressive strength.

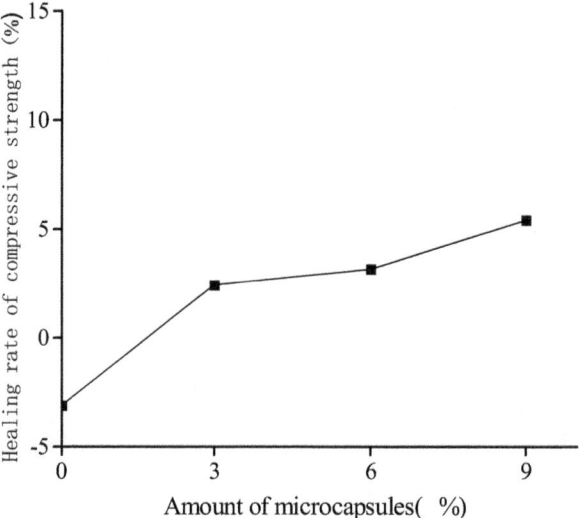

Figure 5. Healing rate (Equation (4)) of compressive strength with amount of microcapsules.

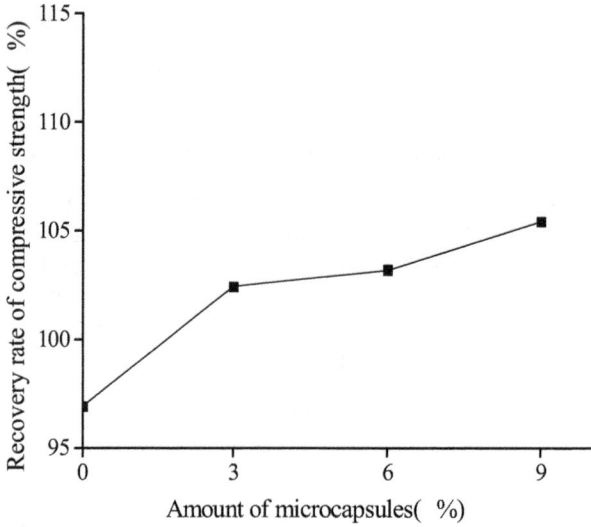

Figure 6. Recovery rate (Equation (3)) of compressive strength with amount of microcapsules.

3.2. DMA Results and Discussions

To investigate the self-healing efficiency of the dynamic modulus of the specimen, we use similar definitions of the recovery rate η_{E-ROC} and healing rate η_{E-HEA} with Equations (3) and (4), respectively, as

$$\eta_{E-ROC} = \frac{E_{healed}}{E_{original}} \times 100\% , \tag{5}$$

$$\eta_{E-HEA} = \frac{E_{healed} - E_{damaged}}{E_{damaged}} \times 100\% , \tag{6}$$

where E_{healed} denotes the dynamic modulus after the healing process (GPa); $E_{original}$ is the original dynamic modulus of the specimen at 28 days (GPa); $E_{damaged}$ is the dynamic modulus of the specimen at the damaged state (GPa).

Figure 7 shows the variation in the dynamic modulus of the specimens at the original, damaged, and healed states. The data are the averaged values of the three specimens. The error bars indicated the standard deviation, wherein the maximum value was 0.061 GPa for 9% of microcapsules after the healing. It is seen that the dynamic modulus of the cement-mortar decreases with the increase in the content of microcapsules. Because of the difference of the elastic modulus between the microcapsules and cement matrix, and the interface between them may also be relatively weak, the stiffness as well as the elastic modulus of the cementitious composite, should be decreased with the content of microcapsules, particularly for the case shown in Figure 7. Moreover, it can be seen that the dynamic modulus of the cementitious composite without microcapsules is considerably higher than that of the composite containing the microcapsules. The dynamic modulus of the specimen after the preloading was lower than that of the original specimens, which indicates that the specimen was damaged after the preloading; the microcracks appeared and the stiffness as well as the elastic modulus of the specimen decreased. The dynamic modulus of the specimen after the healing was higher than that of the damaged sample. In the case of the specimen without microcapsules, the high dynamic modulus could be attributed to the extension of the hydration of the cementitious composite, which self-heals the microcracks. In the case of the specimen containing the microcapsules, the reaction of the healing agent with the catalyst improves the elastic modulus.

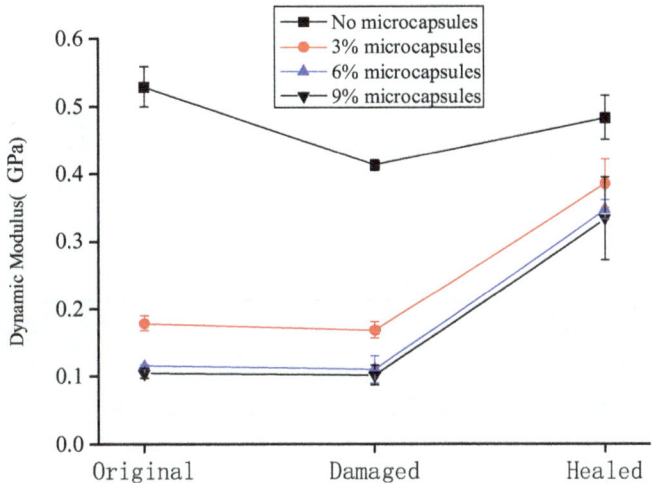

Figure 7. Variation in the dynamic modulus of the specimens.

Figures 8 and 9 show that the healing and recovery rates of the dynamic modulus of the cement-based materials increase with the increase in the amount of microcapsules. The healing and recovery rates of the samples without the microcapsules were 16.95% and 91.30%, respectively. The healing rate was positive, but lower, indicating that the microcracks in the specimen without the microcapsules can achieve self-healing by extending the hydration. However, this effect is limited, as it cannot restore the initial state of the specimens. The healing and recovery rates of the composites with microcapsules were higher than those of the specimens without microcapsules. The results show that the effect of extending the hydration is limited, and the microcapsules have a significant effect on the self-healing process of the cement-based material. It indicates that, the more the amount of microcapsules, the more the healing agent; the self-healing effect of the cement-based material is more evident. It is found that compared to the compressive strength of the specimen, the dynamic modulus has considerably higher healing rate with the microcapsules. However, as the microcapsules decrease the initial value of the elastic modulus and strength, clearly, an appropriate value of the amount of microcapsules should be determined because of the target performance.

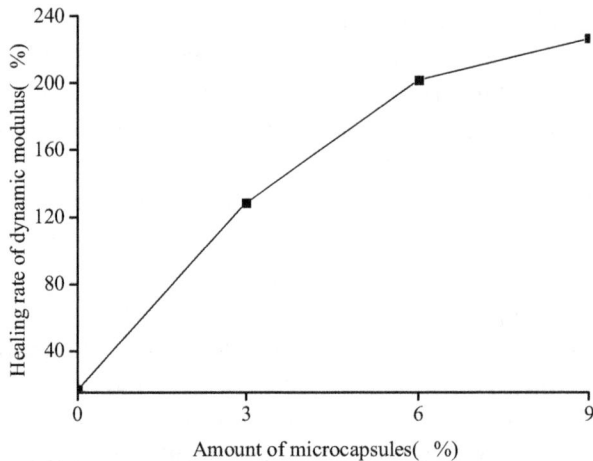

Figure 8. Healing rate (Equation (6)) of the dynamic modulus with the amount of microcapsules.

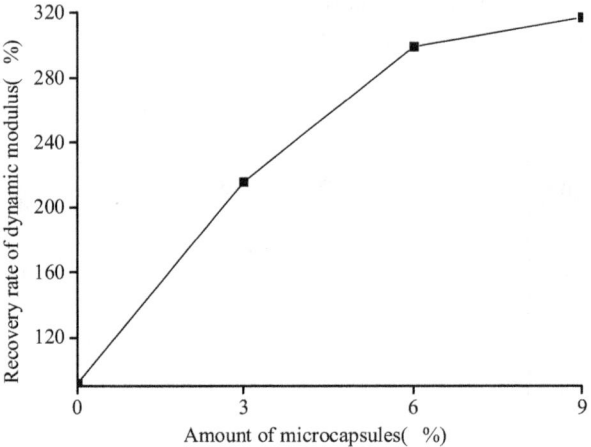

Figure 9. Recovery rate (Equation (5)) of the dynamic modulus with amount of microcapsules.

3.3. MIP Results and Discussions

Based on the MIP test, the pore-structure parameters, such as porosity, pore volume, median value, and the average value of the pore size for the original specimens with different amount of microcapsules are shown in Figures 10–13. It is seen that all the values of the porosity, pore volume, median value, and average value of the pore size increase with the increase in the amount of microcapsules (0%, 3%, 6%, and 9%). However, in the case of 3% microcapsules, the porosity of the specimen is almost the same as the one without the microcapsules. This can also explain the filler effect of the microcapsules when the amount of the mixed microcapsules is small. The variation in the total pore volume of the specimen shows an opposite trend compared to the results of the nitrogen adsorption test (Brunauer–Emmett–Teller (BET)) [34]. The volume of the pores less than 50 nm measured by the BET reduced with the amount of microcapsules. However, it increased when the MIP test was employed, indicating that, both the volume of the pores greater than 50 nm and the number of harmful holes increased. The increase in the porosity, average pore diameter, and the critical diameter indicated an increase in the number of large pores. Then, the increase in the amount of microcapsules increases the large pores and further affects the strength and permeability of the specimen.

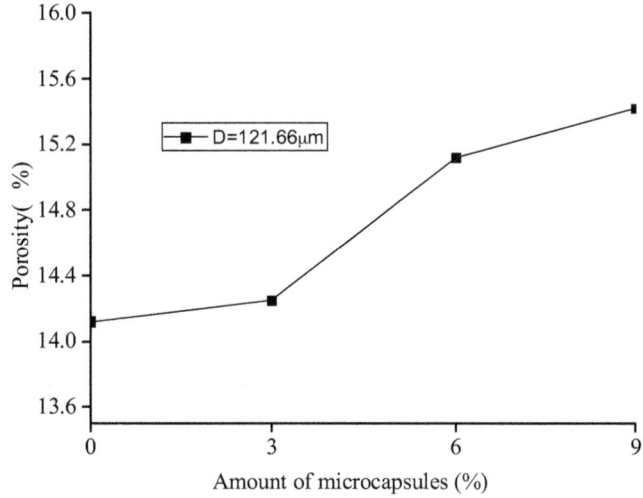

Figure 10. Variation in the porosity with amount of microcapsules.

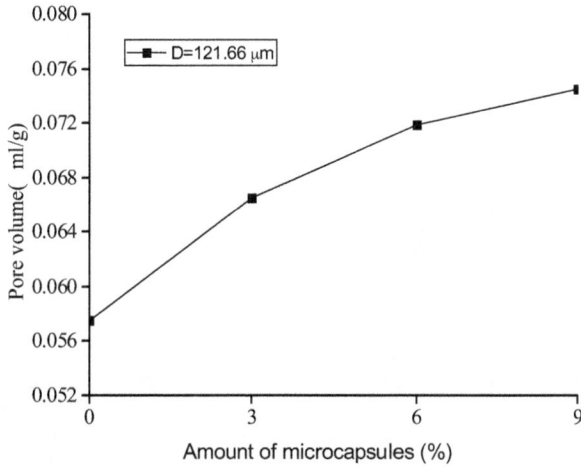

Figure 11. Variation in the pore volume with amount of microcapsules.

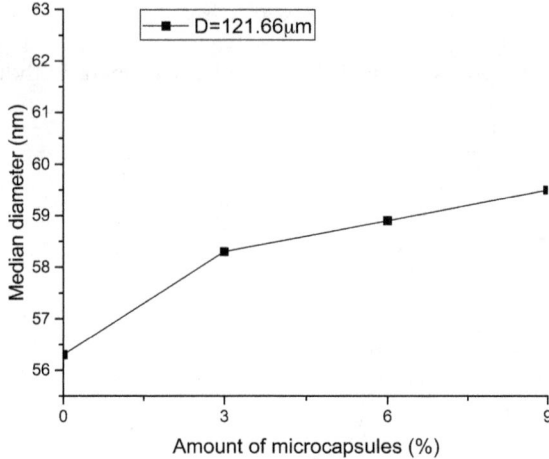

Figure 12. Variation in the median pore diameter microcapsules.

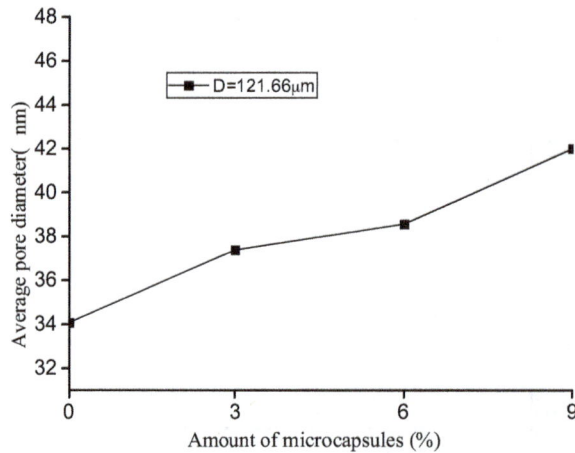

Figure 13. Variation in the average pore diameter with amount of microcapsules.

Figures 14–17 show the pore-size distribution of the specimens before and after the self-healing for the microcapsule size of 121.66 μm for the different amount of microcapsules (0%, 3%, 6%, and 9%), respectively. It can be seen that for all the percentages of microcapsules, the distribution of the cumulative-pore volume after the healing showed lower values than those under damage.

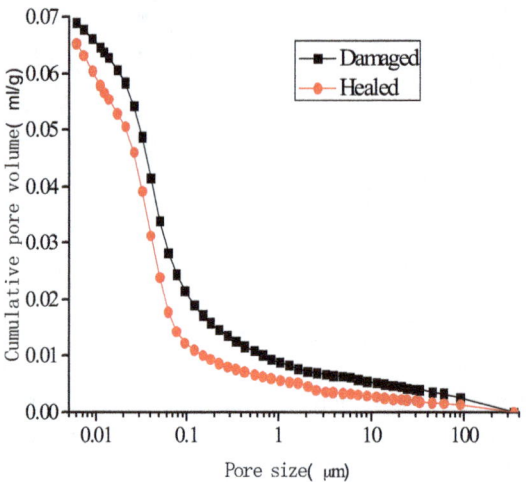

Figure 14. Cumulative-pore volume distribution for specimens without microcapsules.

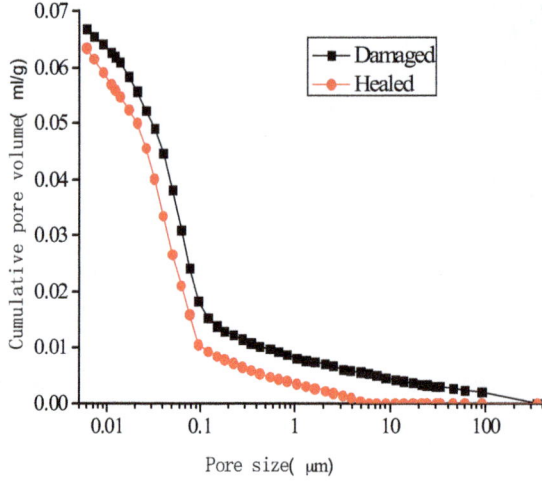

Figure 15. Cumulative-pore volume distribution for specimens with 3% microcapsules.

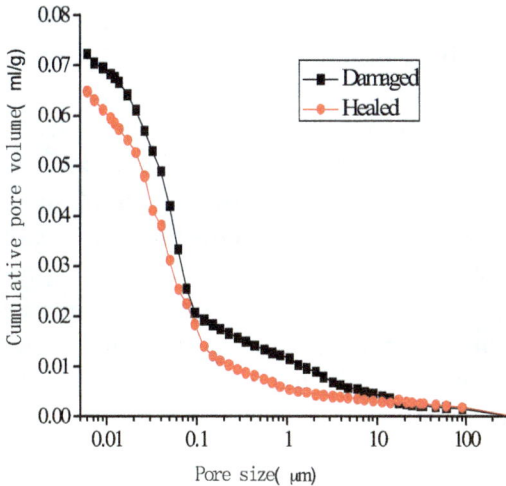

Figure 16. Cumulative-pore volume distribution for specimens with 6% microcapsules.

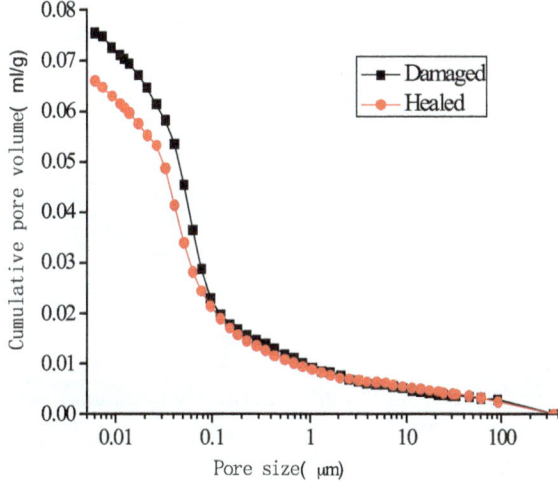

Figure 17. Cumulative-pore volume distribution for specimens with 9% microcapsules.

To investigate the self-healing behavior at the pore-structure level, a series of definitions for the pore structure parameters are given as

$$\eta_P = \frac{P_{damaged} - P_{healed}}{P_{damaged}} \times 100\% \,,\ \eta_{RP} = \frac{P_{original}}{P_{healed}} \times 100\% \,, \tag{7}$$

$$\eta_V = \frac{V_{damaged} - V_{healed}}{V_{damaged}} \times 100\% \,,\ \eta_{RV} = \frac{V_{original}}{V_{healed}} \times 100\% \,, \tag{8}$$

$$\eta_{AVE} = \frac{A_{damaged} - A_{healed}}{A_{damaged}} \times 100\% \,,\ \eta_{RAVE} = \frac{A_{original}}{A_{healed}} \times 100\% \,, \tag{9}$$

where η_P, η_V, and η_{AVE} denote the healing rates of the porosity, total pore volume, and average pore diameter of the specimen, respectively; η_{RP}, η_{RV}, and η_{RAVE} represent the recovery rates of the porosity, total pore volume, and average pore diameter of the specimen, respectively; $P_{original}$, $V_{original}$, and $A_{original}$ are the porosity, total pore volume, and average pore diameter of the original specimen, respectively; $P_{damaged}$, $V_{damaged}$, and $A_{damaged}$ are the porosity, total pore volume, and average pore diameter of the damaged specimen, respectively; P_{healed}, V_{healed}, and A_{healed} are the porosity, total pore volume, and average pore diameter of the healed specimen, respectively.

Figures 18–20 show the healing rates of the porosity, total pore volume, and average pore diameter of the specimen, respectively, with the amount of microcapsules. Figures 21–23 show the recovery rates of the porosity, total pore volume, and average pore diameter of the specimen, respectively, with increasing amount of microcapsules. It is noted that the variation in both the healing and recovery rates of the pore structure parameters, such as the porosity, total pore volume, and average pore diameter of the specimen are consistent; they increase with the amount of microcapsules. The results of the compressive strength and dynamic modulus, given in the previous section, shows that the healing and the recovery rates of the pore structure parameters are consistent with the macro mechanical behaviors. It is validated that the microcapsules can provide self-healing function for the cementitious composite; however, it is shown that the pore structure parameters have important influences on the mechanical behaviors such as compressive strength and dynamic modulus.

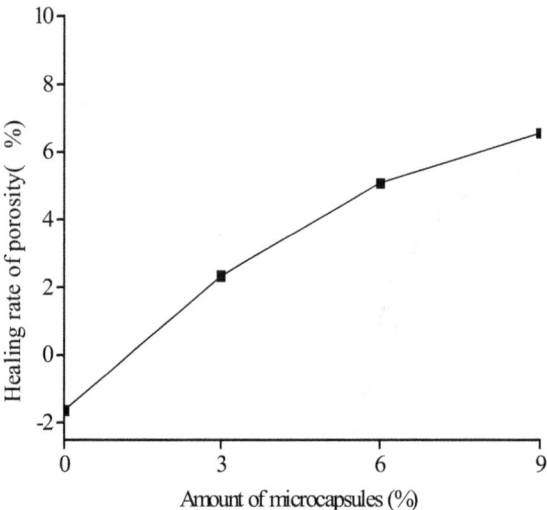

Figure 18. Healing rate (Equation (7): η_P) of porosity with amount of microcapsules.

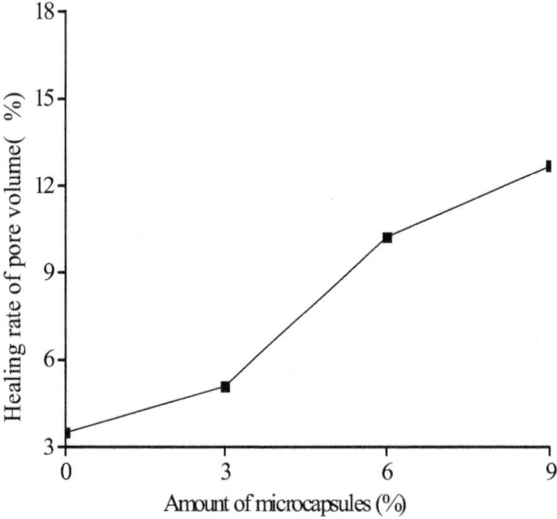

Figure 19. Healing rate (Equation (8): η_V) of pore volume with amount of microcapsules.

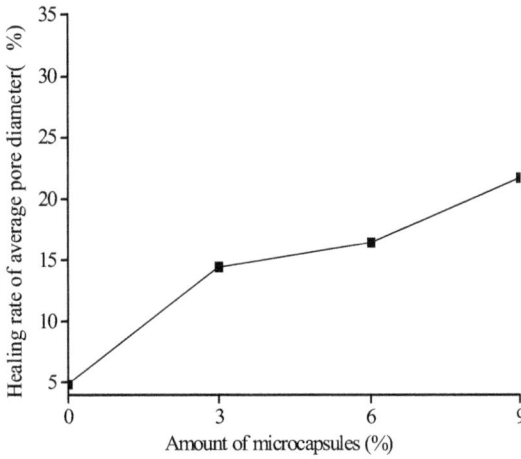

Figure 20. Healing rate (Equation (9): η_{AVE}) of average pore diameter with amount of microcapsules.

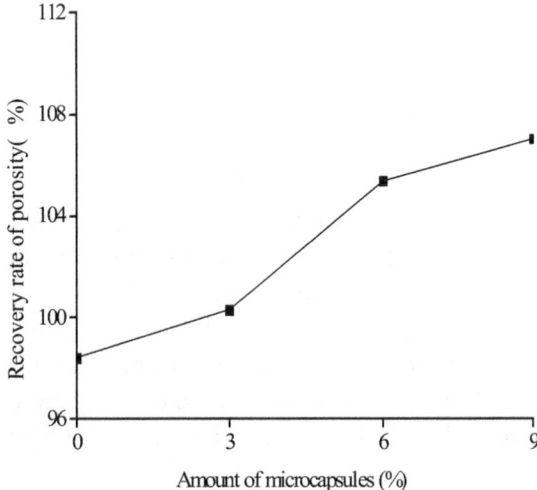

Figure 21. Recovery rate (Equation (7): η_{RP}) of porosity with amount of microcapsules.

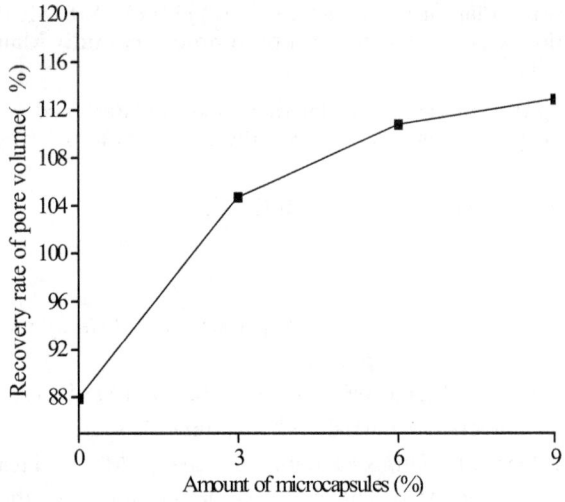

Figure 22. Recovery rate (Equation (8): η_{RV}) of pore volume with amount of microcapsules.

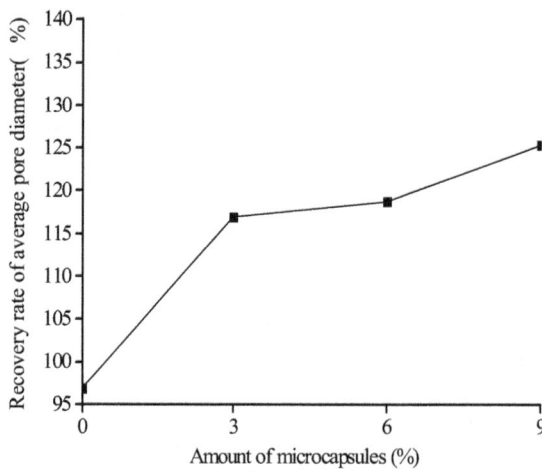

Figure 23. Recovery rate (Equation (9): η_{RAVE}) of average pore diameter with amount of microcapsules.

4. Conclusions

The microcapsules-based self-healing cementitious composite was developed in Guangdong Provincial Key Laboratory of Durability for Marine Civil Engineering, Shenzhen University. Both the macro performance, such as the compressive strength and the dynamic modulus, and the microstructure of the pore-structure parameters were investigated. The results reflect the consistency between the macro behavior and the microstructures. It is concluded that, for the original specimen, both the compressive strength and the dynamic modulus, as well as the pore structure parameters, such as porosity, cumulative pore volume, and average pore diameter, decrease to some extent with increasing amount of microcapsules. However, the self-healing and recovery rates of the specimen performance and the pore structure parameters increase with increasing the amount of microcapsules. Therefore, for the future practical application, there is a need to balance the amount of microcapsules in order to achieve the self-healing function and avoid the drawback of the effect of the microcapsules themselves.

Acknowledgments: The authors gratefully acknowledge the financial support provided by the joint funds of the National Natural Science Foundation and Guangdong Province of China (U1301241), the International Cooperation and Exchange of the National Natural Science Foundation of China (51120185002, 51520105012), the General Program of the National Natural Science Foundation of China (No.51478272); the Science and Technology Foundation for the Basic Research Plan of Shenzhen City (JCYJ20140418182819159, JCYJ20160422095146121), and the Collaborative Innovation Research Center for Environment-Friendly Materials and Structures in Civil Engineering, Southeast University.

Author Contributions: Xianfeng Wang and Ningxu Han developed and designed the experiments; Xianfeng Wang and Peipei Sun performed the experiments and analyzed the data; Xianfeng Wang, Peipei Sun, and Feng Xing wrote and revised the paper.

Conflicts of Interest: The authors declare no conflict of interest.

References

1.   Jonkers, H.M. Self healing concrete: A biological approach. In *Self Healing Materials*; Springer: Dordrecht, The Netherlands, 2007; Volume 100, pp. 195–204.

2.   Zhang, M. A Study on Microcapsule Based Self-Healing Method and Mechanism for Cementitious Composites. Ph.D. Thesis, Central South University, Changsha, China, June 2013.

3.   Xing, F.; Ni, Z.; Han, N.; Dong, B.; Du, X.; Huang, Z.; Zhang, M. Self-Healing Mechanism of a Novel Cementitious Composite Using Microcapsules. In Proceedings of the International Conference on Durability of Concrete Structures, Hangzhou, China, 26–27 November 2008.

4.   Victor, C.L.; Yun, M.L.; Yin, W.C. Feasibility study of a passive smart self-healing cementitious composite. *Compos. Part B Eng.* **1998**, *29*, 819–827.

5.   Xing, F.; Ni, Z. Self-Repairing Concrete Having Polyurethane Polymer Microcapsules and Method for Fabricating the Same. U.S. Patent No. 8362113 B2, 17 July 2012.

6.	Wang, X.F.; Xing, F.; Zhang, M. Experimental study on cementitious composites embedded with organic microcapsules. *Materials* **2013**, *6*, 4064–4081. [CrossRef]

7.	Xiong, W.; Tang, J.; Zhu, G.; Han, N.; Schlangen, E.; Dong, B.; Wang, X.F.; Xing, F. A novel capsule-based self-recovery system with a chloride ion trigger. *Sci. Rep. UK* **2015**, *5*, 10866. [CrossRef] [PubMed]

8.	Wang, X.F.; Xing, F.; Xie, Q.; Han, N.X.; Kishi, T.; Ahn, T.H. Mechanical behavior of a capsule embedded in cementitious matrix-macro model and numerical simulation. *J. Ceram. Process. Res.* **2015**, *16*, 74s–82s.

9.	Han, N.; Xing, F. Intelligent resilience of cementitious materials for marine infrastructures. *J. Ceram. Process. Res.* **2015**, *16*, s14–s21.

10.	Su, J.-F.; Schlangen, E.; Qiu, J. Design and construction of microcapsules containing rejuvenator for asphalt. *Powder Technol.* **2013**, *235*, 563–571. [CrossRef]

11.	Su, J.-F.; Wang, Y.-Y.; Han, N.-X.; Yang, P.; Han, S. Experimental investigation and mechanism analysis of novel multi-self-healing behaviors of bitumen using microcapsules containing rejuvenator. *Constr. Build. Mater.* **2016**, *106*, 317–329. [CrossRef]

12.	Su, J.-F.; Yang, P.; Wang, Y.-Y.; Han, S.; Han, N.-X.; Li, W. Investigation of the Self-Healing Behaviors of Microcapsules/Bitumen Composites by a Repetitive Direct Tension Test. *Materials* **2016**, *9*, 600. [CrossRef]

13.	Dong, B.; Han, N.; Zhang, M.; Wang, X.; Cui, H.; Xing, F. A microcapsule technology based self-healing system for concrete structures. *J. Earthq. Tsunami* **2013**, *7*, 1350014. [CrossRef]

14.	Dong, B.; Fang, G.; Ding, W.; Liu, Y.; Zhang, J.; Han, N.; Xing, F. Self-healing features in cementitious material with urea–formaldehyde/epoxy microcapsules. *Constr. Build. Mater.* **2016**, *106*, 608–617. [CrossRef]

15.	Dong, B.; Wang, Y.; Fang, G.; Han, N.; Xing, F.; Lu, Y. Smart releasing behavior of a chemical self-healing microcapsule in the stimulated concrete pore solution. *Cem. Concr. Compos.* **2015**, *56*, 46–50. [CrossRef]

16.	Lv, L.; Yang, Z.; Chen, G.; Zhu, G.; Han, N.; Schlangen, E.; Xing, F. Synthesis and characterization of a new polymeric microcapsule and feasibility investigation in self-healing cementitious materials. *Constr. Build. Mater.* **2016**, *105*, 487–495. [CrossRef]

17.	Wang, J.Y.; Soens, H.; Verstraete, W.; De Belie, N. Self-healing concrete by use of microencapsulated bacterial spores. *Cem. Concr. Res.* **2014**, *56*, 139–152. [CrossRef]

18.	Van Stappen, J.; Bultreys, T.; Gilabert, F.A.; Hillewaere, X.K.D.; Gómez, D.G.; Van Tittelboom, K.; Dhaene, J.; De Belie, N.; Van Paepegem, W.; Du Prez, F.E.; et al. The microstructure of capsule containing self-healing materials: A micro-computed tomography study. *Mater. Charact.* **2016**, *119*, 99–109. [CrossRef]

19.	Van Breugel, K. Is there a market for self-healing cement-based materials? In Proceedings of the First International Conference on Self-Healing Materials, Noordwijk, The Netherlands, 18–20 April 2007.

20.	Wu, M.; Johannesson, B.; Geiker, M. A review: Self-healing in cementitious materials and engineered cementitious composite as a self-healing material. *Constr. Build. Mater.* **2012**, *28*, 571–583. [CrossRef]

21.	Joseph, C.; Gardner, D.; Jefferson, T.; Isaacs, B.; Lark, B. Self-healing cementitious materials: A review of recent work. *Proc. ICE Constr. Mater.* **2010**, *164*, 29–41. [CrossRef]

22.	Mihashi, H.; Nishiwaki, T. Development of engineered self-healing and self-repairing concrete-state-of-the-art report. *J. Adv. Concr. Technol.* **2012**, *10*, 170–184. [CrossRef]

23.	Van Tittelboom, K.; De Belie, N. Self-Healing in Cementitious Materials—A Review. *Materials* **2013**, *6*, 2182–2217. [CrossRef]

24.	Souradeep, G.; Kua, H. Encapsulation technology and techniques in self-healing concrete. *J. Mater. Civ. Eng.* **2016**, *28*. [CrossRef]

25.	Muhammad, N.Z.; Shafaghat, A.; Keyvanfar, A.; Majid, M.Z.A.; Ghoshal, S.K.; Yasouj, S.E.M.; Ganiyu, A.A.; Kouchaksaraei, M.S.; Kamyab, H.; Taheri, M.M.; et al. Tests and methods of evaluating the self-healing efficiency of concrete: A review. *Constr. Build. Mater.* **2016**, *112*, 1123–1132. [CrossRef]

26.	De Belie, N.; Van der Zwaag, S.; Gruyaert, E.; Van Tittelboom, K.; Debbaut, B. Self-Healing Materials. In Proceedings of the 4th International Conference on Self-Healing Materials, Ghent, Belgium, 16–20 June 2013.

27.	Reichert, M.; Craig, S.; Rubinstein, M.; Genzer, J.; Palmese, G.; White, S.; Li, V.; Van der Zwaag, S. Self-Healing Materials. In Proceedings of the 5th International Conference on Self-Healing Materials, Durham, NC, USA, 22–24 June 2015.

28.	Guo, J.F. The Theoretical Research of Pore Structure and the Strength of Concrete. Master's Thesis, Zhejiang University, Hangzhou, China, June 2004.

29.	Lian, H.Z. *Building Materials Phase Research Foundation*; Tsinghua University Press: Beijing, China, 1996.

30. Abell, A.B.; Willis, K.L.; Lange, D.A. Mercury intrusion porosimetry and image analysis of cement-based materials. *J. Colloid Interface Sci.* **1999**, *211*, 39–44. [CrossRef] [PubMed]

31. Zhao, D.; Liu, F.; Mu, W.; Han, Z.R. Factors Affecting Morphology and Encapsulation Ratio of Chlorpyrifos Microcapsules with UF-resin During Preparation. *Chin. J. Appl. Chem.* **2007**, *24*, 589–592.

32. Standardization Administration of China. *Common Portland Cement*; GB 175-2007; SAC: Beijing, China, 2007.

33. International Standard Organization (ISO). *Method of Testing Cements—Determination of Strength ISO 679*; ISO: Geneva, Switzerland, 1989.

34. Sun, P. Study on Strength and Pore Structure of Microcapsules Based Self-Healing Cementitious Composites. Master's Thesis, Shenzhen University, Shenzhen, China, May 2016.

11

Comparison of Maraging Steel Micro- and Nanostructure Produced Conventionally and by Laser Additive Manufacturing

Eric A. Jägle [1,*], Zhendong Sheng [1,2], Philipp Kürnsteiner [1], Sörn Ocylok [3], Andreas Weisheit [3] and Dierk Raabe [1]

[1] Department Microstructure Physics and Alloy Design, Max-Planck-Institut für Eisenforschung GmbH, Max-Planck-Strasse 1, 40237 Düsseldorf, Germany; zhendong.sheng@iehk.rwth-aachen.de (Z.S.); p.kuernsteiner@mpie.de (P.K.); d.raabe@mpie.de (D.R.)

[2] Institut für Eisenhüttenkunde, Rheinisch-Westfälische Technische Hochschule Aachen, Intzestrasse 1, 52072 Aachen, Germany

[3] Competence Area Additive Manufacturing and Functional Layers, Fraunhofer Institut für Lasertechnik, Steinbachstrasse 15, 52074 Aachen, Germany; soern.ocylok@ilt.fraunhofer.de (S.O.); andreas.weisheit@ilt.fraunhofer.de (A.W.)

* Correspondence: e.jaegle@mpie.de

Academic Editor: Guillermo Requena

Abstract: Maraging steels are used to produce tools by Additive Manufacturing (AM) methods such as Laser Metal Deposition (LMD) and Selective Laser Melting (SLM). Although it is well established that dense parts can be produced by AM, the influence of the AM process on the microstructure—in particular the content of retained and reversed austenite as well as the nanostructure, especially the precipitate density and chemistry, are not yet explored. Here, we study these features using microhardness measurements, Optical Microscopy, Electron Backscatter Diffraction (EBSD), Energy Dispersive Spectroscopy (EDS), and Atom Probe Tomography (APT) in the as-produced state and during ageing heat treatment. We find that due to microsegregation, retained austenite exists in the as-LMD- and as-SLM-produced states but not in the conventionally-produced material. The hardness in the as-LMD-produced state is higher than in the conventionally and SLM-produced materials, however, not in the uppermost layers. By APT, it is confirmed that this is due to early stages of precipitation induced by the cyclic re-heating upon further deposition—i.e., the intrinsic heat treatment associated with LMD. In the peak-aged state, which is reached after a similar time in all materials, the hardness of SLM- and LMD-produced material is slightly lower than in conventionally-produced material due to the presence of retained austenite and reversed austenite formed during ageing.

Keywords: laser metal deposition; additive manufacturing; maraging steel; intrinsic heat treatment; precipitation strengthening; austenite reversion; atom probe tomography

1. Introduction

Maraging steels are materials that combine very high strength, hardness, and toughness [1]. Therefore, they are employed as tool steels in the mold and die making industry, but also for high-performance parts—e.g., in the aerospace industry [2]. They achieve their mechanical properties by a martensitic matrix that contains a high number density of nanometer-sized intermetallic precipitates [3–6]. Different from most tool steels, the martensitic microstructure is not achieved by a relatively high amount of carbon in the alloy composition, but instead by (usually) a high concentration of nickel. The almost complete lack of interstitial alloying elements leads to a good weldability of this class of alloys [7]. This, in turn, makes them amenable to metal additive manufacturing (AM) processes, in particular Laser Metal Deposition (LMD) and Selective Laser Melting (SLM) [8–16]. Since these processes involve a small melt pool generated by a laser beam for the consolidation of powder feedstock to a dense material, they share similarities with micro-welding processes.

One of the main strengths of AM processes is that very complex workpieces can be efficiently generated. In the toolmaking industry, metal AM processes are rapidly becoming the state of the art in the production of tool inserts for (polymer) injection molding processes. The geometrical freedom of AM allows to place cooling channels very close to the tool surface, yielding a very efficient cooling of the injected liquid polymer and avoiding 'hot spots' which would otherwise promote local material damage. Conventionally, cooling channels are produced by deep hole drilling which is only suited to produce (piecewise) straight cooling channels. Thus, cooling channels may not reach all locations in a complex tool and the fluid flow of coolant is hindered by turbulence induced by the rapid change of the channel axis where two bore holes intersect. It has been shown that, using AM-produced tool inserts, the heat removal from the tool can be enhanced such that the cycle time of the process is strongly reduced (by up to 60% [17]) and the productivity of the tool is equally improved.

Maraging steel that is used almost exclusively in AM processes today is the first-generation steel 18Ni-300, also known as 'grade 300 maraging steel' with the material number 1.2709 or slight modifications thereof, such as 'Böhler V720®' (material number 1.6354.9) [18,19]. Conventionally-produced (C-P) material is usually supplied in the solution annealed and quenched condition—i.e., fully martensitic without any precipitates present. These are formed during a subsequent ageing treatment, typically between 480 and 510 °C. Most AM processes used to synthesize metallic materials exhibit a rapid cooling rate during and after solidification (typically ~10^4 K/s in LMD and up to 10^6 K/s in SLM). It is therefore reasonable to assume that the microstructure of AM-produced (AM-P) maraging steel should also consist of martensite without precipitates. Indeed, it has been shown [20] that the microstructure of as-SLM-produced 18Ni-300 maraging steel does not contain any precipitates, however, it does contain a significant amount of retained austenite [13,20]. Normally, AM-produced parts made of maraging steel are not subjected to (thermo-)mechanical treatments—e.g., solution annealing or HIPing—before the final ageing, in contrast to hot-rolled C-P material. The microstructure of C-P and AM-P maraging steel at the start of the ageing treatment can therefore be expected to be quite different.

The aim of this paper is to investigate the difference in crystallography, chemical homogeneity on the micro- and nano-scale and phase distribution of C-P and AM-P 18Ni-300 maraging steel, and to determine the influence of these differences on the microstructural evolution during ageing treatment. For this purpose, we employ optical and electron microscopy including Electron Backscatter Diffraction (EBSD), Energy Dispersive X-ray Spectroscopy (EDS), as well as Atom-Probe Tomography (APT).

2. Materials and Methods

2.1. Additive Manufacturing

The C-P grade 300 maraging steel 'Böhler V720®' was produced by Böhler Edelstahl GmbH (Kapfenberg, Austria) via vacuum induction melting and vacuum arc re-melting. The material was received in form of a rolled bar and analyzed in the solution annealed (0.5 h at 820 °C) and quenched (rapid air quenching) condition. Its composition is given in Table 1.

Table 1. Chemical composition of the 1.6354.9 material used in this study determined by ICP-OES.

Alloying Element	C	Si	Mn	Mo	Ni	Al	Co	Ti	Fe
wt %	0.0018	0.025	0.011	5.03	18.3	0.077	8.74	0.68	Bal.
at %	0.0087	0.052	0.012	3.03	17.95	0.165	8.57	0.82	Bal.

LMD and SLM samples were produced using the parameters given in Table 2. Cuboids are produced by LMD using a simple unidirectional scanning strategy. The LMD samples were produced using a 3 kW diode laser. The powder is fed into the interaction zone of the laser beam and substrate via a coaxial powder feed nozzle. Argon is used as a carrier gas which also provides shielding from the surrounding atmosphere. The samples are 30 mm long (scanning direction, SD), 20 mm wide (transverse direction, TD), and 10 mm high, (build direction, BD). A detailed description of the production process of the SLM samples can be found in reference [13].

Table 2. Processing parameters of two Laser Additive Manufacturing (LAM) methods for 18Ni maraging steels, Laser Metal Deposition (LMD), and Selective Laser Melting (SLM). The energy density is calculated by dividing the laser power by the scan speed, layer thickness, and laser focus diameter.

LAM Process	Laser Power (W)	Scan Speed (mm/s)	Laser Focus Diameter (μm)	Layer Thickness (μm)	Hatch Spacing (μm)	Energy Density (J/mm^3)	Inert Atmosphere
LMD	800	10	1700	420	900	112.0	Ar
SLM Data from [13]	100	150	180	30	112	123.5	N$_2$

2.2. Microstructural Analysis

The AM-P samples were cut in two planes: one is parallel to the scan direction and the build direction (SD-BD) and the other is parallel to the transverse direction and the build direction (TD-BD). The C-P samples were cut parallel to the rolling direction. Standard metallographic sample preparation techniques, including a finishing step of polishing using colloidal silica suspension (OP-S from Stuers ApS, Ballerup, Denmark), were used. A solution of 1% HNO$_3$ in ethanol was used to reveal the microstructure.

EBSD and EDS measurements were carried out in a JEOL 6500F (JEOL, Ltd., Tokyo, Japan) field emitter gun scanning electron microscope (FEG-SEM) equipped with an EDAX Octane Plus EDS detector and a TSL TexSEM DigiView EBSD camera using an acceleration voltage of 15 kV. TSL OIM Analysis™ software (version 7, EDAX, Mahwah, NJ, USA) was used for EBSD data analysis.

APT sample preparation was performed in a FEI Helios NanoLab 600i (FEI, Hillsboro, OR, USA) dual beam device employing the standard liftout process described in reference [21]. Tips were sharpened by applying annular milling patterns with a final step of low kV milling at 5 kV acceleration voltage to minimize Ga contamination at the surface. APT experiments were performed in a Cameca LEAP 3000 X HR (Cameca Instruments, Inc., Madison, WI, USA) in laser mode at a target temperature of 60 K, a laser energy of 0.4 nJ, and a laser pulse frequency of 250 kHz. The target evaporation rate was set to five atoms per 1000 pulses. Data analysis was performed using the IVAS software (version 3.6.6, Cameca Instruments, Inc., Madison, WI, USA).

3. Results

3.1. Ageing Behavior

Figure 1 shows the microhardness of conventionally-produced (C-P), and LMD-produced (LMD-P) maraging steel samples as a function of ageing time at 480 °C. The values are an average of at least six hardness indents placed randomly in the middle of the sample (SD-BD plane) spaced at least 0.5 mm apart. Additionally, the microhardness of SLM-produced (SLM-P) material in the as-produced state and after 480 min ageing are shown (taken from reference [20]). Interestingly, the hardness of LMD-P material is higher than that of C-P and SLM-produced material in the as-produced and as-received states, respectively. The hardness of C-P and SLM-P material is roughly identical. This difference, however, vanishes quickly during ageing and, after 5 to 10 min of ageing, the LMD-P material becomes softer than the C-P material. Peak hardness is reached after 500 to 1000 min of ageing, and in this state the SLM-P material shows a similar hardness than the LMD-P material, both being about 50 HV softer than the C-P material. It is worth noting that the hardness drops after overageing occurs after a slightly longer time in the C-P material than in the LMD-P material. The origins of the differences in hardness, in particular the change in the hardness of the LMD-P material that shifts from being harder than C-P material (as-produced state) to being softer (peak aged state), will be investigated in detail in the remainder of the paper.

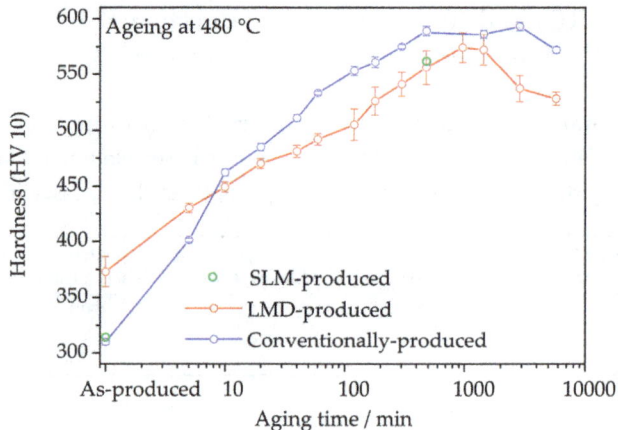

Figure 1. Microhardness of conventionally-produced and LMD-produced material as a function of ageing time at 480 °C. Additionally, the hardness of SLM-produced material in the as-produced state and after 480 min ageing is shown (data from [20]).

3.2. Microstructure in the As-Produced/As-Received State State

Optical micrographs of both C-P (right) and LMD-P material (left) in the as-produced/as-received state are shown in Figure 2. At low magnification, the layer-by-layer structure of the LMD-P material can be seen. The difference in contrast every four to five layers is an artifact from the etching. The build direction of the sample is upwards in the micrographs and the sample was cut in the TD-BD plane. At higher magnifications, the individual melt pools (delineated with dashed white lines in Figure 1), with many solidification dendrites within, are visible. They often cross melt pool boundaries, indicating epitaxial growth of grains between layers. At the highest magnification, it can be seen that the solidification structures are indeed dendrites, albeit with very short (secondary) side arms. The C-P material, on the other hand, does not show any of these solidification structures due to the thermomechanical processing it has experienced after primary synthesis. Instead, etching reveals a fine martensitic microstructure without preferred orientations of the martensite blocks. Note that no features typical for martensitic microstructures such as packets or blocks are visible in the LMD-P material.

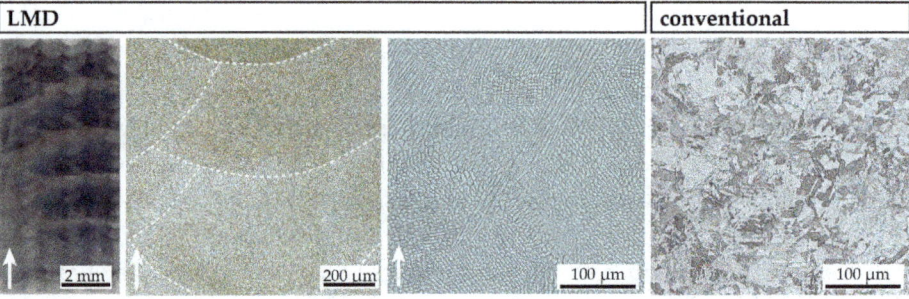

Figure 2. Optical micrograps of LMD-produced and conventionally-produced material in the as-produced/as-received state. Etching with HNO_3 in ethanol reveals the melt pool boundaries and dendrites (LMD-produced material) and the martensite laths (conventionally-produced material). The white arrows indicate the build direction of the LMD-produced sample. LMD samples are cut in the TD-BD plane.

The microstructure is investigated in more detail by EBSD and EDS in Figures 3, 4 and 6. Figure 3a shows the phases detected by EBSD over a relatively large part of the as-LMD-P sample (scan step size: 600 nm). Apparently, a considerable amount of retained austenite is present in the material. It is present in the entire sample, but not distributed entirely homogeneously. There is, for example, a lower apparent austenite fraction present in the areas just below melt pool boundaries. The austenite seems to be located along dendrite boundaries, but the magnification in this figure is not high enough to be certain. In panels (b) and (c), the crystallographic orientation of the martensite and retained austenite phases are shown, respectively. It can be seen that the austenite grains that formed upon solidification are quite large (up to one mm in diameter) and span several deposited layers. Due to this large (prior) austenite grain size and the resulting limited number of grains (and martensite variants) in the EBSD scan, it is not possible to make a statement about the overall crystallographic texture of the material.

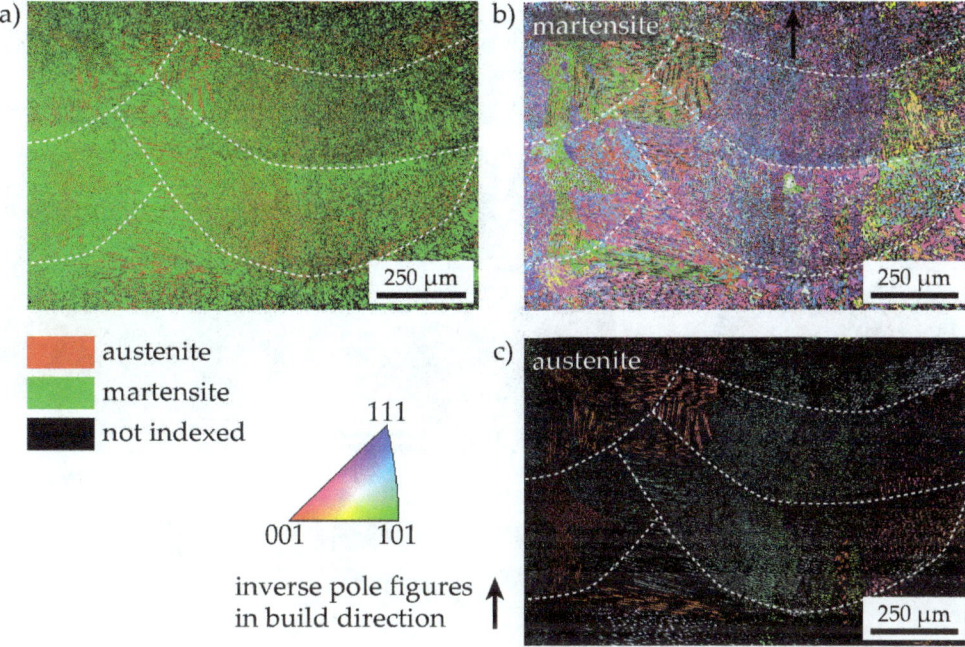

Figure 3. A large-area, low-magnification EBSD scan of as-LMD-produced material. (**a**) Phase map showing the location of the martensite and ferrite phases; (**b**) Inverse pole figure map of martensite only and (**c**) austenite only. Melt pool boundaries are indicated by dashed white lines. The black arrow indicates the build direction.

The exact location of the retained austenite is displayed in Figure 4, where the phase map from a higher resolution EBSD scan (step size: 200 nm) is shown alongside corresponding EDS mappings of the Ti, Mo, Ni, and Co concentrations (the major alloying elements). The retained austenite indeed occurs in the interdendritic regions. Depending on the orientation of the dendrites, the austenite appears as a long needle in the EBSD phase map (when the dendrite axes are in the image plane) or as small circles (when the dendrite axes are perpendicular to the image plane). The interdendritic areas are enriched with Ti, Mo, and Ni. This is due to microsegregation during solidification—i.e., partitioning of solute elements into the remaining liquid during solidification. The enrichment in solutes explains why retained austenite is found only in these locations. Even though Ti and Mo are in general regarded as ferrite stabilizing elements, thermodynamic calculations show that in the interdendritic regions the austenite is stable to lower temperatures than in the matrix. This is depicted in Figure 5, where the relevant part of the phase diagram of the alloy is plotted as calculated by Thermocalc® Version 2016a using database TCFE7 and not considering any phases besides austenite and ferrite in the energy minimization. The compositions of the matrix and interdendritic regions are determined by EDS spot measurements and are noted in the figure. Additionally, empirical equations for the calculation of the martensite start temperature, M_S, of ultra-low carbon steels [22] predict that a higher content of Ti, Mo, and Ni all lower M_S and hence have an austenite-stabilizing effect.

In the EDS maps in Figure 4, also small (<1 μm) Ti-rich particles can be seen. They are most probably oxides formed due to the non-ideal shielding by inert gas during the process (cf. reference [23], wherein it was demonstrated that ODS-materials can be generated by LMD when the shielding gas is turned off).

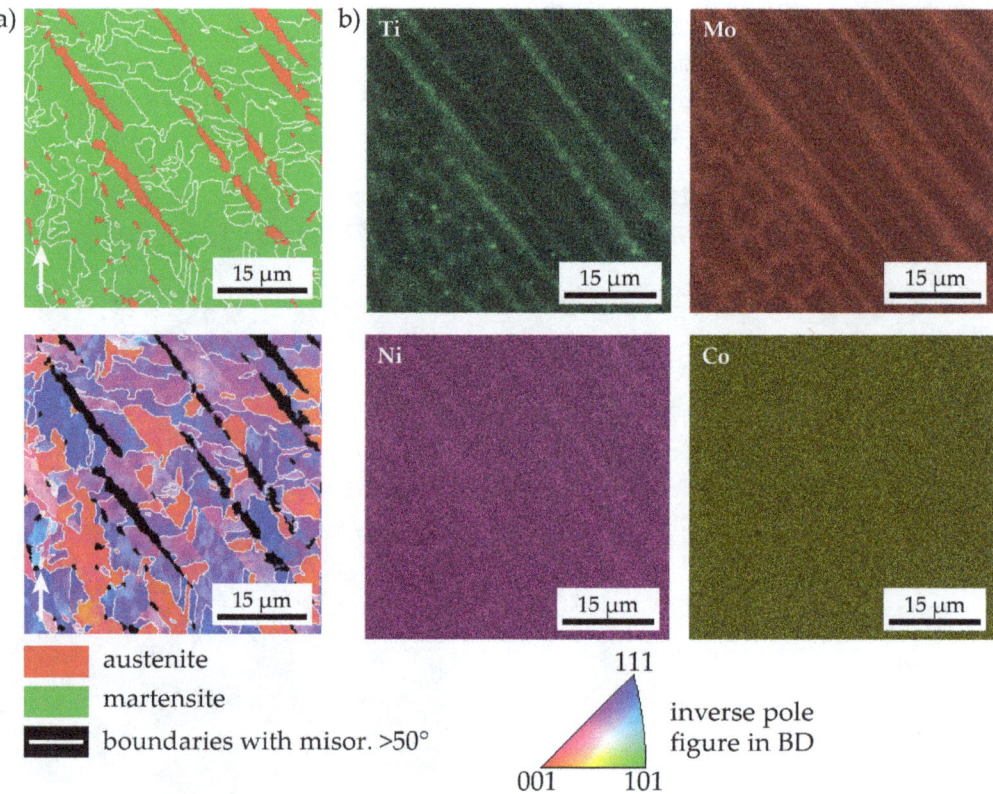

Figure 4. A small-area, high-magnification EBSD scan of as-LMD-produced material together with corresponding EDS element maps of various elements. In the phase map in panel (**a**), the location of the retained austenite in the interdendritic areas can be seen while in panels (**b**), the enrichment of Ti, Mo, and Ni in these regions is apparent.

It is difficult to quantify the exact amount of retained austenite in the LMD-P material. Due to the aforementioned large grain size and unknown crystallographic texture, X-ray diffraction measurements are unreliable. On the other hand, the austenite regions are so small (<5 μm in width) that high-resolution EBSD scans must be performed to correctly capture small austenite grains. Such high-resolution scans, however, only probe a small area and cannot reflect the slightly uneven distribution of austenite within the melt pools (cf. Figure 3). By averaging several small, high-resolution EBSD scans from the bottom to the top of the specimen, we estimate the volume fraction of retained austenite in the as-produced state as 8.5% ± 3.5%.

The microstructure of as-received C-P material is shown in Figure 6. In the lower-resolution scan (step size: 500 nm) of panel (a), it can be seen that the crystallographic texture is random. The prior austenite grains (highlighted as boundaries in the martensite phase with a misorientation between 20° and 50°) are much smaller than in the LMD-P material (cf. Figure 3). A higher-resolution scan (step size: 200 nm, panels (b) and (c)) reveals that there is no detectable retained austenite in this material, in contrast to the LMD-P material. EDS measurements (not shown) confirm the chemical homogeneity of the material.

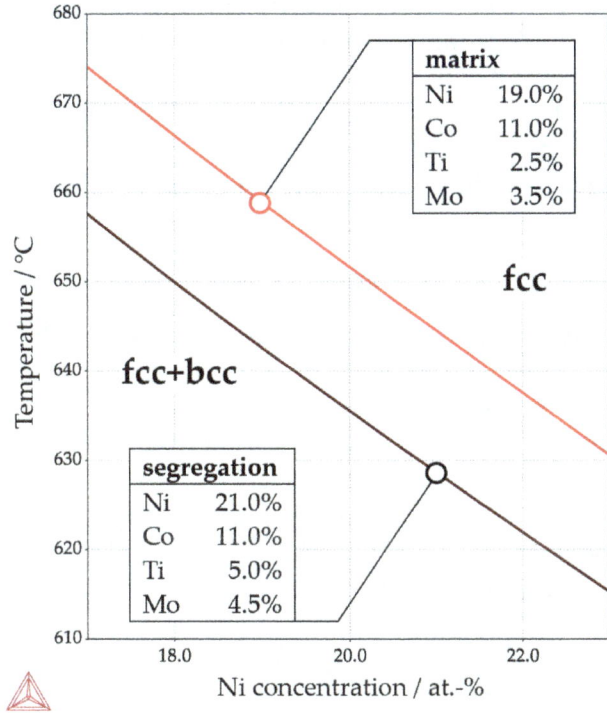

Figure 5. The relevant part of the phase diagram of the maraging steel as calculated by Thermocalc®. The lines separate the austenite and the austenite + ferrit phase fields. The results for two separate calucations are shown: one for the concentration of the interdendritic regions and one for the surrounding matrix. These concentrations were determined by EDS spot measurements. Austenite is stabilized by the solute segregation into the interdendritic regions.

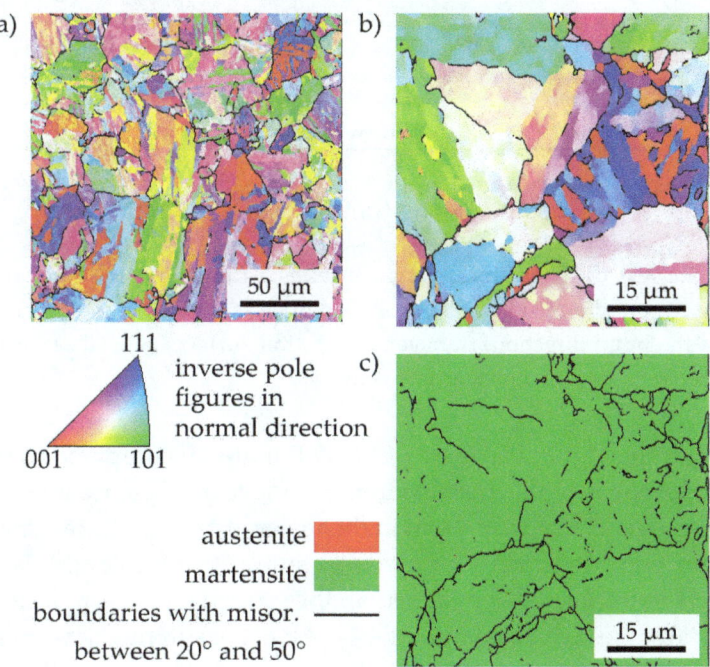

Figure 6. Two EBSD scans at low and high magnification of as-received, conventionally-produced material. Panels (**a**,**b**) show inverse pole figure maps while the phase map in panel (**c**) illustrates the absence of retained austenite.

The appearance of the martensitic microstructure of the LMD-P material is very different from typical martensitic microstructures that contain a hierarchy of prior austenite grains, martensite packets, blocks, and laths. The martensite blocks (delineated by white lines in Figure 4, i.e., boundaries with misorientation above 50°) are in many places confined to a single dendrite, even though all dendrites belonging to the same prior austenite grain (i.e., all dendrites depicted in Figure 4) have only small mutual misorientation. The retained austenite acts as an additional spatial confinement for martensite blocks.

3.3. Hardness Drop in Topmost Layers of As-LMD-Produced Material

Figure 7 shows a peculiar feature of the hardness of the as-LMD-P material. The hardness is constant along the build direction with the exception of the last few layers: At a height of ca. 7.5 mm above the base plate (2.5–3.5 mm below the sample surface), the hardness drops by about 70 HV to values of ca. 310 HV, comparable to the values of C-P and SLM-P material. Note that the scatter in the hardness values of LMD-P material is much higher than that observed for the C-P material (the range of hardness values of C-P materials is shown by the grey bar labelled 'conventional' in the figure). Yet, the observed hardness drop is clearly significant and reproducible. To check this one measurement, a series was performed on the plane build direction/scan direction, the other on the plane build direction/normal direction.

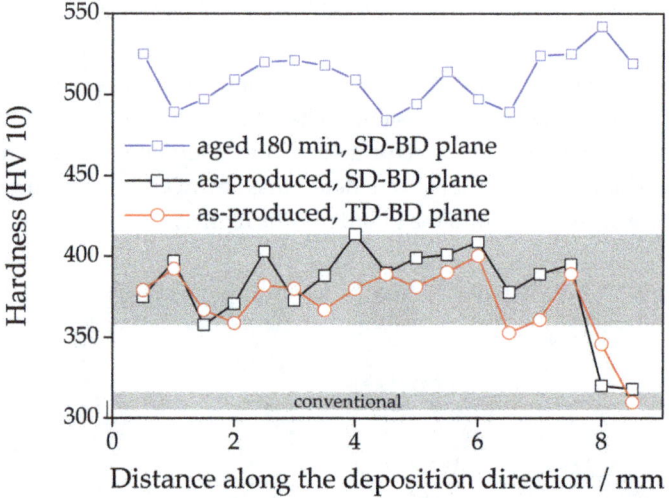

Figure 7. Microhardness of LMD-produced material as a function of distance from the substrate (build plate) along the build direction for as-produced material (on the build direction/scan direction surface as well as the build direction/transverse direction surface) as well as for material aged for 180 min (build direction/scan direction surface only).

It is suspected that the higher hardness of LMD-P material compared to both C-P and SLM-P material in the as-produced/as-received state might be due to a difference in the nanostructure of the material—i.e., whether precipitates are present in the material or not. The fact that the upper layers of LMD-P material do not show this increased hardness supports this hypothesis. These are the layers that have experienced less intrinsic heat treatment in the process (i.e., less re-heating due to deposition of overlying layers). To investigate the hypothesis, APT measurements were performed. Liftouts were performed from an arbitrary position of C-P material, from the middle (in build direction) of both LMD and SLM-produced materials as well as from the topmost layer (middle of a scan track) of LMD-P material. No ageing heat treatment had been done on any of these specimens. The atom maps (not shown here) of all measurements do not show any remarkable features (see reference [20] for atom maps of SLM-produced material).

However, a statistical analysis of the atom positions reveals differences between the datasets. In Figure 8, radial distribution functions (RDFs), computed from these samples are shown. In panel (a), only the Ti-Ti RDFs of the various samples are plotted. The RDFs are normalized by the bulk (average) concentration such that a value of one for a given Ti-Ti-distance indicates that it is equally likely to find a Ti atom at this interatomic distance than it would be to find it in a random solid solution. For C-P and SLM-P material, the RDFs are equal to one (within the error of the measurement) for all Ti-Ti distances. However, for the LMD-P sample taken from the middle of the specimen, i.e., having experienced considerable intrinsic heat treatment, a value larger than one at small Ti-Ti distances is found. This indicates that it is more likely to find two Ti atoms close together than further away from each other, in other words, that clustering of Ti has begun. Note that this deviation from unity cannot be seen for LMD-P material taken from the very top of the specimen (see the black, dashed curve), i.e., for material that has not experienced significant intrinsic heat treatment.

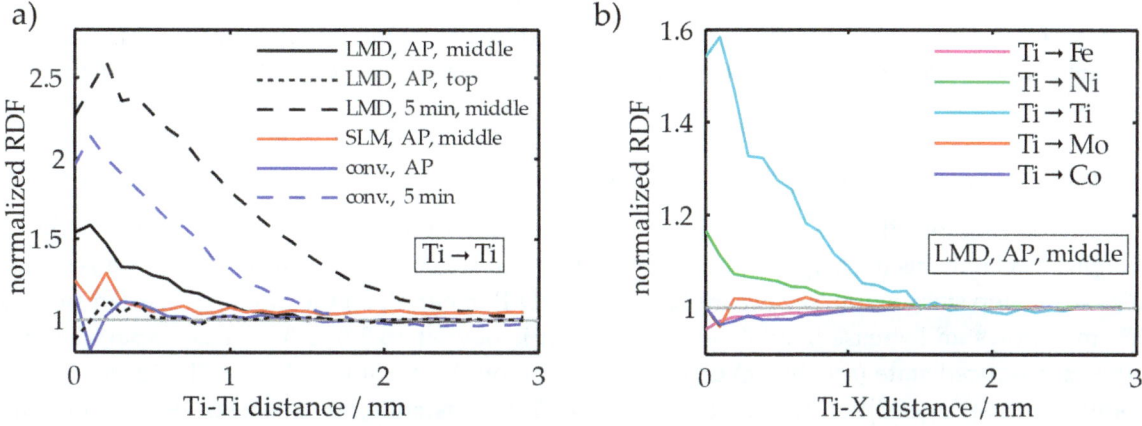

Figure 8. Radial distribution functions of various atom pairs normalized by their bulk average concentrations for different samples as a function of interatomic distance. (**a**) Ti-Ti RDFs for SLM- LMD- and conventionally produced samples in the as-produced (AP) state and after ageing for 5 min. Additionally, the RDF for a sample taken from the very top of as-LMD-produced material is shown; (**b**) Various RDFs (Ti as center atom and Fe, Ni, Ti, Mo, and Co as target atoms) of as-LMD-produced material (taken from the middle of the specimen). Panel (a) reprinted from [24] under the Creative Commons license (http://creativecommons.org/licenses/by/4.0/).

3.4. Microstructure after Short Term (5 min) Ageing

After a short ageing treatment (5 min at 480 °C), both in LMD-P and in C-P material, strong Ti-Ti clustering occurs. This is indicated by the high values of the RDFs at low interatomic distances as determined by APT and as shown in Figure 8a. In panel (b), the type of clustering is shown: Apart from Ti-Ti RDF values being larger than one, also the Ti-Ni RDF is increased at small interatomic distances. This indicates that the clusters contain both Ti and Ni atoms and are possibly very small Ni_3Ti precipitates, a phase that is expected to form in this steel. Corresponding atom maps for C-P and LMD-P material are depicted in Figure 9. Even though clustering is definitely detected in the statistical analysis and the hardness is already significantly increased compared to the as-produced state, the arrangement of the atoms visually still appears random with the possible exception of the Ti atoms (cf. the enlarged inset in the Ti atom map).

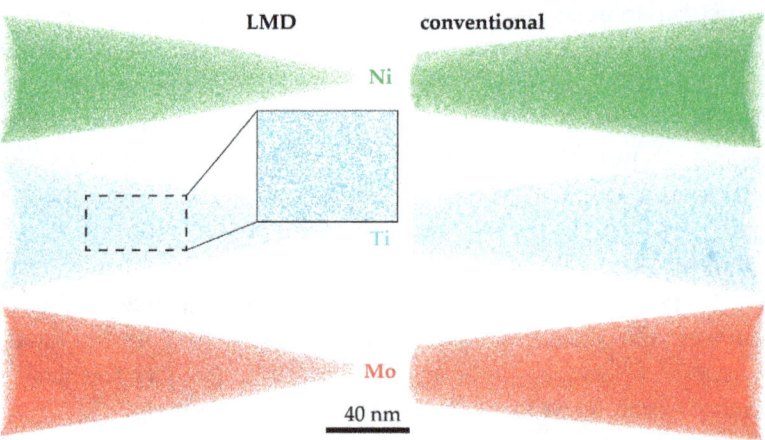

Figure 9. Atom maps obtained by APT of LMD-produced and conventionally-produced material after 5 min of ageing at 480 °C. In the Ti atom maps, early stages of precipitation are beginning to become visible.

3.5. Microstructure in the Peak Aged (480 min) State

After ageing for 480 min—i.e., nearly in the peak-aged condition—the microstructure of the C-P maraging steel remains unchanged. In particular, still no austenite can be detected by a high-resolution EBSD scan (step size: 50 nm, see Figure 10a. The LMD-P material, on the other hand (scan step size: 200 nm), shows an increase in austenite fraction. It is now at 16.5% ± 3.5% as compared to 8.5% in the as-produced state (see, however, the discussion on the accuracy of this value in Section 3.2). This means that, in addition to the presence of retained austenite, reversed austenite is now also present in the microstructure. The same holds for aged SLM-P material (Figure 10c). Due to the very small dendrite width in this material, no reliable mapping of phases by EBSD was possible. Instead, an SEM micrograph of a lightly etched specimen is displayed that depicts very fine austenite films along dendrite boundaries as well as unidentified linear and spherical structures (possibly precipitates). The fcc (Cu type) crystal structure of the interdendritic films is confirmed in multiple spots by EBSD point measurements.

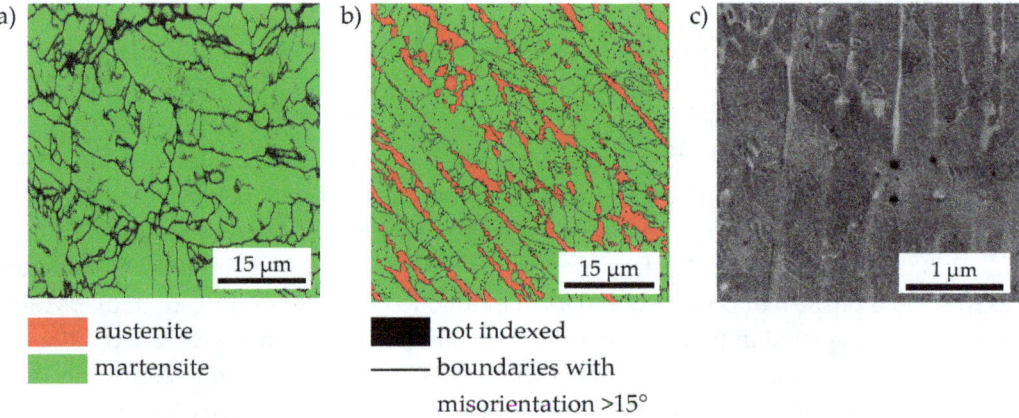

Figure 10. Small-area, high-magnification EBSD scans of conventionally-produced (**a**) and LMD-produced material (**b**) after ageing for 480 min at 480 °C. There is no reversed austenite in the conventionally-produced material while the increased fraction of austenite in the LMD-produced material as compared to the as-produced state indicates that austenite reversion has occurred; (**c**) an SEM micrograph of SLM-produced material aged at 480 °C for 480 min. Very fine austenite films along dendrite boundaries as well as fine precipitates are visible.

In the nanostructure, as revealed by APT, a strong change has taken place upon ageing: A very high number density of precipitates emerges (see Figure 11). There are three different kinds of precipitates present: $(Fe,Ni,Co)_3(Ti,Mo)$, $(Fe,Ni,Co)_3(Mo,Ti)$, and $(Fe,Ni,Co)_7Mo_6$. The different precipitates are delineated in Figure 11 by three different iso-concentration surfaces: c(Mo) > 25 at %, c(Mo) > 10 at % and c(Ti) > c(Mo). In Table 3, the compositions and number densities of the various precipitates are compiled as determined by the constant concentration in the middle of precipitates found in proximity histograms based on the three previously mentioned iso-concentration surfaces. In our previous work [20], we analyzed SLM-produced material annealed for the same time and found the same kinds of precipitates. We speculated that the $(Fe,Ni,Co)_3(Ti,Mo)$-precipitates were probably formed first, which we can now confirm by analyzing the APT measurements after 5 min annealing. Interestingly, the chemistry, number density, and sizes of precipitates in all three differently produced materials are very similar. Note that due to the interconnected nature of the $(Fe,Ni,Co)_7Mo_6$-precipitates, their number density cannot be determined. Due to the limited sample size and the chemical inhomogeneity of the material, only an approximation of the average number density values may be given.

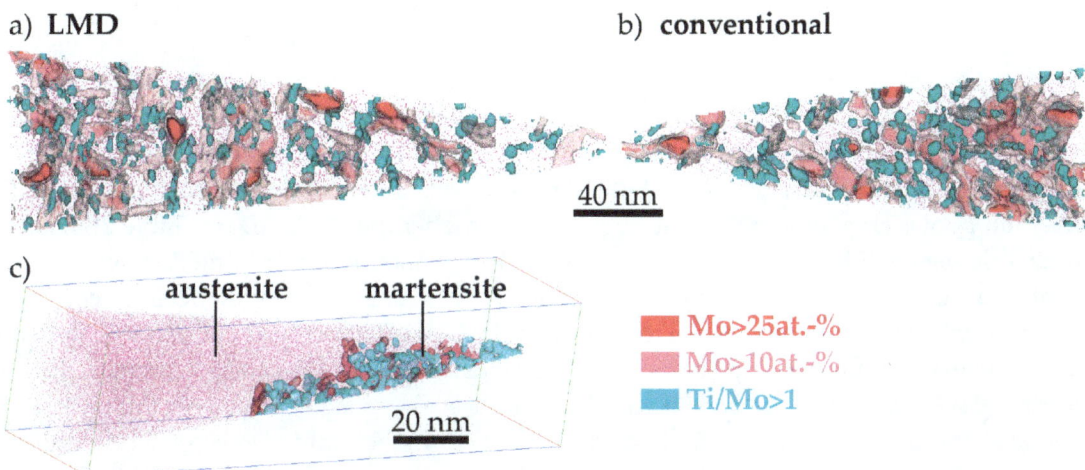

a) **LMD**   b) **conventional**

40 nm

c)

austenite   martensite

20 nm

Mo>25at.-%
Mo>10at.-%
Ti/Mo>1

Figure 11. APT datasets of LMD-produced (**a**) and conventionally-produced material (**b**). Three different kinds of precipitates are present in both materials, as delineated by three different kinds of iso-concentration surfaces (panels (**a,b**) reprinted in modified form from [24] under the Creative Commons license (http://creativecommons.org/licenses/by/4.0/)). In panel (**c**), a measurement including both precipitate-containing martensite and precipitate-free austenite is shown.

Table 3. Chemical composition (in at %) of the precipitates in material produced by the three different processes (conventionally produced, C-P; Laser Metal Deposition-produced, LMD-P; and Selective Laser Melting-produced, SLM-P). The compositions and approximate number densities are determined from proximity histograms based on the three different iso-concentration surfaces (see text).

Precipitate	Material	Fe	Ni	Co	Mo	Ti	Number Density
		(at %)					(m^{-3})
$(Fe,Ni,Co)_3(Ti,Mo)$	C-P	10	60	6	4	20	$\sim 4 \times 10^{23}$
	LMD-P	15	61	2	3	20	$\sim 3 \times 10^{23}$
	SLM-P [20]	12	60	4	5	18	–
$(Fe,Ni,Co)_3(Mo,Ti)$	C-P	21	50	5	19	5	–
	LMD-P	22	52	2	22	2	–
	SLM-P [20]	22	52	5	16	5	–
$(Fe,Ni,Co)_7Mo_6$	C-P	40	17	4	39	0	$\sim 8 \times 10^{22}$
	LMD-P	38	20	2	40	0	$\sim 5 \times 10^{22}$
	SLM-P [20]	37	17	4	38	0	–

An additional APT data set of LMD-P material in aged condition is displayed in panel (c) of Figure 11. In it, the two phases present in the maraging steel can be discerned: The martensite phase contains a high density of precipitates while the austenite phase is completely devoid of precipitates. The composition of the austenite is equal to the average composition of the alloy. Note that here, for LMD-P material, we did not find a Ni-enriched shell around the retained austenite indicative of austenite reversion as we did in the case of the SLM-produced material [20]. The EBSD measurements (cf. Figure 10), however, prove that austenite reversion does occur. The interface between austenite and martensite appears faceted, however the crystallographic orientation of the phases could not be determined from this particular APT measurement.

4. Discussion

The most striking differences between AM-produced and conventionally-produced 18Ni-300 maraging steel are summarized by the hardness versus ageing time curve (Figure 1). There are three notable effects:

(i).   The hardness of as-LMD-produced material is higher than both as-SLM-produced and conventional as-received material. The reason for this lies in the early stages of precipitation that are detected by APT in the (and only in the) LMD-P material. Apparently, already the very small clusters have a significant strengthening effect. In principle, the precipitation (clustering) could occur either during the cooling down just after deposition and solidification of material in the LMD-process or during the pulse-like re-heating (intrinsic heat treatment) upon deposition of adjacent tracks and overlying layers. Both of these effects differ between SLM and LMD. In SLM, the melt pool is smaller and the scanning speed is higher than in LMD (cf. Table 2), leading to both a slower cooling rate after deposition and a less pronounced reheating during the intrinsic heat treatment. However, the results of Figure 7 allow to separate the effects of cooling rate and intrinsic heat treatment. Since the hardness of LMD-P material is comparable to C-P and SLM-P material in the very top layers, clearly the clustering in the as-produced state is due to the intrinsic heat treatment (that does not apply or applies less strongly to the very top layers). If the cooling after deposition were the origin of the clustering, it could also be observed in the top layers (cf. also the discussion in reference [24]). The fact that the intrinsic heat treatment is strong enough to induce (early stages of) precipitation in the present maraging steel, a material that needs several hours to reach a peak aged stage, suggests that the intrinsic heat treatment might be exploited to design a maraging steel that is fully in-situ precipitation strengthened—i.e., that does not need an additional heat treatment after AM production. We are currently optimizing the LMD-process and designing a model maraging steel to achieve this goal. First results with a Fe-Ni-Al alloy show promising results, including a very high number density of precipitates in the as-LMD-P state (publication in preparation).

(ii).   The peak hardness is lower in both AM-produced materials compared to the C-P material. This is most likely due to the significant amount of comparatively soft retained (and reversed) austenite in the AM-produced materials (cf. Figure 3, Figure 4 and Figure 10). The austenite, in turn, is present because of the chemical inhomogeneity due to microsegregation during solidification. This microsegregation is present also in SLM-P material [13,20] that has been solidified at a higher rate than LMD-P material. Despite often being referred to as a process inducing rapid solidification, obviously neither LMD nor SLM enable effective trapping of solutes in the solidifying material in this particular alloy. Even though C-P material initially—i.e., after casting—surely also contained such inhomogeneity, these had been homogenized during the subsequent standard downstream thermomechanical processing such as hot rolling and annealing. Due to the near-net shape nature of AM processes this is not a viable option for AM-P material. Potentially, a prolonged solution annealing before ageing might remove the segregation and hence the retained austenite in AM-produced material, yet, it would also add an undesired additional processing step and additional cost to the AM production process. Note that the interdendritic spacing and thus also the width of the austenite regions is smaller in SLM-P material than in LMD-P material. Also the volume fraction of retained and reversed austenite—namely, 5.8% and 9.4% in the as-produced and peak aged states, respectively [13]—could be smaller, but this is not certain given the errors in the determination of the austenite content. The presence of reversed austenite in (over-)aged maraging steels is well documented (see e.g., [6,25,26]). Even though austenite lowers the strength of the material, it may be a desired microstructure constituent because it allows tailoring the ductility and toughness [27,28]. A slightly overaged condition may therefore be ideal for certain applications [29]. The exact influence of the fraction of retained and reversed austenite on the hardness of the materials is beyond the scope of this study due to the difficulty in separating the effects of grain size (prior austenite grain size, martensite block size, and morphology) and crystallographic texture from the austenite fraction.

(iii). The kinetics of precipitation is not noticeably different between C-P and AM-P material. It could have been expected that the presence of a high density of lattice defects in the AM-P material, originating due to the high residual stress imposed in the AM processes, leads to a quicker nucleation and growth of precipitates in the AM-produced materials or to a different morphology. This is, however, not the case. After 480 min of ageing, all differences in the nanostructure that have been present in the as-produced state are of no significance any more, as evidenced by the very similar composition, distribution, and sizes of the observed three types of precipitates (cf. Figure 11 and Table 3). Another point to note is a certain inhomogeneity in the LMD-P material. This is visible in the slightly uneven distribution of austenite and in the high scatter of the hardness values compared to the C-P material. We performed sets of microhardness indents with decreased load across a melt pool but did not find systematic changes in hardness. Hence, the observed variation in the hardness values seems to be truly random. A homogenizing heat treatment might alleviate this effect, too.

5. Conclusions

We studied the same nominal 18Ni-300 maraging steel alloy produced by three different processes: conventionally synthesized—i.e., by vacuum induction melting, vacuum arc re-melting, and hot rolling, Selective Laser Melting (SLM) and Laser Metal Deposition (LMD), two additive manufacturing methods. We find that the intrinsic heat treatment inherent to LMD (i.e., the heat input by adjacent tracks and overlying layers) is sufficient to induce early stages of precipitation in as-LMD-produced material, however, not in the topmost layers. This in turn leads to a higher hardness of as-LMD-produced material compared to as-received conventionally-produced material and as-SLM-produced material (except for the topmost layers). Upon ageing, however, the effect of the intrinsic heat treatment is superseded by the hardness increase due to precipitation. Precipitation kinetics and precipitate chemistry, size, and morphology is practically identical in all three studied materials. In the peak aged condition, the hardness of the two AM-produced materials is lower than that of the conventionally-produced material. This is due to the presence of both retained and reversed austenite in these materials, while there is no austenite at all in the conventionally-produced material. The reason for the austenite formation lies in the chemical inhomogeneity caused by microsegregation upon solidification that is suppressed neither in SLM nor in LMD, despite relatively quick cooling and solidification rates. This work illustrates that more microstructure and spatially resolving composition investigations need to be conducted to reveal and understand the micro- and nanostructures of allegedly well-known materials after synthesizing and processing them via the various novel AM methods. Additionally, our study demonstrates the possibility of designing materials that are in-situ precipitation hardened in the LMD process.

Acknowledgments: This work was carried out in the scope of the AProLAM project, funded jointly by the Fraunhofer society and the Max Planck Society in their strategic cooperation framework. The authors gratefully acknowledge the help of Benjamin Breitbach with XRD experiments. The authors would like to thank S. Kleber

(Böhler Edelstahl GmbH) for providing the conventionally-produced material and Jan van Humbeeck for providing the SLM-produced material.

Author Contributions: Eric A. Jägle and Dierk Raabe designed the project. Sörn Ocylok and Andreas Weisheit produced the samples used in this study. Zhendong Sheng performed the microstructural analyses. All authors contributed to the interpretation of the results and the writing of the manuscript.

Conflicts of Interest: The authors declare no conflict of interest.

References

1. Sha, W.; Guo, Z. *Maraging Steels: Modelling of Microstructure, Properties and Applications*; Elsevier: Amsterdam, The Netherlands, 2009.
2. Kundig, K.J.A. Copper and copper alloys. In *Handbook of Materials Selection*; Myer, K., Ed.; John Wiley & Sons: New York, NY, USA, 2002; pp. 135–200.
3. Vasudevan, V.; Kim, S.; Wayman, C. Precipitation reactions and strengthening behavior in 18 wt pct nickel maraging steels. *Metall. Trans. A* **1990**, *21*, 2655–2668. [CrossRef]
4. Sha, W.; Cerezo, A.; Smith, G.D.W. Atom Probe Studies of Early Stages of Precipitation in Maraging Steels. *Scr. Mater.* **1992**, *26*, 517–522. [CrossRef]
5. Tewari, R.; Mazumder, S.; Batra, I.S.; Dey, G.K.; Banerjee, S. Precipitation in 18 wt% Ni maraging steel of grade 350. *Acta Mater.* **2000**, *48*, 1187–1200. [CrossRef]
6. Rao, M.N. Progress in understanding the metallurgy of 18% nickel maraging steels. *Int. J. Mater. Res.* **2006**, *97*, 1594–1607. [CrossRef]
7. Lang, F.H.; Kenyon, N. *Welding of Maraging Steels*; Welding Research Council: Shaker Heights, OH, USA, 1971.
8. Stanford, M.; Kibble, K.; Lindop, M.; Mynors, D.; Durnall, C. An investigation into fully melting a maraging steel using direct metal laser sintering (DMLS). *Steel Res. Int.* **2008**, *2*, 847–852.
9. Cabeza, M.; Castro, G.; Merino, P.; Pena, G.; Román, M. Laser surface melting: A suitable technique to repair damaged surfaces made in 14 Ni (200 grade) maraging steel. *Surf. Coat. Technol.* **2012**, *212*, 159–168. [CrossRef]
10. Grum, J.; Slabe, J.M. The State of Differently Heat-Treated 12% Ni Maraging Steel after Laser Remelting. *Mater. Sci. Forum* **2007**, *537–538*, 647–654. [CrossRef]
11. Thijs, L.; Humbeeck, J. Van Investigation on the inclusions in maraging steel produced by Selective Laser Melting. In *Innovative Developments in Virtual and Physical Prototyping*; Bártolo, P.J., Ed.; CRC Press: Boca Raton, FL, USA, 2011; pp. 297–304.
12. Yasa, E.; Kempen, K.; Kruth, J. Microstructure and mechanical properties of Maraging Steel 300 after selective laser melting. In Proceedings of the 21st International Solid Freeform Fabrication Symposium, Austin, TX, USA, 9–11 August 2010; pp. 383–396.
13. Kempen, K.; Yasa, E.; Thijs, L.; Kruth, J.-P.; Van Humbeeck, J. Microstructure and mechanical properties of Selective Laser Melted 18Ni-300 steel. *Phys. Procedia* **2011**, *12*, 255–263. [CrossRef]
14. Becker, T.; Dimitrov, D. The achievable mechanical properties of SLM produced Maraging Steel 300 components. *Rapid Prototyp. J.* **2016**, *22*, 487–494. [CrossRef]
15. Casalino, G.; Campanelli, S.L.; Contuzzi, N.; Ludovico, A.D. Experimental investigation and statistical optimisation of the selective laser melting process of a maraging steel. *Opt. Laser Technol.* **2015**, *65*, 151–158. [CrossRef]
16. Casati, R.; Lemke, J.N.; Tuissi, A.; Vedani, M. Aging behavior and mechanical performance of 18-Ni 300 steel processed by selective laser melting. *Metals* **2016**, *6*. [CrossRef]
17. LBC Engineering "Leistungsspektrum Lasergenerieren". Available online: http://www.lbc-engineering.de/de/lasergenerierung.php (accessed on 23 November 2016).
18. Decker, R.F. *Source Book on Maraging Steels*; American Society for Metals: Metals Park, OH, USA, 1979.
19. Decker, R.F.; Eash, J.T.; Goldman, A.J. 18% Nickel maraging steel. *Trans. ASM* **1962**, *55*, 58–76.
20. Jägle, E.A.; Choi, P.; van Humbeeck, J.; Raabe, D. Precipitation and austenite reversion behavior of a maraging steel produced by selective laser melting. *J. Mater. Res.* **2014**, *29*, 2072–2079.
21. Larson, D.; Prosa, T.; Kelly, T. *Local Electrode Atom Probe Tomography—A User's Guide*; Springer: Berlin/Heidelberg, Germany, 2013.

22. Liu, C.; Zhao, Z.; Northwood, D.O.; Liu, Y. A new emperical formula for the calculation of ms temperatures in pure iron and super low carbon alloy steels. *J. Mater. Process. Technol.* **2001**, *113*, 556–562. [CrossRef]

23. Springer, H.; Baron, C.; Szczepaniak, A.; Jägle, E.A.; Wilms, M.B.; Weisheit, A.; Raabe, D. Efficient additive manufacturing production of oxide- and nitride-dispersion-strengthened materials through atmospheric reactions in liquid metal deposition. *Mater. Des.* **2016**, *111*, 60–69. [CrossRef]

24. Jägle, E.A.; Sheng, Z.; Wu, L.; Lu, L.; Risse, J.; Weisheit, A.; Raabe, D.; Wu, L.; Risse, J.; Weisheit, A.; et al. Precipitation Reactions in Age-Hardenable Alloys During Laser Additive Manufacturing. *JOM* **2016**, *68*, 943–949.

25. Shekhter, A.; Aaronson, H.; Miller, M. Effect of aging and deformation on the microstructure and properties of Fe-Ni-Ti maraging steel. *Metall. Mater. Trans. A* **2004**, *35*, 973–983.

26. Galindo-Nava, E.I.; Rainforth, W.M.; Rivera-Díaz-del-Castillo, P.E.J. Predicting microstructure and strength of maraging steels: Elemental optimisation. *Acta Mater.* **2016**, *117*, 270–285. [CrossRef]

27. Raabe, D.; Ponge, D.; Dmitrieva, O.; Sander, B. Designing Ultrahigh Strength Steels with Good Ductility by Combining Transformation Induced Plasticity and Martensite Aging. *Adv. Eng. Mater.* **2009**, *11*, 547–555. [CrossRef]

28. Raabe, D.; Ponge, D.; Dmitrieva, O.; Sander, B. Nanoprecipitate-hardened 1.5 GPa steels with unexpected high ductility. *Scr. Mater.* **2009**, *60*, 1141–1144. [CrossRef]

29. Viswanathan, U.K.; Dey, G.K.; Sethumadhavan, V. Effects of austenite reversion during overageing on the mechanical properties of 18 Ni (350) maraging steel. *Mater. Sci. Eng. A* **2005**, *398*, 367–372. [CrossRef]

12

Scaling-Up Techniques for the Nanofabrication of Cell Culture Substrates via Two-Photon Polymerization for Industrial-Scale Expansion of Stem Cells

Davide Ricci [1], Michele M. Nava [1,*], Tommaso Zandrini [2], Giulio Cerullo [2], Manuela T. Raimondi [1] and Roberto Osellame [2]

[1]   Department of Chemistry, Materials and Chemical Engineering "Giulio Natta", Politecnico di Milano, 20133 Milano, Italy; davide2.ricci@mail.polimi.it (D.R.); manuela.raimondi@polimi.it (M.T.R.)

[2]   Istituto di Fotonica e Nanotecnologie (IFN)-CNR and Department of Physics, Politecnico di Milano, 20133 Milano, Italy; tommaso.zandrini@polimi.it (T.Z.); giulio.cerullo@polimi.it (G.C.); roberto.osellame@polimi.it (R.O.)

*   Correspondence: michele.nava@polimi.it

Academic Editor: Alina Maria Holban

Abstract: Stem-cell-based therapies require a high number (10^6–10^9) of cells, therefore in vitro expansion is needed because of the initially low amount of stem cells obtainable from human tissues. Standard protocols for stem cell expansion are currently based on chemically-defined culture media and animal-derived feeder-cell layers, which expose cells to additives and to xenogeneic compounds, resulting in potential issues when used in clinics. The two-photon laser polymerization technique enables three-dimensional micro-structures to be fabricated, which we named synthetic nichoids. Here we review our activity on the technological improvements in manufacturing biomimetic synthetic nichoids and, in particular on the optimization of the laser-material interaction to increase the patterned area and the percentage of cell culture surface covered by such synthetic nichoids, from a low initial value of 10% up to 88% with an optimized micromachining time. These results establish two-photon laser polymerization as a promising tool to fabricate substrates for stem cell expansion, without any chemical supplement and in feeder-free conditions for potential therapeutic uses.

Keywords: two-photon laser polymerization; microfabrication; synthetic nichoids; stem cell expansion; pluripotency maintenance; biomimetics

1. Introduction

1.1. Rationale Underlying Industrial-Scale Expansion of Stem Cells and Limitation of Feeder-Cell Layers and Exogenous Conditioning

Stem cell-based therapies represent the most challenging and, potentially, the most successful applications for stem cells (SCs) [1]. Multipotent adult stem cells, including mesenchymal stem cells (MSCs) and adipose stem cells, are likely to be important sources for such therapies both because of the ease of access and the autologous derivation, which involves a low risk of infection and low immunoresponse from the host [2,3]. In addition, induced pluripotent stem cells (iPSCs), which can be obtained from the genetic reprogramming of somatic cells to their pluripotent stage and then induced towards a specific cell phenotype [2,4], represent a further promising source because they allow ethical issues related to embryonic stem cells (ESCs) to be overcome. However, several limitations need to be overcome prior to the successful exploitation of SC-based therapies in clinics. For example, the low efficiency of iPSC reprogramming [5] and the average number of cells needed for cell therapies and regenerative strategies is typically in the order of 10^6 to 10^9 cells, whereas the number of SCs from 4 mL-bone-marrow aspirate is around 700 cells [6]. Therefore, a (minimal) in vitro cell manipulation, including expansion in order to achieve high cell densities by ensuring either multi/pluripotency maintenance or differentiation towards the correct lineage to meet clinical demands is necessary.

In this context, efforts have focused on optimized culture media with a well-defined composition. Such media include additives and small molecules that inhibit signaling pathways associated with cell death and differentiation. For example, Valamehr and colleagues [7] reported the maintenance of a homogeneous population of undifferentiated human iPSCs by supplementing the standard culture media with "SMC4" containing the cell pathway inhibitors Glycogen Synthase Kinase 3, MAPK/ERK Kinase, Rho-Associated Protein Kinase, and Transcription Growth Factor-beta. The same research group derived a culture medium, namely a fate maintenance medium, containing SMC4 and leukemia inhibitory factor (LIF), and basic Fibroblast Growth Factor (bFGF) [8]. They reported that (mouse) ESCs had a lower expression of genes related to three germ layers differentiated with respect to those cells expanded on conventional feeder layers. Another example relevant to the development of protocols of chemically-defined media for the maintenance and expansion of (mouse) ESCs consisted in growing cells in suspension as spheroids in LIF-bFGF-conditioned medium to ensure long-term pluripotency and very few differentiated cells [9].

Examples of commercially available chemically-defined media for a feeder-free (FF) culture of human iPSCs are PluriSTEM (Merck Millipore, Darmstadt, Germany), StemPro (Invitrogen, Carlsbad, CA, USA), mTeSR1 (STEMCELL Technologies, Vancouver, BC, Canada), Pluripro (Cell Guidance Systems, St. Louis, MO, USA), Stemline (Sigma, St. Louis, MO, USA), and Essential-8 (Invitrogen, Carlsbad, CA, USA) [10]. Despite all the advantages in terms of expansion efficiency, the long-term clinical side-effects on human beings in the case of SC transplantation needs to be assessed in advance. For example, LIF has been proven to increase cancer expansion and metastasis in human osteosarcoma and carcinoma, enhancing the phosphorylation of signal transducers and activators of transcription-3 [11], as well as mediating the proinvasive activation of stromal fibroblasts [12]. In addition, such culture media typically may contain animal-derived components, (e.g., fetal bovine serum and/or bovine serum albumin) to enhance SC growth. The serum composition as well as the presence of differentiation factors are generally undefined, so that the serum lots need to be screened and certified for SC use, resulting in high direct/indirect costs [9]. For these reasons, serum-free and additive-free culture conditions are preferable.

For human pluripotent stem cells, an animal feeder cell layer (e.g., mouse embryonic fibroblasts) is usually employed to support cell adhesion, survival, and self-renewal [3,13]. However, the biological products secreted by such xenogeneic feeder-cells are also a source of pathogens and mycoplasma contaminations which might induce an acute immune response in the host upon SC transplantation. To prevent such issues, human feeder cell layers have been introduced, such as recombinant E8 fragments of laminin isoforms [14], recombinant human laminin 511 [15,16], and clinical grade human foreskin fibroblasts [17,18]. Nevertheless, there are still several drawbacks related to the cost, the high lab-to-lab variability, high batch variability, as well as a limited scalability. An additional issue is the persistence of viral and non-viral infections from allogenic materials, the complexity in the maintenance of cells, the isolation and separation of a pure population of SCs with respect to feeder-cells. Finally, feeder-cells release a not-well defined variety of factors which may affect the interpretation of the biological results [3] and lead to safety issues when the SCs are transplanted. An example of FF culture consists in different extracellular matrix (ECM) extracts such as Matrigel® [19] derived from mouse sarcoma cell basement membrane. Matrigel® is currently the gold standard among ECM culture substrates for expanding SCs in feeder-free conditions. Nevertheless, exposure to animal-derived pathogens, xenogeneic components, and immunogenic epitopes would make these cells unsafe for clinical applications in humans [3]. Recently, a serum- and xeno-free substrate composed of conditioned medium from human dermal fibroblasts for long-term expansion of human ESCs and human iPSCs has been reported, named "RoGel" [20]. Despite decreasing the risk of animal-derived pathogen contamination, this system may have limitations, including high batch variability, costs, and the risks associated with human-derived pathogens and allogenic epitopes which severely limit potential therapeutic applications in humans.

Besides, the development of well-defined culture media, feeder-free and xeno-free substrates for cell expansion in vitro, researchers have focused on culture substrates that mimic at least one of the features of the physiological microenvironment surrounding cells. Increasing evidence has shown that, despite their extensive usage, conventional substrates such as culture polystyrene or glass dishes do not resemble the in vivo milieu. Thus, all the complex interactions that occur in vivo are impaired [21–23]. Various strategies have been developed to overcome these limitations and improve substrates for cell culture [13]. In order to recreate feeder free substrates for expansion and pluripotency maintenance of human stem cells, Park and colleagues developed a chemically-defined coating (i.e., polydopamine-mediated oligovitronectin) for conventional polystyrene culture plates, thereby promoting human ESC self-renewal and pluripotency maintenance [24]. Another approach is to mimic the physical and mechanical features of the natural microenvironment, including the nanotopography and the three-dimensional (3-D) geometry of such substrates [25–27]. A "smart" 3-D substrate, may be able to instruct cells towards the right fate, limiting and, whenever possible, avoiding any medium additive or supplement and/or xenogeneic-factor. There would be enormous benefits in terms of safety and risk mitigation during in vitro manipulation, thus leading to a potential industrial-scale expansion of cells for therapeutic purposes [10].

1.2. State of the Art in the Use of 3-D Substrates for the Expansion of Stem Cells

Several studies have proposed natural and synthetic polymers to recreate 3-D culture conditions, such as gelatin [28], collagen-I [29], fibronectin [30], Polyethylene-glycol (PEG) [31], Poli-L-Glycolic Acid [32,33], hyaluronic acid [34], vitronectin [35], poly(N-isopropylacrylamide)-PEG [36,37], and carboxymethyl-hexanoyl chitosan [38] in the form of hydrogels, nanofiber scaffolds and other 3-D structures [21,26]. These studies have shown that SCs could maintain a proliferative and undifferentiated state only by exploiting the biophysical cues offered by the surrounding environment. Besides established fabrication processes (e.g., solvent casting-particulate leaching, gas foaming, electrospinning, and fiber bonding), rapid prototyping methods (e.g., 3-D printing, fused deposition modelling, selective laser sintering, and electron beam lithography) have been increasingly used in scaffold fabrication because of the high resolution that can be achieved [39]. However these techniques, based on a layer-by-layer approach, require the use of (multiple) masks, and the complex 3-D architectures are difficult to manufacture [40]. A recent work proposed a fabrication method that produces poly(ε-caprolactone) (PCL) nanofibers over a large culture surface, by pressing the PCL substrate against a femtosecond laser fabricated glass mold to study the effects of the mechanics on cell fate [41].

1.3. State of the Art in the Use of 2PP for the Nanofabrication of Substrates for Cell Culture

An alternative scaffolding fabrication method is two-photon laser polymerization (2PP), a direct laser writing technique with three-dimensional (3-D) capabilities and a spatial resolution down to 100 nm [42–44]. 2PP employs a photosensitized resist, where simultaneous absorption of two photons from a near-infrared ultrashort laser pulse triggers a photochemical process which results in the cross-linking of monomers and oligomers. The two-photon absorption mechanism enables selective polymerization in a small volume around the focus, known as volume element or voxel. This enables sub-diffraction limited resolution in both the lateral and axial directions as well as 3-D fabrication capabilities at different depths by adjusting the laser focus with little or no collateral effects. Examples of photopolymers used in 2PP are organically modified ceramics, epoxides (e.g., SU-8), and acrylic monomers whose biocompatibility has been assessed [40,45]. Recent studies have shown that 3-D geometries can be manufactured by 2PP in PEG-diacrylate, poly lactic acid (PLA) and PLA-poly-ε-caprolactone copolymer [46–48]. In our previous work, we manufactured 3-D "structurally" biomimetic synthetic niches for MSC culture [49–52]. Evidence of spontaneous lineage commitment was observed in the monolayer culture surrounding the structural niches, but not inside the niches, thus suggesting that structural niches were able to direct SC homing, proliferation, and multipotency maintenance. However, the quantitative biological measurements were inevitably diluted because of the limited percentage (10%) of cell culture surface covered by the structural niches [52].

In this paper, we review our activity on the technological improvements in manufacturing the 3-D "structurally" biomimetic synthetic niche system for SC culture. In particular, more than on the biological results, we focus here on the optimization of the laser-material interaction that we performed in the last years to increase the patterned area and the percentage of cell culture surface covered by such 3-D synthetic niches, reaching the most recent result of 88% surface coverage with an optimized micromachining time. The aim of this study is to obtain a greater number of cells experiencing the 3-D microenvironment provided by the increased number of niches and, therefore to obtain quantitative, reliable, biological data to conclusively demonstrate the effect of the 3-D architecture on the SC fate.

2. Materials and Methods

2.1. Description of Photosensitive Material and Photo-Initiator

Synthetic niches were directly fabricated by 2PP in the commercially available SZ2080 photoresist, composed of a sol-gel-synthetized silicon (S)-zirconium (Z) hybrid inorganic-organic resin (hereafter called SZ2080) (Maria Farsari, IELS-FORTH, Heraklion, Greece) [53]. The main components of SZ2080 are methacryloxypropil trimethoxysilane and zirconium propoxide with the addition of 1% concentration of Irg photoiniziator (Irgacure 369, 2-Benzyl-2-dimethylamino-1-(4-morpholinophenyl)-butanone-1) [50]. SZ2080 has many advantages, such as biocompatibility [45,49], chemical and electrochemical inertia, long-term stability, good optical transmission, and mechanical stability after polymerization due to low shrinkage compared to other commercial photoresists [45].

2.2. The Laser Fabrication Set-Up

Synthetic niches were directly written onto circular glass cover slips with a 150-μm thickness and 12-mm diameter (Bio-Optica, Milan, Italy). Initially, we used a home-built Ti:Sapphire femtosecond laser (87 MHz repetition rate, 40 fs pulse duration, up to 400 mW average power) and a 3-D piezo positioning system (P-611.3 NanoCube, Physik Instrumente, Karlsruhe, Germany) with a travel range of 100 μm in each direction. Subsequently, we changed both the femtosecond laser and the positioning system. We used a home-built cavity-dumped Yb:KYW mode-locked system with a 1030 nm wavelength, generating pulses with 300-fs duration, energy up to 1 μJ and 1-MHz repetition rate, corresponding to 1 W of average power. The laser beam was focused on samples with a 1.4 numerical aperture 100$\times$ oil immersion objective (Plan-Apochromat, Carl Zeiss, Oberkochen, Germany). A power control stage, consisting of a polarizing beam splitter and a rotating waveplate, was used to tune the laser output power, while a mechanical shutter (LS Series, Uniblitz Electronics, Rochester, NY, USA), was used to switch the laser beam impinging on the sample on and off. A computer-controlled three-axis brushless motion stage (ANT130, Aerotech, Hanover, MD, USA) was used to control the position of the sample with respect to the laser focus in the two plane dimensions (X and Y), and to move the vertical position of the laser focus (Z) with respect to the sample plane, via specific motion controller software (Automation 3200 CNC Operation Interface, Aerotech, Hanover, MD, USA). This fabrication system has a very high precision and resolution down to $\approx$5 nm, maximum displacement in the order of tens of cm, and a very high speed up to 150 mm$\cdot$s^{-1}. Its characteristics are thus ideal for patterning large areas through 2PP. A two-axis stage moves the sample in the X-Y plane, while a third axis, moving the focusing objective, scans along the Z direction. The system controller enables the three axes to move simultaneously, thus enabling 3-D writing at a constant speed. A gimbal mechanical system (Gimbal Mounts GM100/M, Thorlabs Inc., Newton, NJ, USA) fixed to the X-Y motion stages, was used to tilt the sample and to set it perpendicular with respect to the laser beam. A CMOS camera (DCC1545M, Thorlabs Scientific Imaging (TSI), Austin, TX, USA) and a beam splitter were used to visualize on computer software (uEye Cockpit, 4.71, IDS Imaging Development Systems GmbH, Obersulm, Germany) the image of the working field, back-lit by a red-light LED. The on-line vision of the polymerized lines enabled the structure fabrication to be accurately positioned in X-Y and Z. Prior to laser exposure, the SZ2080 photoresist was manually placed on glass coverslips baked at 105 °C (ramp starting from room temperature) for 1 h. Samples with the photoresist were mounted on a parallelepiped metallic custom-made holder and exposed to the laser in order to identify the microstructures by 2PP. To remove the unpolymerized regions, the samples were immersed in a 50% (v/v) 3-pentanone, 50% (v/v) isopropyl alcohol solution (Sigma-Aldrich, St. Louis, MO, USA). The fabricated 3-D niche substrates were then imaged first by optical microscopy (Eclipse ME600, Nikon, Tokyo, Japan), and then by scanning electron microscopy (SEM, Phenom Pro, Phenom World, Eindhoven, The Netherlands). All SEM observations were carried out at 5 kV.

2.3. Description of the Up-Scaling Techniques

In order to obtain quantitative and reliable biological data to assess the effect of the 3-D architecture on the stem cell fate, we increased the surface of the culture substrate covered by the 3-D synthetic niches to obtain a greater number of cells experiencing the 3-D microenvironment provided by the niches. We thus adopted the following strategy. Firstly, we initially optimized the fabrication process of the elementary 3-D niches, which we named 'nichoids'. Each nichoid was 30 μm high and 90 × 90 μm in transverse dimensions, and consisted of a lattice of interconnected lines, with a graded spacing between 10 and 30 μm transversely and a uniform spacing of 15 μm vertically. Each nichoid was surrounded by four outer confinement walls formed by horizontal lines spaced by 5 μm, resulting in gaps of 1 μm (Figure 1a,b). Secondly, the nichoids were arranged at the vertexes of an equilateral triangle with 200 μm sides (Figure 1d) [49]. Then, to increase the substrate coverage to the whole slide surface, the nichoids were arranged at the vertexes and at the center of a hexagon with a 300 μm side (Figure 1e) and the total number of elementary nichoids was increased to 367 μm [49,52]. However, the increased number of nichoids led to an increased manufacturing time, thus the manufacturing parameters (laser power and scan speed) needed to be optimized (see Section 3.1). This last configuration however covered only a 10% fraction of the cell culture surface with nichoids, resulting in a large percentage of SCs experiencing the 2-D environment of the glass slide surface. Thus, to further increase the coverage, we designed a very large scaffold that covered a circular area with a 3 mm radius, composed of continuously-packed nichoids (CPN) with about 3500 elements, where each nichoid shared its external walls with the adjacent ones (Figure 1f).

Despite the large surface covered with this approach, there were several drawbacks, including the low structural stability (see Section 3.2). To overcome these issues, we fractioned the large scaffolds into submatrixes with 5 × 5 nichoids (Figure 1c) with a small spacing of 30 μm between adjacent matrixes, resulting in a fractionated supermatrix of nichoids (FSN) including 218 matrixes (Figure 1g). In this last configuration, the entire structure was reinforced by an external 4-μm high base wall, which anchored the structure to the underlying glass surface and forced the cells to enter the nichoids from the top. In addition, to ensure the robustness of the structures, the pillars were reinforced by a double scan irradiation with a lateral shift of 0.5 μm. To prevent mechanical damage to the shutter, due to frequent opening and closing, our control software quickly relocated the laser focus inside the glass substrate to avoid photopolymerization when positioning the writing beam in the X-Y plane.

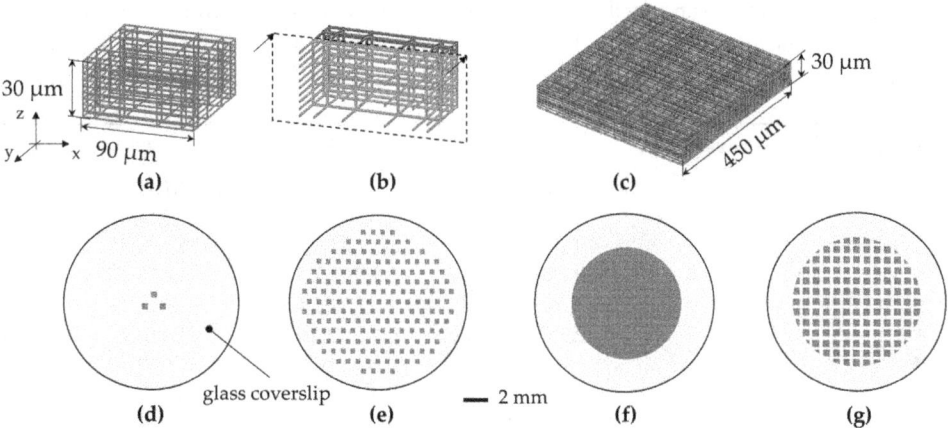

Figure 1. The nichoid culture substrate. (**a**) CAD of the nichoid elementary unit; (**b**) Cross-section of the nichoid; (**c**) CAD of the matrix of nichoids, consisting of 5 × 5 elementary nichoids. Scale-up approach: (**d**) the culture substrate composed of three elementary nichoids; (**e**) The substrate composed of approximately 367 elementary nichoids; (**f**) The culture substrate covered with continuously-packed nichoids (CPN), resulting in 3500 adjacent elementary nichoids; (**g**) The culture substrate covered with a fractionated supermatrix of nichoids (FSN) including 218 matrixes at a distance of 30 μm.

3. Results and Discussion

3.1. Parameter Optimization of 2PP-Engineered Elementary Nichoids

Up-scaling the synthetic nichoid culture system to obtain more nichoid-cultured SCs that could be compatible with clinical demands had various limitations, including the long machining time required to fabricate an elementary nichoid. Initially, we were only able to manufacture a few (e.g., three) elementary nichoids for each culture substrate. Using the Ti:Sapphire laser and the piezo positioning system, we were limited to a processing speed of 10 μm·s^{-1}, thus requiring a long manufacturing time of 30 min per nichoid (Table 1, column A).

Table 1. Technical parameters regarding the various nichoid layouts and upscaling processes. A = three elementary nichoids, B = 367 elementary nichoids, C = continuously-packed nichoids (CPN), D = fractionated supermatrix of nichoids (FSN).

Parameters	A	B	C	D
Total machining time	1.5 h	3 h	17 h	12 h
Power–scan speed (mW–mm·s^{-1})	15 [1]–0.01	12 [2]–1.5	12 [2]–1	13 [2]–3
Elementary nichoid writing time	30 min	30 s	18 s	7 s
Surface of cell culture (mm^2)	0.24	28.27	28.27	50.26
% of surface covered by the nichoids	10%	10%	100%	88%
Number of nichoids	3	367	3500	5450
Estimated nichoid-cultured cells/sample	60	8000	7 × 10^4	10.9 × 10^4

[1] Ti:Sapphire laser. [2] Yb:KYW laser.

To speed up the fabrication, we used a different femtosecond laser and translation stages (see Section 2.2). In this new configuration we were able to work in new processing conditions with writing speeds increased by two orders of magnitude, thus fabricating an elementary nichoid in about 30 s (Table 1, column B). In these new processing windows, we optimized the laser manufacturing parameters (laser power and scan speed) to identify the processing windows that would provide stable structures (Table 2). We tested several combinations of laser power and scan speed ranging from 12–15 mW and 1–10 mm·s^{-1}, respectively.

Table 2. Optimization of the scan speed (mm·s^{-1}) and laser power (mW) resulting in the process window for the microfabrication of niches. STABLE = structurally stable niches; UNSTABLE = structurally unstable niches; DAMAGED = structurally damaged niches; Ø = no polymerization occurred.

Scan Speed (mm·s^{-1})	Power (mW)	12	13	14	15
1		STABLE	DAMAGED	DAMAGED	DAMAGED
2		STABLE	DAMAGED	DAMAGED	DAMAGED
3		UNSTABLE	STABLE	STABLE	DAMAGED
4		UNSTABLE	STABLE	STABLE	STABLE
5		UNSTABLE	STABLE	STABLE	STABLE
6		UNSTABLE	UNSTABLE	STABLE	STABLE
7		Ø	UNSTABLE	STABLE	STABLE
8		Ø	UNSTABLE	UNSTABLE	UNSTABLE
9		Ø	Ø	UNSTABLE	UNSTABLE
10		Ø	Ø	UNSTABLE	UNSTABLE

As shown in Table 2, for a high average laser power, the fabricated structures were visibly damaged, with interrupted or warped polymeric lines (DAMAGED in Table 2). Conversely, for high scan speeds, the amount of energy per unit of time delivered to the material was too low so that 2PP did not occur (Ø in Table 2). Some of the tested writing parameters led to unstable microstructures which partially or completely collapsed (UNSTABLE in Table 2, Figure 2), while others resulted in stable nichoids (STABLE in Table 2, Figure 3), thus enabling us to identify several useful combinations of fabrication parameters. However the reproducibility in the fabrication of the structures decreased with the increasing scan speed when fabricating a large number of structures for a long time. In addition, it is worth noting that 2PP is a threshold nonlinear phenomenon. Indeed, the radical polymerization in the resist may occur only if the irradiated energy per focal volume is above a certain threshold and below a damage threshold. This explains why both laser average power and scan speed influence the damage/non-polymerization of the photosensitive material. Examples of non-optimized fabrication outcomes are shown in Figure 2.

Referring to the initial configurations with few elementary nichoids (Figure 1d,e), the main drawbacks were the long micromachining time and the detachment and misalignment of grids which led to a structural collapse of the nichoids and, thus, an inability to accurately control the 3-D microgeometry (Figure 2a,b).

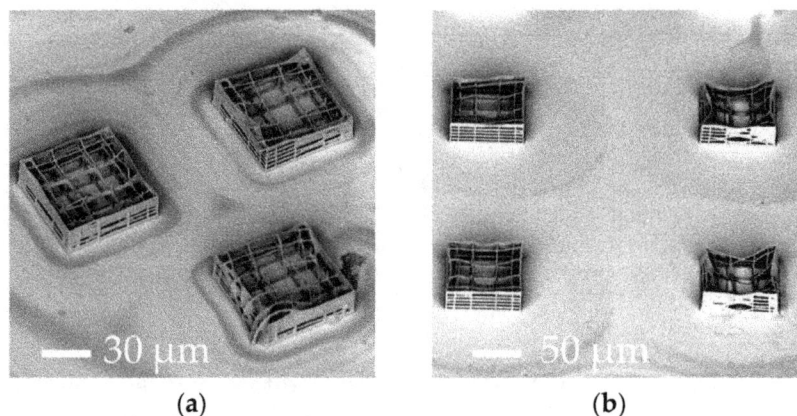

(a)   (b)

Figure 2. *Cont.*

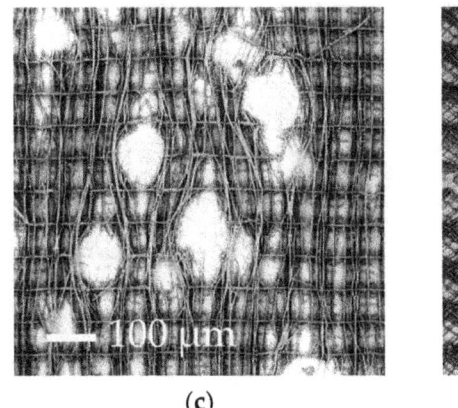

(c)   (d)

Figure 2. SEM images of the micro-fabrication before the optimization. Instability and deformation on (**a**) three elementary nichoids; (**b**) on the substrate composed of 367 elementary nichoids; (**c**) Collapse of the culture substrate covered with continuously-packed nichoids (CPN), more likely due to cavitation phenomena and vibration during the manufacturing process; (**d**) Instability issues in the culture substrate covered with a fractionated supermatrix of nichoids (FSN), including 218 matrixes at a distance of 30 µm.

A mathematical model was developed [54] to predict the dependence of the transversal diameter d and the height h of the polymerized voxel on the average laser power P and the exposure time t:

$$d = K \times [\ln(P^2 \times t \times E_{th}^{-1})]^{1/2} \tag{1}$$

$$h = K \times [(P^2 \times t \times E_{th}^{-1})] \tag{2}$$

where E_{th} is the threshold energy for the 2PP process and K is a constant that depends on the properties of the material and the experimental setup. However, this model requires the empirical determination of the parameter K, and its predictive ability lies within a narrow and specific processing window (i.e., above the 2PP threshold energy and below the damage threshold) [55]. In this work, we used a trial and error procedure, which was systematic, simple, and enabled us to quickly obtain reliable results.

3.2. Increasing the Percentage of Glass Surface Covered by the Nichoids

Using the optimized parameters, we were able to structurally fabricate stable nichoids with a reduced machining time. The first configurations with elementary nichoids arranged in a triangular layout (relative distance 200 µm) (Figure 1d) and in hexagonal distribution (300 µm side) (Figure 1e) resulted in a good architectural stability and an optimal control of the 3-D geometry of the nichoids (Figure 3a,b respectively).

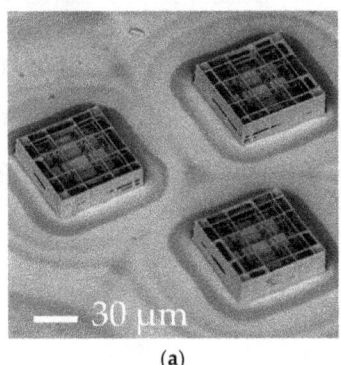
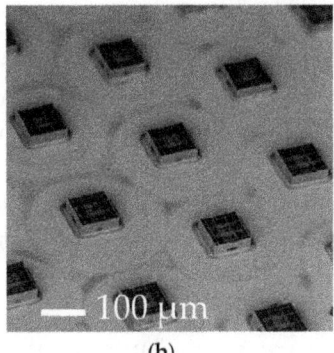

(a)   (b)

Figure 3. *Cont.*

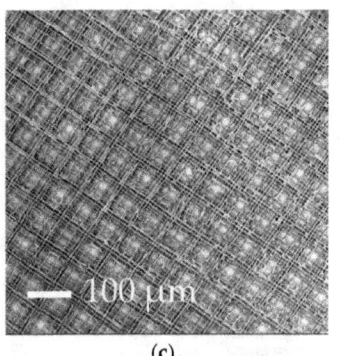

(c)   (d)

Figure 3. SEM images of the micro-fabrication after the optimization. (**a**) The culture substrate composed of three elementary nichoids patterned in a 200-μm side triangle; (**b**) The culture substrate of elementary nichoids in a hexagonal layout (300 μm side); (**c**) The culture substrate covered with continuously-packed nichoids (CPN), which share external walls with the adjacent ones; (**d**) The culture substrate covered with a fractionated supermatrix of nichoids (FSN), including 218 matrixes at a distance of 30 μm.

Having reduced the elementary nichoid manufacturing time to 30 s, we were able to fabricate almost 400 nichoids onto the whole glass slide, covering approximately 10% of the cell culture surface in 3 h (Table 1, column B). Considering a typical occupation of about 20 cells/nichoid [49], we could theoretically obtain around 8000 nichoid-cultured cells per sample in this configuration (Table 1, column B).

Despite such technological improvements, the percentage surface coverage of the 3-D structures was still too low, so that biological measurements were inevitably diluted by the cells deposited on the flat glass surface surrounding the elementary nichoids, which did not experience the 3-D nichoid environment. Thus, in order to decrease the 2-D unpatterned surface between the nichoids, we designed a continuous scaffold CPN (Figures 1f and 3c) by fabricating longer and continuous lines instead of fabricating the nichoids one by one. This new design reduced the manufacturing time for an elementary nichoid to 18 s. Therefore, by using the following parameters 12 mW–1 mm·s^{-1}, we fabricated almost 3500 adjacent nichoids in a machining time of 17 h, covering 100% of the treated culture surface (Table 1, column C). However, the CPN structure (Figure 1f) also showed structural instabilities (Figure 2c). In fact, even a single defect inside the structure has negative effects on the stability of the whole scaffold because of its packed configuration. Moreover, a manufacturing time of 17 h was incompatible with a potential industrial exploitation, and it was difficult to remove the unexposed photoresist from the internal lattice of the nichoids as the solvent can only be accessed from the top. Therefore, although this configuration produced very nice samples (Figure 3c) which could theoretically provide 70,000 nichoid-cultured cells (Table 1, column C), these drawbacks caused us to abandon the CPN design. To overcome the above issues, we developed the FSN configuration with 168 matrixes, each composed of 5 × 5 nichoids, spaced by 80 μm [56]. With this layout, the residual tensions due to the volumetric shrinkage and local defects in the microstructures did not affect the whole 3-D architecture. This was due to the 80 μm spacing that made each matrix a stand-alone entity, not subject to the problems of the neighboring matrixes. Initially, the FSN structure was fabricated in 12 h of machining time with the following parameters 12 mW, 1.5 mm·s^{-1} [56]. Then, by further increasing the fabrication speed to 3 mm·s^{-1}, with a laser power of 13 mW, we were able to reduce the fabrication time for an elementary nichoid to 7 s (Table 1, column D). This was possible because, by tracing longer continuous lines, the translation stages have more time to accelerate, thus reaching a higher velocity. In order to increase the number of nichoids, we reduced the distance between the matrixes of nichoids to 30 μm. Thus, thanks to these technological improvements, we were able to fabricate culture substrates by including more nichoid matrixes (i.e., 218) (Figure 1g) than the previous 168 matrixes in the first FSN configuration [56] with the same machining time of 12 h (Table 1, column D). The results were also very good in terms of structural stability and precise control on the 3-D geometry of the 2PP structures (Figure 3d). We were able to extend the laser-treated surface of cell culture from a circle with a 3-mm radius to one with a 4-mm radius. Thus, this last FSN configuration enabled us to cover up to the 88% of the surface of cell culture with the nichoids and to obtain an estimated number of 10.90×10^4 cells experiencing the nichoid environment (Table 1, column D). The larger number of nichoid-cultured cells will enable us to achieve a more homogeneous cell population and thus obtain a more reliable sample for further quantitative analyses including the investigation of gene expression. However, this latter configuration also initially had some instability issues due to focus variations onto the non-perfectly planar glass surface. We observed differences in the Z-position of the glass surface ranging between 6 μm to 10 μm from the center to the edges of the substrate, which were not negligible with respect to the structure height (30 μm) and thus caused the nichoids to detach from the glass samples or collapse on themselves (Figure 2d). We pre-characterized the sample surface and modified our control software by implementing a point-by-point compensation of the focus in the vertical direction to balance out the glass concavity. In addition, we increased the structure height and initiated the irradiation deeper inside the glass by 8 μm, to ensure that the manufactured blocks anchored on the substrate surface. All these actions enabled us to achieve stable structures (Figure 3d). The optimal processing window was rather narrow and was sensitive to uncontrolled variations in the laser characteristics (pulse duration, wavelength, and spot size) and in the photosensitive resin (different batches, ageing). Therefore, after the initial optimization, we had to periodically re-optimize the processing window by finding optimal values that changed slightly but always came close to the ones here presented. Finally, to confirm the cell adhesion on all the nichoid culture substrates, human bone marrow-derived MSCs were seeded and stained with DAPI (Figure 4a–d) [49,50].

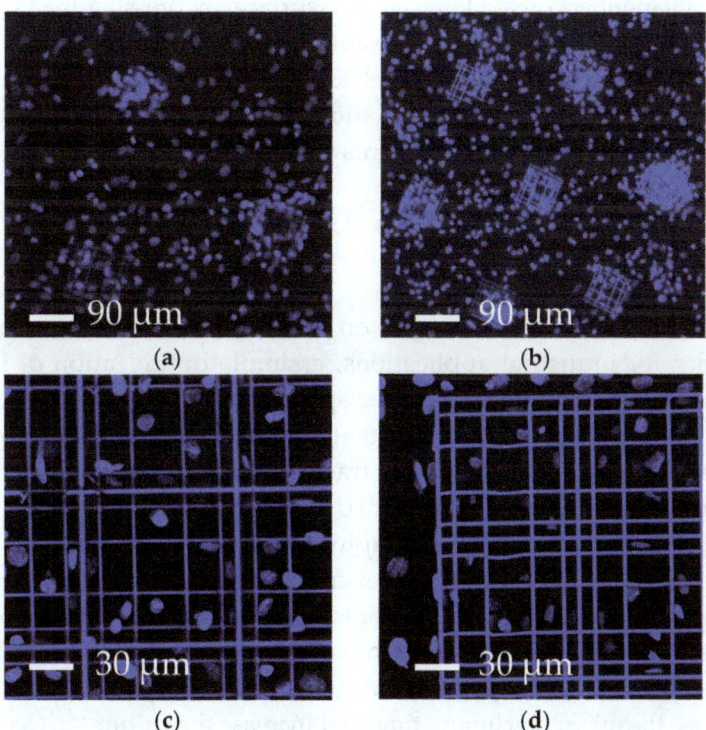

Figure 4. Fluorescence images of cell-populated niche substrates. (**a**) Elementary nichoids patterned in a 200-μm side triangle; (**b**) The culture substrate of elementary nichoids in a hexagonal layout (300 μm side); (**c,d**) The culture substrate covered with a fractionated supermatrix of nichoids (FSN), including 218 matrixes at a distance of 30 μm. Nuclei are stained in DAPI (blue).

3.3. Comparison with Results from the Literature

Cells in our engineered nichoids can adhere to a 3-D environment, experiencing a high surface-to-volume ratio (Figure 4a–d), can be easily extracted/detached by standard culture protocols (e.g., trypsin detachment), and can be imaged both live and fixed by (inverted) optical and confocal microscopes. In fact, the nichoids were laser written directly on standard glass coverslips (150 μm-thick). There are studies in the literature dealing with 3-D culture systems for SC expansion, including gelatin [28], collagen-I [29], fibronectin [30], Polyethylene-glycol (PEG) [31], Poli-L-Glycolic Acid [32,33], hyaluronic acid [34], vitronectin [35], poly(N-isopropylacrylamide)-PEG [36,37], and carboxymethyl-hexanoyl chitosan [38] in the form of hydrogels, nanofiber scaffolds, and other 3-D structures [21,26]. However, since cells were typically embedded in packed cross-linked matrices, imaging them by optical microscopy or collecting viable cells either for further analysis or for potential therapeutic applications would be difficult and, in some cases, almost impossible. In addition, our nichoid-based culture substrate is made of a chemically stable and biocompatible photoresist, so that cell-material surface interactions can be neglected.

The evidence of multipotency maintenance without any chemical supplements reported in [52] was attributed to the interaction between cells and the 3-D nichoid structure. This is an interesting issue from an industrial and clinical perspective, because of the minimization of cell manipulation and associated risks for the host. This is a great advantage over the most recent literature in which chemical exogenous factors were used for this purpose [8,9]. Finally, the FSN culture substrate was demonstrated to preserve the pluripotency of mouse ESCs without either soluble factors (i.e., LIF) or a feeder layer [56]. These results are in agreement with other studies in which 3-D polymer-based scaffold systems for stem cell culture made up of biodegradable matrices have been used [3,20,33], but with the advantages previously outlined. A recent work proposed a fabrication method that produces poly (ε-caprolactone) (PCL) nanofibers over a large culture surface, by pressing the PCL substrate against a femtosecond laser fabricated glass mold [41]. The authors reported pluripotency maintenance and cell detachment by trypsin. However, the low geometrical control of the microstructured surface did not allow a fine study of the effects of the mechanical stimuli exerted on the cells. Indeed, cells grow along the bent nanofibers, instead of being inside a regular 3-D scaffold matrix.

3.4. Future Prospectives

3.4.1. Further Reduction in the Machining Time

2PP is intrinsically a serial micro-fabrication technique. In order to massively increase the production of samples for industrial applications, a simple optimization of laser manufacturing parameters would not be effective because of the speed and acceleration limitations of the current high-precision translation stages. We observed that at higher scan speeds, the machining time was not proportionally reduced due to the significant fraction of time occupied by repositioning the focus between writing steps, which is already performed at the highest speed. A solution to improve the production rate might consist in a soft lithography technique to replicate microscopic structures, namely micro transfer molding (μTM) [57]. However, the internal network of the structures would be irreversibly damaged when extracted from the mold. There are studies in the literature dealing with the resolution of this problem, such as membrane-assisted micro-transfer molding (MA-μTM) [58], but the high complexity of 3-D geometry of the nichoids limits the use of this technique.

To further decrease the micro-machining time and increase the culture surface covered by nichoids, an attractive solution consists in parallel processing, in which multiple foci are created from a single laser beam, using optical systems such as a microarray of lenses [59], a holographic spatial light modulator, and digital micro-mirrors [60]. With these systems, the multiple spots created would be focused into the same objective, allowing the polymerization of multiple lines simultaneously, and decreasing the fabrication time by a factor equal to the number of spots. All the spots would be localized within the field of view of the objective, which in our case is larger than one nichoid, and translating the sample, lines of any length could be polymerized. Another interesting method to decrease the fabrication time could be to integrate a galvanometer scanning mirror system which uses moving magnets for fast and precise positioning of mirrors for the deflection of laser beams, leading to a rapid and accurate scanning of the spot in the focal plane. Closed-loop galvanometer systems have a frequency response up to 10 kHz, and can provide a constant velocity of beam deflection and fast-step response times in the 100 µs range [61]. This could bypass the problem of the limited acceleration of the stages and thus reduce the machining time, while maintaining the same high resolution of our current 2PP apparatus. Galvanometer mirrors are routinely used for beam scanning in the focal plane in high speed confocal microscopy, in combination with high numerical aperture objectives. The drawback of this approach would be the limited field of view of our high numerical aperture objective that would compel us to stitch different parts together to fabricate structures of several hundreds of micrometers, and the impossibility of galvanometric scanners to trace vertical lines (along the Z-axis) unless coupled with mechanical stages or piezoelectric systems for the vertical displacement. Indeed, galvanometric scanners allow the focus position only in the X-Y plane.

3.4.2. Decreasing Cell Adhesion in the Nichoid Surroundings

So far, we have neglected the fact that the surface of the glass substrate is not completely treated by the laser writing process. In fact, as shown in Figure 1f,g, an annular untreated region is left at the edge of the substrate. This region is not structured to enable the sample to be fixed to the holder during the irradiation process, and afterwards to enable sample manipulation with tweezers without damaging the scaffolds. To avoid cell adhesion to this surface, before culturing the cells, we applied a PDMS ring with the same shape. This operation is however rather complicated as the ring is positioned manually with the risk of ruining the edges of the scaffolds. In the future, we plan to reduce the fraction of the cells not experiencing the 3-D nichoid environment using substrate glasses with a hydrophobic coating. These substrates will prevent cell adhesion not only on the external annular region, but on the whole flat substrate, thus favoring homing and 3-D suspension of the cells in the nichoid microstructures.

4. Conclusions

In this work, we have reviewed our activity on the upscaling of 2PP-fabricated 3-D substrates for the expansion of stem cells, by increasing the surface coverage and reducing the machining time. We have discussed how, by suitably tailoring the irradiation parameters, we could reduce the fabrication time of a single nichoid from 30 min to 7 s, achieving the most recent result of 88% of the surface of cell culture covered by nichoids. This result will enable the majority of the stem cells to grow in the 3-D scaffold. We thus expect to obtain more evident biological data to confirm our previous hypothesis that a truly 3-D culture substrate, allowing 3-D isotropic cell adhesion, can drive stem cell responses to allow cell multi/pluripotency without supplementing any chemical media and/or feeder. We plan to further improve our culture system by increasing the surface patterned with nichoids by a parallelization of the manufacturing process, and by removing the 2-D areas by depositing a non-fouling (e.g., hydrophobic) coating.

If successful, this system could be used for the expansion of patient-specific stem cells for clinical treatments, such as customized cell therapies for degenerative diseases such as Alzheimer's and Parkinson's diseases.

Acknowledgments: This study received funding from the European Research Council (ERC) under the European Union's Horizon 2020 research and innovation program (grant agreement No. 646990-NICHOID). These results reflect only the authors' views and the Agency is not responsible for any use that may be made of the information contained.

Author Contributions: R.O., G.C. and M.T.R. conceived the study, provided scientific supervision on the nanofabrication of the culture substrates and critically reviewed the manuscript. D.R. fabricated the culture substrates and contributed to the manuscript. T.Z. contributed and critically reviewed the manuscript. M.M.N. conceived and generated all the figures and wrote the largest part of the manuscript. All authors read and approved the manuscript.

Conflicts of Interest: G.C., M.T.R., and R.O. are listed as inventors on a patent application (102015000048704) entitled "Matrici di nicchioidi sintetiche per la coltivazione di cellule staminali" which deals with the implementation of a surface coating by 2PP for stem cell culturing, according to the results presented in this paper. All the other authors declare no competing interests.

References

1. Trounson, A.; McDonald, C. Stem Cell Therapies in Clinical Trials: Progress and Challenges. *Cell Stem Cell* **2015**, *17*, 11–22. [CrossRef] [PubMed]

2. Herberts, C.A.; Kwa, M.S.G.; Hermsen, H.P.H. Risk factors in the development of stem cell therapy. *J. Transl. Med.* **2011**, *9*, 29. [CrossRef] [PubMed]

3. Villa-Diaz, L.G.; Ross, A.M.; Lahann, J.; Krebsbach, P.H. The evolution of human pluripotent stem cell culture: From feeder cells to synthetic coatings. *Stem Cells* **2013**, *31*, 1–7. [CrossRef] [PubMed]

4. Miki, T.; Yasuda, S.Y.; Kahn, M. Wnt/β-catenin signaling in embryonic stem cell self-renewal and somatic cell reprogramming. *Stem Cell Rev. Rep.* **2011**, *7*, 836–846. [CrossRef] [PubMed]

5. Ebrahimi, B. Reprogramming barriers and enhancers: Strategies to enhance the efficiency and kinetics of induced pluripotency. *Cell Regen.* **2015**, *4–10*. [CrossRef] [PubMed]

6. Hernigou, P.; Homma, Y.; Lachaniette, C.H.F.; Poignard, A.; Allain, J.; Chevallier, N.; Rouard, H. Benefits of small volume and small syringe for bone marrow aspirations of mesenchymal stem cells. *Int. Orthop.* **2013**, *37*, 2279–2287. [CrossRef] [PubMed]

7. Valamehr, B.; Abujarour, R.; Robinson, M.; Le, T.; Robbins, D.; Shoemaker, D.; Flynn, P. A novel platform to enable the high-throughput derivation and characterization of feeder-free human iPSCs. *Sci. Rep.* **2012**, *2*, 213. [CrossRef] [PubMed]

8. Valamehr, B.; Robinson, M.; Abujarour, R.; Rezner, B.; Vranceanu, F.; Le, T.; Medcalf, A.; Lee, T.T.; Fitch, M.; Robbins, D.; et al. Platform for Induction and Maintenance of Transgene-free hiPSCs Resembling Ground State Pluripotent Stem Cells. *Stem Cell Rep.* **2014**, *2*, 366–381. [CrossRef] [PubMed]

9. Tamm, C.; Galito, S.P.; Annere, C.A. Comparative Study of Protocols for Mouse Embryonic Stem Cell Culturing. *PLoS ONE* **2013**, *8*, e81156. [CrossRef] [PubMed]

10. Viswanathan, P.; Gaskell, T.; Moens, N.; Culley, O.J.; Hansen, D.; Gervasio, M.K.R.; Yeap, Y.J.; Danovi, D. Human pluripotent stem cells on artificial microenvironments: A high content perspective. *Front. Pharmacol.* **2014**, *5*, 150. [CrossRef] [PubMed]

11. Liu, B.; Lu, Y.; Li, J.; Liu, Y.; Liu, J.; Wang, W. Leukemia inhibitory factor promotes tumor growth and metastasis in human osteosarcoma via activating STAT3. *APMIS* **2015**, *123*, 837–846. [CrossRef] [PubMed]

12. Albrengues, J.; Bourget, I.; Pons, C.; Butet, V.; Hofman, P.; Tartare-Deckert, S.; Feral, C.C.; Meneguzzi, G.; Gaggioli, C. LIF Mediates Proinvasive Activation of Stromal Fibroblasts in Cancer. *Cell Rep.* **2014**, *7*, 1664–1678. [CrossRef] [PubMed]

13. Joddar, B.; Ito, Y. Artificial niche substrates for embryonic and induced pluripotent stem cell cultures. *J. Biotechnol.* **2013**, *168*, 218–228. [CrossRef] [PubMed]

14. Miyazaki, T.; Futaki, S.; Suemori, H.; Taniguchi, Y.; Yamada, M.; Kawasaki, M.; Hayashi, M.; Kumagai, H.; Nakatsuji, N.; Sekiguchi, K.; et al. Laminin E8 fragments support efficient adhesion and expansion of dissociated human pluripotent stem cells. *Nat. Commun.* **2012**, *3*. [CrossRef] [PubMed]

15. Rodin, S.; Domogatskaya, A.; Ström, S.; Hansson, E.M.; Chien, K.R.; Inzunza, J.; Hovatta, O.; Tryggvason, K. Long-term self-renewal of human pluripotent stem cells on human recombinant laminin-511. *Nat. Biotechnol.* **2010**, *28*, 611–615. [CrossRef] [PubMed]

16. Hongisto, H.; Vuoristo, S.; Mikhailova, A.; Suuronen, R.; Virtanen, I.; Otonkoski, T.; Skottman, H. Laminin-511 expression is associated with the functionality of feeder cells in human embryonic stem cell culture. *Stem Cell Res.* **2012**, *8*, 97–108. [CrossRef] [PubMed]

17. Meng, G.; Liu, S.; Li, X.; Krawetz, R.; Rancourt, D.E. Extracellular matrix isolated from foreskin fibroblasts supports long-term xeno-free human embryonic stem cell culture. *Stem Cells Dev.* **2010**, *19*, 547–556. [CrossRef] [PubMed]

18. Yang, H.; Qiu, Y.; Zeng, X.; Ding, Y.; Zeng, J.; Lu, K.; Li, D. Effect of a feeder layer composed of mouse embryonic and human foreskin fibroblasts on the proliferation of human embryonic stem cells. *Exp. Ther. Med.* **2016**, *11*, 2321–2328. [CrossRef] [PubMed]

19. Totonchi, M.; Taei, A.; Seifinejad, A.; Tabebordbar, M.; Rassouli, H.; Farrokhi, A.; Gourabi, H.; Aghdami, N.; Hosseini-Salekdeh, G.; Baharvand, H. Feeder- and serum-free establishment and expansion of human induced pluripotent stem cells. *Int. J. Dev. Biol.* **2010**, *54*, 877–886. [CrossRef] [PubMed]

20. Pakzad, M.; Ashtiani, M.K.; Gargari, S.L.M.; Baharvand, H. Development of a simple, repeatable, and cost-effective extracellular matrix for long-term xeno-free and feeder-free self-renewal of human pluripotent stem cells. *Histochem. Cell Biol.* **2013**, *140*, 635–648. [CrossRef] [PubMed]

21. Tibbitt, M.W.; Anseth, K.S. Hydrogels as Extracellular Matrix Mimics for 3D Cell Culture. *Biotechnol. Bioeng.* **2009**, *103*, 655–663. [CrossRef] [PubMed]

22. Discher, D.E.; Janmey, P.; Wang, Y. Tissue Cells Feel and Respond to the Stiffness of Their Substrate. *Science* **2005**, *310*, 1139–1143. [CrossRef] [PubMed]

23. Discher, D.E.; Mooney, D.J.; Zandstra, P.W. Growth factors, matrices, and forces combine and control stem cells. *Science* **2009**, *324*, 1673–1677. [CrossRef] [PubMed]

24. Park, H.-J.; Yang, K.; Kim, M.-J.; Jang, J.; Lee, M.; Kim, D.-W.; Lee, H.; Cho, S.-W. Bio-inspired oligovitronectin-grafted surface for enhanced self-renewal and long-term maintenance of human pluripotent stem cells under feeder-free conditions. *Biomaterials* **2015**, *50*, 127–139. [CrossRef] [PubMed]

25. Nava, M.M.; Raimondi, M.T.; Pietrabissa, R. Controlling Self-Renewal and Differentiation of Stem Cells via Mechanical Cues. *J. Biomed. Biotechnol.* **2012**, *2012*. [CrossRef] [PubMed]

26. Kress, S.; Neumann, A.; Weyand, B.; Kasper, C. Stem cell differentiation depending on different surfaces. *Adv. Biochem. Eng. Biotechnol.* **2012**, *126*, 263–283. [CrossRef] [PubMed]

27. Peerani, R.; Zandstra, P.W. Enabling stem cell therapies through synthetic stem cell–niche engineering. *J. Clin. Investig.* **2010**, *120*, 60–70. [CrossRef] [PubMed]

28. Zhao, G.; Liu, F.; Lan, S.; Li, P.; Wang, L.; Kou, J.; Qi, X.; Fan, R.; Hao, D.; Wu, C.; et al. Large-scale expansion of Wharton's jelly-derived mesenchymal stem cells on gelatin microbeads, with retention of self-renewal and multipotency characteristics and the capacity for enhancing skin wound healing. *Stem Cell Res. Ther.* **2015**, *6*, 38. [CrossRef] [PubMed]

29. Leisten, I.; Kramann, R.; Ferreira, M.S.V.; Bovi, M.; Neuss, S.; Ziegler, P.; Wagner, W.; Knüchel, R.; Schneider, R.K. 3D co-culture of hematopoietic stem and progenitor cells and mesenchymal stem cells in collagen scaffolds as a model of the hematopoietic niche. *Biomaterials* **2012**, *33*, 1736–1747. [CrossRef] [PubMed]

30. Zhou, X.; Rowe, R.G.; Hiraoka, N.; George, J.P.; Wirtz, D.; Mosher, D.F.; Virtanen, I.; Chernousov, M.A.; Weiss, S.J. Fibronectin fibrillogenesis regulates three-dimensional neovessel formation. *Genes Dev.* **2008**, *22*, 1231–1243. [CrossRef] [PubMed]

31. Hassan, W.; Dong, Y.; Wang, W. Encapsulation and 3D culture of human adipose-derived stem cells in an in-situ crosslinked hybrid hydrogel composed of PEG-based hyperbranched copolymer and hyaluronic acid. *Stem Cell Res. Ther.* **2013**, *4*, 32. [CrossRef] [PubMed]

32. Alamein, M.A.; Stephens, S.; Liu, Q.; Skabo, S.; Warnke, P.H. Mass Production of Nanofibrous Extracellular Matrix with Controlled 3D Morphology for Large-Scale Soft Tissue Regeneration. *Tissue Eng. Part C Methods* **2013**, *19*, 458–472. [CrossRef] [PubMed]

33. Alamein, M.A.; Wolvetang, E.J.; Ovchinnikov, D.A.; Stephens, S.; Sanders, K.; Warnke, P.H. Polymeric nanofibrous substrates stimulate pluripotent stem cells to form three-dimensional multilayered patty-like spheroids in feeder-free culture and maintain their pluripotency. *J. Tissue Eng. Regen. Med.* **2015**, *9*, 1078–1083. [CrossRef] [PubMed]

34. Chung, C.B.S.; Burdick, J.A. Influence of 3D Hyaluronic Acid Microenvironments on Mesenchymal Stem Cell Chondrogenesis. *Tissue Eng. Part A* **2009**, *15*, 243–254. [CrossRef] [PubMed]

35. Rowland, T.J.; Miller, L.M.; Blaschke, A.J.; Doss, E.L.; Bonham, A.J.; Hikita, S.T.; Johnson, L.V.; Clegg, D.O. Roles of integrins in human induced pluripotent stem cell growth on Matrigel and vitronectin. *Stem Cells Dev.* **2010**, *19*, 1231–1240. [CrossRef] [PubMed]

36. Lei, Y.; Schaffer, D.V. A fully defined and scalable 3D culture system for human pluripotent stem cell expansion and differentiation. *Proc. Natl. Acad. Sci. USA* **2013**, *110*, 5039–5048. [CrossRef] [PubMed]

37. Lei, Y.; Jeong, D.; Xiao, J.; Schaffer, D.V. Developing defined and scalable 3d culture systems for culturing human pluripotent stem cells at high densities. *Cell Mol. Bioeng.* **2014**, *7*, 172–183. [CrossRef] [PubMed]

38. Chien, Y.; Liao, Y.-W.; Liud, D.-M.; Lin, H.-L.; Chen, S.-J.; Chen, H.-L.; Peng, C.-H.; Lian, C.-M.; Mou, C.-Y.; Chiou, S.-H. Corneal repair by human corneal keratocyte-reprogrammed iPSCs and amphiphatic carboxymethyl-hexanoyl chitosan hydrogel. *Biomaterials* **2012**, *33*, 8003–8016. [CrossRef] [PubMed]

39. Abdelaal, O.A.M.; Darwish, S.M.H. Review of Rapid Prototyping Techniques for Tissue Engineering Scaffolds Fabrication. In *Characterization and Development of Biosystems and Biomaterials*; Springer: Berlin/Heidelberg, Germany, 2013; pp. 33–54. [CrossRef]

40. Maruo, S.; Fourkas, J.T. Recent progress in multiphoton microfabrication. *Laser Photon. Rev.* **2008**, *2*, 100–111. [CrossRef]

41. Hofmeister, L.H.; Costa, L.; Balikov, D.A.; Crowder, S.W.; Terekhov, A.; Sung, H.J.; Hofmeister, W.H. Patterned polymer matrix promotes stemness and cell-cell interaction of adult stem cells. *J. Biol. Eng.* **2015**, *9*, 18. [CrossRef] [PubMed]

42. Kawata, S.; Sun, H.-B.; Tanaka, T.; Takada, K. Finer features for functional microdevices. *Nature* **2001**, *412*, 697–698. [CrossRef] [PubMed]

43. Narayan, R.J.; Doraiswamy, A.; Chrisey, D.B.; Chichkov, B.N. Medical prototyping using two photon polymerization. *Materialstoday* **2010**, *13*, 42–48. [CrossRef]

44. Malinauskas, M.; Danilevičius, P.; Baltriukienė, D.; Rutkauskas, M.; Žukauskas, A.; Kairytė, Ž.; Bičkauskaitė, G.; Purlys, V.; Paipulas, D.; Bukelskienė, V.; et al. 3D artificial polymeric scaffolds for stem cell growth fabricated by femtosecond laser. *Lith. J. Phys.* **2010**, *50*, 75–82. [CrossRef]

45. Danilevičius, P.; Rekštytė, S.; Balčiūnas, E.; Kraniauskas, A.; Širmenis, R.; Baltriukienė, D.; Bukelskienė, V.; Gadonas, R.; Sirvydis, V.; Piskarskas, A.; et al. Laser 3D micro/nanofabrication of polymers for tissue engineering applications. *Optlastec* **2013**, *45*, 518–524. [CrossRef]

46. Ovsianikov, A.; Malinauskas, M.; Schlie, S.; Chichkov, B.; Gittard, S.; Narayan, R.; Löbler, M.; Sternberg, K.; Schmitz, K.P.; Haverich, A. Three-dimensional laser micro- and nano-structuring of acrylated poly(ethylene glycol) materials and evaluation of their cytoxicity for tissue engineering applications. *Acta Biomater.* **2011**, *7*, 967–974. [CrossRef] [PubMed]

47. Melissinaki, V.; Gill, A.A.; Ortega, I.; Vamvakaki, M.; Ranella, A.; Haycock, J.W.; Fotakis, C.; Farsari, M.; Claeyssens, F. Direct laser writing of 3D scaffolds for neural tissue engineering applications. *Biofabrication* **2011**, *3*, 045005. [CrossRef] [PubMed]

48. Felfel, R.M.; Poocza, L.; Gimeno-Fabra, M.; Milde, T.; Hildebrand, G.; Ahmed, I.; Scotchford, C.; Sottile, V.; Grant, D.M.; Liefeith, K. In vitro degradation and mechanical properties of PLA-PCL copolymer unit cell scaffolds generated by two-photon polymerization. *Biomed. Mater.* **2016**, *11*, 015011. [CrossRef] [PubMed]

49. Raimondi, M.T.; Eaton, S.M.; Laganà, M.; Aprile, V.; Nava, M.M.; Cerullo, G.; Osellame, R. Three-dimensional structural niches engineered via two-photon laser polymerization promote stem cell homing. *Acta Biomater.* **2013**, *9*, 4579–4584. [CrossRef] [PubMed]

50. Raimondi, M.T.; Nava, M.M.; Eaton, S.M.; Bernasconi, A.; Vishnubhatla, K.C.; Cerullo, G.; Osellame, R. Optimization of Femtosecond Laser Polymerized Structural Niches to Control Mesenchymal Stromal Cell Fate in Culture. *Micromachines* **2014**, *5*, 341–358. [CrossRef]

51. Nava, M.M.; Raimondi, M.T.; Credi, C.; de Marco, C.; Turri, S.; Cerullo, G.; Osellame, R. Interactions between structural and chemical biomimetism in synthetic stem cell niches. *Biomed. Mater.* **2015**, *10*. [CrossRef] [PubMed]

52. Nava, M.M.; Di Maggio, N.; Zandrini, T.; Cerullo, G.; Osellame, R.; Martin, I.; Raimondi, M.T. Synthetic niche substrates engineered via two-photon laser polymerization for the expansion of human mesenchymal stromal cells. *J. Tissue Eng. Regen. Med.* **2016**. [CrossRef] [PubMed]

53. Ovsianikov, A.; Mironov, V.; Stampf, J.; Liska, R. Engineering 3D cell-culture matrices: Multiphoton processing technologies for biological and tissue engineering applications. *Expert Rev. Med. Devices* **2012**, *9*, 613–633. [CrossRef] [PubMed]

54. Lee, K.-S.; Yang, D.-Y.; Park, S.H.; Kim, R.H. Recent developments in the use of two-photon polymerization in precise 2D and 3D microfabrications. *Polym. Adv. Technol.* **2006**, *17*, 72–82. [CrossRef]

55. Sun, H.-B.; Takada, K.; Kim, M.-S.; Lee, K.-S.; Kawata, S. Scaling laws of voxels in two-photon photopolymerization nanofabrication. *Appl. Phys. Lett.* **2003**, *83*, 1104. [CrossRef]

56. Nava, M.M.; Piuma, A.; Figliuzzi, M.; Cattaneo, I.; Bonandrini, B.; Zandrini, T.; Cerullo, G.; Osellame, R.; Remuzzi, A.; Raimondi, M.T. Two-photon polymerized "nichoid" substrates maintain function of pluripotent stem cells when expanded under feeder-free conditions. *Stem Cell Res. Ther.* **2016**, *7*, 132. [CrossRef] [PubMed]

57. Uppal, N. A Mathematical Model Development and Sensitivity Analysis of Two Photon Polymerization for 3d Micro/Nano Fabrication. Ph.D. Thesis, University of Texas, Arlington, TX, USA, August 2008.

58. LaFratta, C.N.; Li, L.; Fourkas, J.T. Soft-lithographic replication of 3D microstructures with closed loops. *Proc. Natl. Acad. Sci. USA* **2006**, *103*, 8589–8594. [CrossRef] [PubMed]

59. Formanek, F.; Takeyasu, N.; Tanaka, T.; Chiyoda, K.; Ishikawa, A.; Kawata, S. Three-dimensional fabrication of metallic nanostructures over large areas by two-photon polymerization. *Opt. Express* **2006**, *14*, 800–809. [CrossRef] [PubMed]

60. Yang, L.; El-Tamer, A.; Hinze, U.; Li, J.; Hu, Y.; Huang, W.; Chu, J.; Chichkov, B. Parallel direct laser writing of micro-optical and photonic structures using spatial light modulator. *Opt. Lasers Eng.* **2015**, *7*, 26–32. [CrossRef]

61. Li, Z.; Pucher, N.; Cicha, K.; Torgersen, J.; Ligon, S.C.; Ajami, A.; Husinsky, W.; Rosspeintner, A.; Vauthey, E.; Naumov, S.; et al. A Straightforward Synthesis and Structure—Activity Relationship of Highly Efficient Initiators for Two-Photon Polymerization. *Macromolecules* **2013**, *46*, 352–361. [CrossRef]

13

Monitoring the Damage State of Fiber Reinforced Composites Using an FBG Network for Failure Prediction

Esat Selim Kocaman [1], **Erdem Akay** [2], **Cagatay Yilmaz** [1,3,4], **Halit Suleyman Turkmen** [2], **Ibrahim Burc Misirlioglu** [1,3,4], **Afzal Suleman** [5] **and Mehmet Yildiz** [1,3,4,*]

[1] Faculty of Engineering and Natural Sciences, Sabanci University, Tuzla, 34956 Istanbul, Turkey; esatselim@sabanciuniv.edu (E.S.K.); cagatayyilmaz@sabanciuniv.edu (C.Y.); burc@sabanciuniv.edu (I.B.M.)

[2] Faculty of Aeronautics and Astronautics, Ayazaga Campus, Istanbul Technical University, Maslak, 34469 Istanbul, Turkey; erdemakay@itu.edu.tr (E.A.); halit@itu.edu.tr (H.S.T.)

[3] Integrated Manufacturing Technologies Research and Application Center, Sabanci University, Tuzla, 34956 Istanbul, Turkey

[4] Sabanci University-Kordsa Global, Composite Technologies Center of Excellence, Istanbul Technology Development Zone, Sanayi Mah. Teknopark Blvd. No: 1/1B, Pendik, 34906 Istanbul, Turkey

[5] Mechanical Engineering Department, University of Victoria, Victoria, BC V8W 2Y2, Canada; suleman@uvic.ca

[*] Correspondence: meyildiz@sabanciuniv.edu

Academic Editors: M.H. Ferri Aliabadi and Zahra Sharif Khodaei

Abstract: A structural health monitoring (SHM) study of biaxial glass fibre-reinforced epoxy matrix composites under a constant, high strain uniaxial fatigue loading is performed using fibre Bragg grating (FBG) optical sensors embedded in composites at various locations to monitor the evolution of local strains, thereby understanding the damage mechanisms. Concurrently, the temperature changes of the samples during the fatigue test have also been monitored at the same locations. Close to fracture, significant variations in local temperatures and strains are observed, and it is shown that the variations in temperature and strain can be used to predict imminent fracture. It is noted that the latter information cannot be obtained using external strain gages, which underlines the importance of the tracking of local strains internally.

Keywords: polymer-matrix composites; fatigue; mechanical testing; damage monitoring; fibre Bragg grating

1. Introduction

Among various classes of composite materials, fibre-reinforced polymeric composites are frequently utilized as structural components in a variety of industries ranging from aeronautics and automotive to civil infrastructure owing to their high specific stiffness and strength. Despite the care shown in the design and manufacturing of these structures using high-tech equipment, composites have certain drawbacks that need to be addressed for their reliable usage at prolonged time scales. Unlike conventional isotropic and homogeneous materials, such as metals and alloys, their mechanical behaviour is complex to model due to their heterogeneous internal structure. Hence, continuous monitoring of the strain state of composites under different environmental conditions and mechanical loads, especially cyclic loads, is particularly important given that such loading is the most common cause inflicting catastrophic damage on composite structures during service. Therefore, it is crucial to develop techniques to monitor or "sense" the strain state of the composite during service and use these data in situ to predict the onset of failure [1].

One of the most recent and precise sensing techniques for strain and structural health monitoring (SHM) of structures is to use fibre Bragg grating (FBG) sensors. FBGs are optical sensors that are sensitive to both strain and temperature (via thermal strains). They are fabricated by periodic modulation of the refractive index of a single mode optical fibre core using Ultraviolet (UV) light [2]. Upon exposure to a broad band (near IR) of light sent through the optical fibre, an FBG sensor acts like a mirror reflecting a narrow light signal centred at a particular wavelength, referred to as Bragg wavelength λ_B. The Bragg wavelength is a function of the grating pitch (Λ) (spacing between periodic variation of refractive index) and effective refractive index (n_{eff}) of the sensor and can be formulated as $\lambda_B = 2n_{eff}\Lambda$. Strain and temperature can be measured based on the shift in the wavelength using the following relation $\frac{\Delta\lambda_B}{\lambda_B} = (1 - p_e)\varepsilon + (\alpha + \xi)\Delta T$, where $\Delta\lambda$ is the change in the wavelength, p_e is the photo-elastic coefficient and α and ξ are the thermal expansion and thermo-optic coefficient of fibres, respectively. ε is the strain, and ΔT represents the temperature change of the sensor [3]. The sensitivity of strain and temperature of a bare FBG is around 1.2 pm/$\mu\varepsilon$ and 13.7 pm/$^\circ$C, respectively; however, it is crucial to measure the strain and temperature sensitivity of every embedded FBG sensor to account for the strain transfer between the sensor and the host material, since factors, such as the variation in material properties and manufacturing conditions, may alter the sensitivity [4].

Being small and flexible, FBG sensors can be embedded discretely into composites at locations of interest, thereby allowing for the tracking of local strain distribution and evolution without compromising the structural integrity of the host material [2]. In [5], we demonstrated the reliability of FBG sensors in fatigue experiments where they could outlast the sample life time, providing real-time strain data for almost more than 5.5 million fatigue cycles. FBG sensors possess several other important attributes, such as immunity to the electromagnetic interference, light weight, multiplexing, absolute measurement capability and high corrosion resistance. Moreover, the same set of embedded FBG sensors can be used for process monitoring of composites, such as resin flow [6], cure monitoring [7,8] and the detection of cure-induced residual strains [7,9], thus assuring the high quality of the manufactured components [10,11]. Furthermore, FBG sensors were successfully used to monitor composite structures under real-time operating conditions, as can be seen in [12,13].

There have been several efforts reported in the open literature proposing to utilize FBG sensors to monitor internal strain states of composite structures [14–23]. For example, Takeda et al. introduced a new method to predict damage patterns and strains of notched composite laminates with embedded FBG sensors utilizing a layer-wise finite element model [24]. Doyle et al. performed in situ processing and condition monitoring to evaluate different fibre optic sensor systems and obtained a good correlation between strains acquired by surface-mounted optical sensors and external strain gages. They also demonstrated the feasibility of these sensor systems to monitor the stiffness degradation in composites due to fatigue [25]. Baere et al. evaluated the performance of embedded and bonded FBG sensors with organic modified ceramic coating through measuring strains in thermoplastic composites under fatigue loading conditions, wherein it was shown that the two strain quantities agree well with each other, indicating the feasibility of the sensing system [26]. Shin et al. used embedded FBG sensors to monitor fatigue damage evolution in graphite/epoxy composite specimens with a central circular hole and demonstrated the potential of such sensors to detect the occurrence and progress of damage [27]. In Takeda's study, embedded FBG sensors were used to sense local strain distribution due to transverse crack formation by measuring the power of the reflected light from sensors [28]. In another work, both spectrum and wavelength information received from FBG sensors were utilized for the durability tests (drop-weight impact and two periodic fatigue tests) and for the condition monitoring of a composite wing structure [29].

One important problem with the prediction of failure in a composite structure is that, due to their heterogeneous structure, damage can initiate and progress at multiple locations within the loaded section, unlike metals, where the failure is often caused by the propagation of a single microscopic crack that turns into a macroscopic one later on. Matrix cracking, matrix-fibre debonding, delamination, transverse-ply cracking and fibre breakage are some of the well-known damage forms observed in fibre-reinforced composites exposed to fatigue loads. Moreover, various damage modes can occur independently, concurrently or interactively, making time-based estimates of a potential failure very difficult. Since the damage formation is distributed and progressive, the reductions in strength and stiffness can start at the early stages of the fatigue loading [30]. It is now understood that stiffness degradation of fibre-reinforced polymer matrix composites in response to cyclic loads is characterized by three distinct stages. In the first stage comprising the initial 15%–20% of fatigue life, the rapid formation and interconnection of matrix cracking causes a sharp, non-linear decrease in stiffness. The second stage accounts for 15%–20% up to 90% of the fatigue life where there is a gradual and linear decrease in stiffness, which is attributed to the propagation of several cracks, fibre debonding and delamination. The final stage is distinguished by a sharp nonlinear decrease in stiffness due to the plurality of fibre breakages [31]. Note that this sequence is identical to that of homogeneous metals and alloys, but accompanied by rather different damage mechanisms.

Despite the recent interest in SHM, the number of such studies, particularly about the usage of FBG and similar sensor networks to monitor and understand the governing processes of fatigue, is scarce, and the reported studies are limited to low cycle and low strain amplitude fatigue, fatigue monitored with surface bonded sensors and/or the fatigue of thermoplastic materials with embedded FBG sensors. Correlating the local internal strain readings to external gage data under low-cycle fatigue conditions is important in terms of predicting an approaching failure that would allow time for removing the composite from service. It should also be kept in mind that the stability of the internal sensor network is crucial; as they are exposed to repetitive high strain amplitude dynamic loads, the composite endures [32]. With this motivation, we use local strain data provided by FBG sensors embedded in glass-reinforced composites under constant high strain and low-cycle fatigue loading conditions to focus on the distribution and evolution of strain along the composite specimens, allowing thus to demonstrate the onset of an approaching failure. It is noted that the FBG network provides a useful means to identify local strains, which start to deviate significantly as the fatigue progresses, a sign of an approaching failure, as is shown in this study. In addition, when a specimen is subjected to cyclic loading, a portion of the mechanical energy is dissipated as heat (also referred to as autogenous heating) causing a rise in the temperature of the specimen [33–42]. This was already demonstrated for metallic materials during the fatigue loading to predict the remaining usable life of the material [41,42]. Upon experimental analysis on the possible mechanisms of heat generation in vibrothermography, several reasons were proposed in the literature, including viscoelasticity/material damping, heating due to the plastic deformation at the end of the crack tips that are formed due to the repeated loading and, finally, the friction between the internal surfaces of the cracks [43,44].

In this work, combined with the data gathered from the FBG network inside several composite specimens composed of biaxial glass fibres and epoxy matrix, we present a hypothesis that the heat generated due to rubbing of the surfaces during the fatigue process is dominant and demonstrate that both the local strain data and rise in temperature can be used to monitor the internal damage state of a fibre glass epoxy composite. It is also shown that the FBG network can remain useful and collect data to track local strains all the way until the fracture of the specimen. Practical issues in relation to performing fatigue testing under constant displacement and strains achieved by using a linear variable differential transformer (LVDT) and extensometer, respectively, as control sensors are also addressed.

2. Experimental Section

2.1. Specimen Preparation and Testing Equipment

In the course of this study, a total of five flat composite panels were manufactured. Four of them were produced using a resin transfer moulding (RTM) method, while the fifth one with a vacuum infusion (VI) technique. Each panel contained two embedded FBG networks (each network includes

3 FBG sensors at different locations), leading to in total 10 test specimens with FBG sensors. Two of the ten specimens were discarded due to FBG sensor failure/breakage at ingress/egress locations. A laboratory-scale RTM apparatus with the capability to embed optical fibres into composite parts was used to produce panels with dimensions of 620 mm × 320 mm × 3.5 mm. Composite laminates consisted of E-glass fibre and epoxy resin and had a stacking sequence of $[90/0]_{6S}$. Metyx LT300 E10A 0/90 biaxial E-glass stitched fabric (Metyx Composites, Istanbul, Turkey) was used as the reinforcement, which has an area density of 161 g/m^2 in the 0° orientation, which is aligned along the resin flow direction in the mould, and 142 g/m^2 in the 90° orientation, leading to the total area density of 313 g/m^2. The selected resin system is Araldite LY 564 epoxy resin mixed with XB 3403 hardener (manufactured by Huntsman Corporation, The Woodlands, TX, USA) with the ratio of 100 and 36 parts by weight. The composite panels underwent an initial cure at 65 °C for 24 h with a post cure at 80 °C for 24 h. Three 1 mm-long FBG sensors with Bragg wavelengths of 1540, 1550 and 1560 nm that are written on the same fibre optic cable with 4-cm or 6-cm intervals were purchased from Technica SA (Beijing, China). Prior to manufacturing, the fibre optic cable was fixed onto the 0° surface of a ply through passing it under fibre stitches. The plies were stacked such that the fibre optic cable was between the 6th and 7th layers of the laminate, as shown in Figure 1a. Mechanical test specimens with and without FBG sensors were cut out from the composite panels using a water-cooled diamond circular blade saw into dimensions of 250 mm × 25 mm × 3.7 mm with a 150-mm gage length. The length (also the loading axis) of the specimen was aligned with the 0° fibre orientation. In specimens with FBG sensors, the middle FBG (1550 nm) was positioned at the centre of the gage length of the specimens, and the remaining two sensors were located towards the grips of specimens, as shown in Figure 1b. All three sensors were oriented along the loading direction. To avoid damage and in turn the breakage of test specimens at grip locations, both ends of specimens were tabbed with an aluminium tab having dimensions of 50 mm × 25 mm × 1 mm using a two-component room temperature curing epoxy system (Araldite 2011, Huntsman Corporation, The Woodlands, TX, USA).

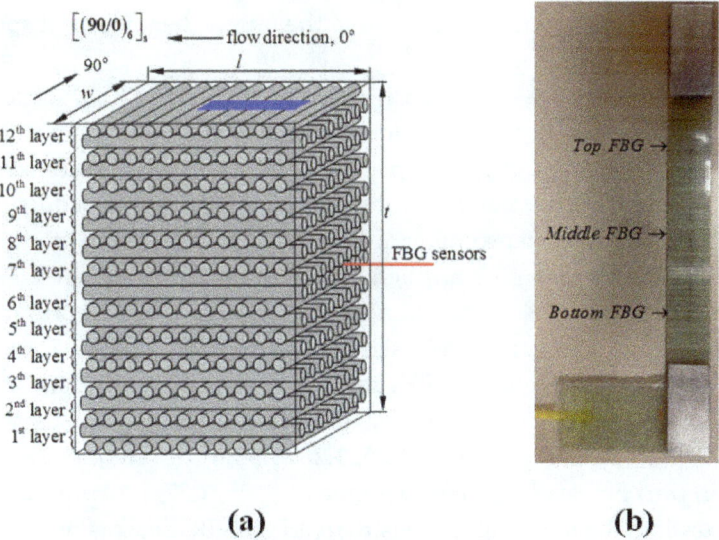

(a)   **(b)**

Figure 1. (a) The schematic drawing for stacking sequences together with the placement of FBG sensors and also the orientation of the cut test specimen indicated by the blue region where l, w and t indicate the length, width and the thickness of the manufactured composite plate; (b) L-shaped specimen that enables easy gripping of the test specimen by the testing system.

All tests were performed on an MTS 322 test frame (MTS Systems Corporation, Eden Prairie, MN, USA) with MTS 647 hydraulic wedge grips using an MTS FlexTest GT digital controller with MTS Station Manager software (FlexTest GT Station Manager Version 3.5C, MTS Systems Corporation, Eden Prairie, MN, USA). Load and displacement data were collected with a built-in load cell (MTS 661.20F-03,

MTS Systems Corporation, Eden Prairie, MN, USA) and linear variable differential transformer (LVDT) (MTS Systems Corporation, Eden Prairie, MN, USA), respectively. Strain was collected using an axial extensometer (MTS 634.25F-24, MTS Systems Corporation, Eden Prairie, MN, USA). In some tests, a second axial extensometer (Epsilon 3542, Epsilon Technology Corp., Jackson, WY, USA) was used simultaneously to study the effect of gage length on the strain measurement. The Micron Optics SM130-700 interrogator (Micron Optics, Inc., Atlanta, GA, USA) with Micron Optics Enlight software was used to collect FBG data. K-type thermocouples (OMEGA Engineering, Inc., Norwalk, CT, USA) were used to measure temperature, and corresponding data were collected using a National Instruments NI SCXI-1314 DAQ card (National Instruments Corporation, Austin, TX, USA) in an NI SCXI-1000 chassis with Signal Express software. All data were acquired at a sampling rate of 100 Hz.

2.2. Test Procedure

The baseline parameters for fatigue tests were determined by performing eleven static tests whereby the average ultimate tensile stress and strain of the composite specimens were measured to be 320 MPa and 16.31 $\mu\varepsilon$, respectively. In this study, eight FBG sensor embedded specimens were extracted intact from five flat composite panels and then were subjected to constant amplitude strain and tension-tension sine wave tests at various strain ratios ($\varepsilon_{max}/\varepsilon_{ult}$) varying between 0.5 and 0.6, where the maximum fatigue loading to be imposed on these test specimens was determined based on the strain ratio of interest. To ensure that fatigue tests were performed in tension-tension mode, all specimens were subjected to a minimum stress of 27.6 MPa. Autogenous heating of test specimens becomes a concern during fatigue testing of glass fibre-reinforced composites since the generated heat is not transferred to the environment as fast as in metallic materials. Knowing that fatigue properties of composites are especially sensitive to heat, tests were performed at a frequency of 4 Hz to prevent extensive overheating of samples. Of the eight experiments on composite specimens with embedded FBG sensor networks, three of them were rejected due to experimental errors, such as power outage, unexpected failure and slipping of specimens at grip locations. However, the collected and processed data in these experiments were still useful to observe the fatigue behaviour of specimens at the strain ratio of interest.

For each fatigue specimen, the strain sensitivities of embedded FBG sensors were determined as follows. First, the minimum and maximum loads were calculated using minimum and maximum stress and the area of the specimen. A given specimen was statically tensioned up to the calculated maximum load and then unloaded down to the minimum load while acquiring the displacement data by LVDT, strain data by the extensometer and the Bragg wavelength by the optical interrogator. The loading and unloading procedure is applied for the second time. Then, all of the collected data were processed such that the extensometer data corresponding to the second ramp were plotted as a function of pertinent Bragg wavelength for each FBG and LVDT, and linear regression was used to determine strain calibration coefficients for FBG and LVDT sensors. The strain sensitivities of bottom, middle and top FBGs for specimens L1 , L2, L3, E1 and E2 were determined respectively as (1.24, 1.25, 1.24 in pm/$\mu\varepsilon$), (1.27, 1.25, 1.31 in pm/$\mu\varepsilon$), (1.25, 1.28, 1.30 in pm/$\mu\varepsilon$), (1.27, 1.30, 1.24 in pm/$\mu\varepsilon$) and (1.23, 1.26, 1.26 in pm/$\mu\varepsilon$), leading to the average of (1.25, 1.27, 1.27 pm/$\mu\varepsilon$).

Prior to fatigue testing, the temperature sensitivity of the FBG sensors was determined in order to account for the temperature variation in the specimens due to autogenous heating. To this end, FBG sensors were placed in a furnace. The temperature in the furnace was ramped from 30 °C up to 60 °C and allowed to soak at each 10 °C temperature increment for one hour before the temperature and wavelength were recorded. A plot of wavelength vs. temperature was constructed for each FBG, and linear regression was used to extract an average temperature sensitivity of 0.010 nm/°C for the FBG sensors. Before fatigue tests, three thermocouples were fastened to the surface of specimens, such that each one was located just above one of the FBG sensors in order to monitor the temperature increase due to autogenous heating in the vicinity of the corresponding FBG sensor. Then, the surface temperature data of each thermocouple were converted into the wavelength shift using the previously-determined temperature sensitivity coefficient, and the wavelength changes due to the increase in temperature were subtracted from the corresponding FBG wavelength data at each data point. Here, it should be noted that the surface temperature may not be the same as that sensed by the embedded FBG; however, it provides temperature information very close to the exact values, thereby allowing for higher accuracy in the measured strain.

The behaviour of FBG sensors under different experimental conditions was also investigated by utilizing two different experimental procedures, namely fatigue experiments controlled by the LVDT and the extensometer, as tabulated in Table 1. Three of the five presented fatigue experiments (i.e., L1, L2 and L3) were performed under LVDT control, whereas the remaining two experiments (E1 and E2) were conducted under the extensometer control. For experiments with LVDT control, the specimen of interest was subjected to a constant displacement corresponding to the desired strain ratio through using the LVDT sensor of the fatigue testing system. It is recalled that the displacement recorded by the LVDT was related to the strain acquired by the extensometer through a calibration coefficient as described previously. Prior to fatigue experiments controlled by the LVDT sensor, the extensometer was dismounted from the specimens, and the fatigue experiments were performed between minimum and maximum displacements imposed by LVDT. As for the experiments with extensometer control, the extensometer was mounted onto the specimens and kept thereon throughout all of the fatigue tests. The specimens were strained between the minimum and maximum strains corresponding to desired strain ratios. After fatigue tests, the relevant sections of the broken fatigue specimens (Figure 2c) were used to cut out cross-section samples using a circular saw blade for microscope examination of the microscopic details of the failed samples, particularly the interfaces between FBG and host composite material. Their pertinent surfaces were polished using different grits of abrasive papers and inspected under optical microscope and a Leo Supra 35VP Field Emission Scanning Electron Microscope (SEM).

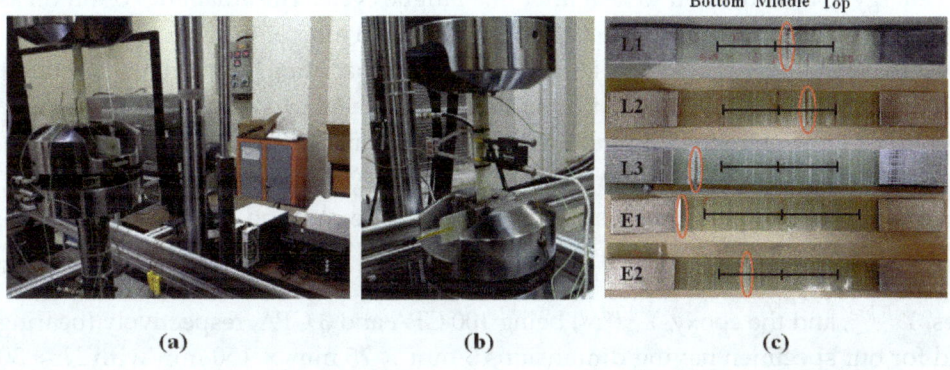

Figure 2. (**a**) Fatigue testing system with the data acquisition set-up; (**b**) the specimen E2 with double axial extensometers; (**c**) failed specimens where failure locations are marked with the red circles and sensor locations are indicated by black vertical ticks.

Table 1. Test parameters for fatigue experiments. LVDT, linear variable differential transformer.

Specimen Code	Strain Ratio	Fatigue Method	Fatigue Life (Cycle)
L1	0.60	LVDT	20,951
L2	0.55	LVDT	30,966
L3	0.60	LVDT	19,701
E1	0.57	Extensometer	19,385
E2	0.50	Extensometer	13,811

3. Results and Discussion

Data acquired from relevant sensors throughout the fatigue tests of specimens were processed and presented as temperature, maximum strains at each cycle and strain energy density variations as a function of cycle numbers, as shown in Figures 3–7. The global strain energy density F in the specimen was calculated, using $F = 0.5\sigma\varepsilon$ where $\sigma = S/A$ with S being the force measured by the testing machine, A the cross-sectional area of the specimen normal to the force and ε the longitudinal strain, either measured using external gages (global) or local FBGs. Here, it should be noted that for all specimens given in Table 1, the intervals between the subsequent FBG sensors are 4 cm, except for the specimen E1, for which it is 6 cm, as shown in Figure 2c.

3.1. Specimen L1

Figure 3a shows the variation of surface temperatures for the specimen L1 at three different FBG sensor locations, which were recorded using a K-type thermocouple at a sampling rate of 100 Hz. As can be observed, the temperature evolution reveals three distinct stages: an initial increase called Thermal Stage I, followed by a second phase (Thermal Stage II) for which the rate of temperature variation (in this case temperature increase) is smaller than Thermal Stage I, and finally, a notable increase prior to failure, which is identified by Thermal Stage III. There could be several main contributing factors to the autogenous heating as stated previously, namely viscoelastic/damping effects, plasticity and frictional rubbing of the surfaces of the internal cracks that formed during the test. Intuitively speaking, for specimen L1, the temperature is expected to be noticeably higher for the top FBG sensor, which measures visibly larger local strain, since the higher the strain, the larger the local strain energy in the specimen (as seen in Figure 3c); therefore, one would expect more heat dissipation. This type of behaviour reveals that the heat generated at locations where the FBG sensor reads low strain is not of viscoelastic or plastic origin. If one assumes a fully viscoelastically behaving specimen, the total strain energy accumulated in the volume $F_\Sigma = 0.5 \int_V \sigma\varepsilon dV$ is fully converted to heat upon removal of the force on the system under adiabatic conditions using the first law of thermodynamics $dU = F_\Sigma$ where dU is the internal energy to be converted to heat after one fatigue cycle. The adiabatic condition assumption is the most conservative one that indicates the heat generated after N number of cycles will remain in the specimen that can be written as $\Sigma_N dU = C_p \Delta T$ stating that after N cycles, the temperature of the specimen will increase by ΔT in proportion to the average heat capacity, C_p. We take C_p as a linear combination in proportion to the volume fractions of the constituents $C_p = \alpha C_p^{Fibre} + (1-\alpha)C_p^{Matrix}$ where α is the volume fraction of the fibres, and C_p^{Fibre} and C_p^{Matrix} are the heat capacities of the fibres and the matrix at constant pressure, respectively. Thus, one can find the final temperature after N cycles as $\Delta T = \dfrac{1}{2C_p} \sum_N \int_V \sigma\varepsilon dV$. For $C_p^{Fibre} = 0.810$ J/g/°C and $C_p^{Matrix} = 0.12$ J/g/°C, elastic modulus of the fibres, Y^{Fibre}, and the epoxy, Y^{Epoxy}, being 100 GPA and 3 GPA, respectively (bearing the same strain), and for our specimen having dimensions 3 mm $\times$ 25 mm $\times$ 150 mm with $N \approx 30{,}000$ until failure, as in our experiments, we find that the final temperature of the specimen will differ from the ambient test environment by less than a degree, a significantly lower value than observed in our tests. Moreover, that viscoelastic heating should increase with increased local strains is just the opposite of what we observe in our specimen: temperatures are in general the highest where the strain is the lowest. Henceforth, the contribution of the viscoelasticity to temperature increase in the specimens can be ruled out, leaving us with plasticity effects and rubbing of internal crack surfaces, which form with progressing fatigue.

Eliminating the viscoelastic-induced heating as a possible mechanism that gives rise to the observed sample temperatures and noting that the static tensile tests do not give rise to any detectable heating of the specimen, despite the strain being several times more than that of the fatigue test, it can be concluded that heating due to plasticity effects is also negligible, leaving us with only the crack surface rubbing as the main contributor to autogeneous heating. We again remind here that the highest temperatures are locally reached where strain is relaxed. Indeed, upon inspection of Figure 3, the middle FBG sensor region has almost the same temperature value, possibly differing by a degree or so, compared to the top FBG region despite that strain being visibly higher in the middle FBG region, again helping us rule out that global heating during the test is due to plasticity effects. An identical argument was also put forth in the work of Lang and Manson [45], where what is called here "heating due to surface rubbing" is termed as "frictional heating" in that work.

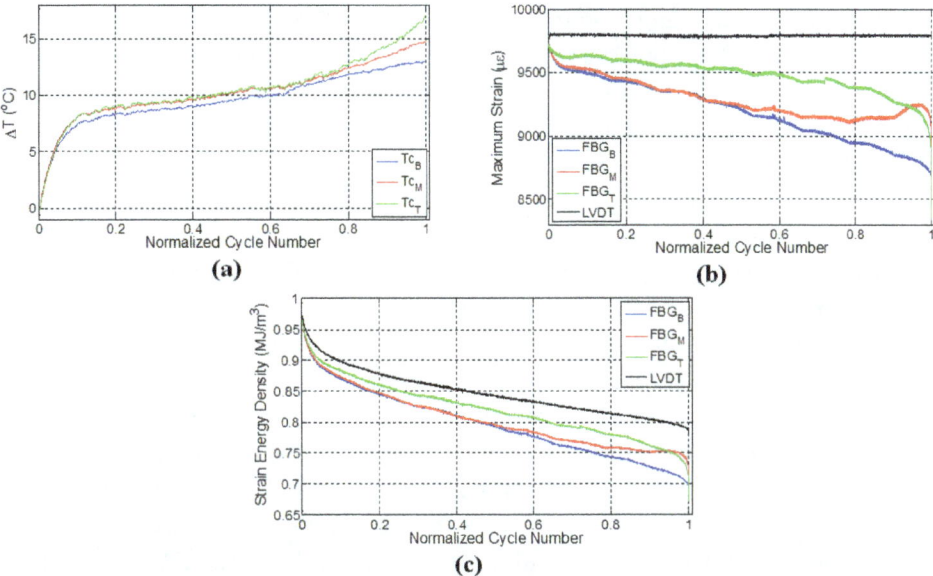

Figure 3. Evolution of (**a**) temperature; (**b**) strain and (**c**) strain energy density for specimen L1 where temperature is monitored by thermocouples, while the strain data are obtained using both FBG and LVDT sensors. The letters B, M and T in subscripts refer to the sensor locations: bottom, middle and top, respectively.

It has been shown in a previous work [5] that, after the initial fast rises in the temperature (Thermal Stage I), there is a gradual and linear decrease in the temperature, which is followed by a level off behaviour (Thermal Stage II). This decrease in temperature is possibly due to the fact that the rate of heat generation is smaller than the rate of heat given off to the environment (as a combination of conduction, convection and radiation, albeit being small), since that test was done at low strain. Thus, the heat generation rate due to damage accumulation is relatively low. In this work, the almost linear rise in temperature of the specimen at the second stage in Figure 3a is attributed to the fact that the damage sites and new surfaces forming in the course of this fatigue experiment are able to generate heat at a rate higher than the heat removal rate. This implies that the higher the crack damage density in the structure, the higher the heat generation, since the temperature rise occurs even though strain energy input to the specimen drops down in accordance with the decline in force due to the damage evolution in the material. At a later stage, around 75% of the fatigue life (1.6×10^4 cycles), there is a noticeable deviation in temperature from the linear region for the top and middle thermocouples, such that the temperature rises are augmented, indicating the onset of Thermal Stage III. It is worth mentioning that the temperature regimes observed in this study correlate quite well with the first, second and third phases of the strain energy density versus cycle number curves for the composite material in Figure 3c. However, it is interesting to see that at the locations where temperature rises towards the end of Stage II and beginning of Stage III, the strain data read from FBGs at these locations are changing significantly, implying that portions of the matrix into which the FBG is embedded have a tendency to "detach" from the load-bearing portions still adhered to the fibres.

Figure 3b shows the evolution of maximum strains (i.e., peak strains in the sinusoidal strain form) as a function of cycle number, which was recorded by LVDT and FBG sensors. Recalling that the fatigue test on this specimen was conducted under constant displacement using the LVDT sensor, one may at first expect that FBG sensors should also give constant strain values. However, maximum strains recorded by FBGs can be significantly different in comparison to the global strain of the specimen. As the fatigue experiment progresses, the local strains measured by FBG sensors drop down such that the trend has three separate regions consisting of an initial sharp decrease superseded by a gradual and almost linear decline followed by a final sharp variation after which failure occurs. The drop in the FBG recorded strain as the fatigue experiment continues is due to the damage formed within the specimen, in turn leading to an elongation in the gage length of the specimen. Hence, it is expected that the fatigue equipment should apply less force to induce the desired maximum displacement, whereby the specimen effectively experiences less local strain. Interestingly, these three stages are in agreement with the fatigue phases observed in temperature and strain energy density (based on LVDT) versus cycle number plots in Figure 3a,c, respectively. Besides, each FBG sensor reads notably different local strains, and the relative difference in the FBG measured strains further increases as the fatigue test continues, thereby demonstrating the clear existence of the non-uniform strain distribution due to the local differences in the damage type, density and evolution along the specimen gage length. Moreover, near failure, the strain of the middle FBG sensor starts to increase, while the top one decreases notably, again a sign of inhomogeneous damage accumulation. Note that the corresponding temperatures for these two sensors' locations increased drastically in the third thermal stage, as well. Such sudden changes in the strain values may signify the possible formation of major deformation other than fibre-matrix debonding and delamination and can be used as an alert for an approaching catastrophic failure. It is interesting to note that the specimen failed at a location close to the middle FBG between the middle and top FBG sensors. These findings also indicate that discrete embedded FBG sensors are reliable in predicting an approaching failure, which would not have been possible otherwise with externally-attached strain gages, especially at high strain fatigue, as strain gages mounted on the surfaces of the specimen can experience debonding from the specimen surface with progressing loading cycles and lose their performance at earlier stages of fatigue experiments.

Figure 3c presents a plot of the strain energy density versus cycle number for all sensors. One can clearly notice that strain energy density calculated using LVDT-based strain possesses three distinct phases. The sudden drop in the strain energy density towards the end is due to the fibre breakage and is caused by the relaxation of applied force on the specimen. As the strains acquired from FBG and LVDT sensors differ, so do their corresponding strain energy density, and FBG sensors experience a larger decrease in their respective strain energy density compared to the one calculated based on LVDT. The variation of strain as a function of cycle number for FBG sensors in Figure 3b resembles that of strain energy density in terms of having initial sudden change followed by a linear region. The difference in the duration of the first phase detected based on the FBG strain and LVDT-based strain energy density can be associated with the fact that the strain field of the sensor with a larger gage length, such as the extensometer and LVDT, is affected by all of the matrix cracks along the gage length, whereas FBG strains are influenced only by the local cracks in the vicinity of the sensor having a much shorter gage length. Therefore, the first phase demarcated based on the FBG strain is slightly shorter. Please note that the strain energy curves given here are obtained by taking the product of the global stress and local strain. Considering the fact that the local strains are read from the FBGs embedded inside the epoxy matrix and that we cannot measure local stress, one should use $F = 0.5Y^{Epoxy}\varepsilon^2$ where Y^{Epoxy} is the Young's modulus of the matrix is and ε is the local strain. Therefore, the differences in the strain energy curves computed based on $F = 0.5Y^{Epoxy}\varepsilon^2$ will be more considerable than what has been given in Figure 3c. Going back to the viscoelastic heating, if one uses the relation $\Delta T = \frac{1}{2C_p^{Epoxy}} \sum_N \int_V Y^{Epoxy}\varepsilon^2 dV$, which stands for local heating in a local epoxy volume under adiabatic conditions, one will find the local ΔT, supposing that the local temperature rise occurs only in proportion with the heat capacity of the epoxy only (that later on dissipates in accordance with the local heat transfer coefficients), still to be much smaller than what is observed globally, proving that viscoelastic heating effects are indeed negligible. In another conservative scenario where the high strain values read by the local FBG are representative of the fibre strain (due to the assumption that the epoxy is perfectly adhered to a local fibre and carries the same strain as that of the fibre), and one modifies the above relation such that C_p^{Epoxy} and Y^{Epoxy} are replaced by that of the fibre; ΔT is calculated as 1 °C at most, approaching the values calculated from the global stress and strain values obtained from $\Delta T = \frac{1}{2C_p} \sum_N \int_V \sigma \varepsilon dV$, meaning that fibre viscoelasticity is also not the cause of the temperature.

3.2. Specimen L2

A second experiment was performed on the specimen L2 under displacement control with the strain ratio of 0.55 and was designed to consolidate the repeatability of the previous experiment. In this experiment, after 25,000 cycles (80% of the fatigue life), the test was paused; the specimen was kept unloaded for 30 min; and then, the fatigue test was reinitiated while keeping the experimental conditions the same. The specimen failed close to the middle FBG sensor denoted by the red curve in Figure 4b where the local strain has reached a maximum just before the failure. Figure 4a presents the variation of surface temperatures at three different thermocouple locations for specimen L2. For the first fatigue loading, similar to the previous case (Figure 3a), temperatures of all three locations increase sharply and then follow a gradual linear increase, which corresponds to the second fatigue stage. Upon terminating the loading, all temperature values start to decrease as expected. A rather important observation that deserves special consideration is the temperature trend for secondary fatigue loading. Upon re-initiating the fatigue loading, it was noted that the rate of change of temperature was higher than that corresponding to the initial fatigue loading. Moreover, the temperature curve after the pausing-restarting action very rapidly catches up with the curve before the pause, confirming the extensive damage, hence "new surface density" presence in the sample compared to its virgin state. For the latter case, since the specimen is expected to possess much higher damage and related crack density throughout the specimen, friction between the newly-formed crack or damage surfaces act as sources for heat generation in the specimen, thereby increasing the temperature faster compared to the beginning of the test, just like the experiments on the previously-discussed L1 sample.

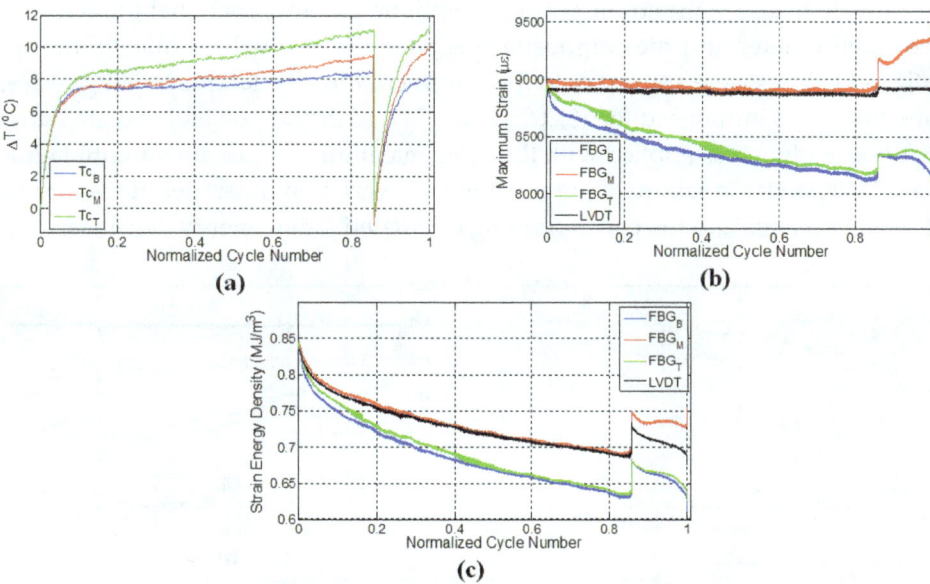

Figure 4. Evolution of (a) temperature; (b) strain and (c) strain energy density for specimen L2.

FBG strains and strain energy densities for the specimen L2 are respectively given in Figure 4b,c. Similar to the results of the previous experiment, FBG strains have a decreasing trend throughout the initial fatigue loading. When the fatigue loading was reinitiated, maximum strains measured by FBG sensors experienced a sudden jump compared to the maximum strain of the last cycle belonging to the first fatigue loading, even though the applied maximum displacement was kept the same. This is due to the thermal strain. It is seen that the specimen has a higher temperature at the end of the first fatigue loading than at the beginning of the second fatigue loading, which implies that a certain portion of the applied strain in the former case is contributed by the thermal strains associated with the thermal expansion of the specimen. When the specimens cool down and the predefined displacement is applied again for the second fatigue loading, the contribution from the thermal strain diminishes, and more force is required to induce the desired displacement onto the specimen, resulting in an upward jump in the measured force. This in turn influences the FBG strains along the specimen, causing it to experience a sudden increase.

After the jump in the strain, the maximum strains start to drop down again until about 28,000 cycles (90% of the fatigue life) for all of the sensors. The rate of these decreases in strain is significantly different compared to that corresponding to the end of the first fatigue loading for all of the respective sensors. At this stage, the strains recorded by top and bottom FBG sensors continue to decrease, whereas the strain of the middle FBG starts to increase, pointing to significant deformations in the vicinity of the middle FBG sensor (recall that a similar behaviour was also noted for the middle FBG sensor of specimen L1). After the restart of the fatigue test, specimen L2 withstands an additional 6000 cycles of fatigue loading before the failure, leading to total of 30,966 cycles to failure. Both L1 and L2 specimens failed at a location close to the middle FBG in the upper part of the specimen (section above the middle FBG), as shown in Figure 2c. The positions at which the specimens have failed are consistent with the abrupt variations in the strain fields presented in both Figures 3b and 4b. Moreover, the failure locations of L1 and L2 specimens coincide with the vicinity of the pair of FBG sensors recording higher local strains on average over the duration of the experiment.

3.3. Specimen L3

For the sake of completeness and to validate the possible mechanisms effective in the second experiment, another specimen L3 was prepared and tested. In this test, fatigue loading was applied using LVDT as the control sensor, and an extensometer with the wavelength of 50 mm was mounted onto the specimen to measure the strain during the fatigue loading, such that the centres of the gage length of the extensometer and the composite specimen are aligned. Similar to the specimen L2, two stage fatigue loadings were applied onto the specimen. Initially, fatigue loading corresponding to the strain ratio of 0.6 was introduced for 15,000 cycles (76% of the fatigue life) and after around 30 min; the second fatigue loading was applied with the same maximum displacement until failure, resulting in an additional 4694 cycles to failure. Figure 5 shows the variation of temperatures, strains measured by different sensor systems and the corresponding calculated strain energy densities.

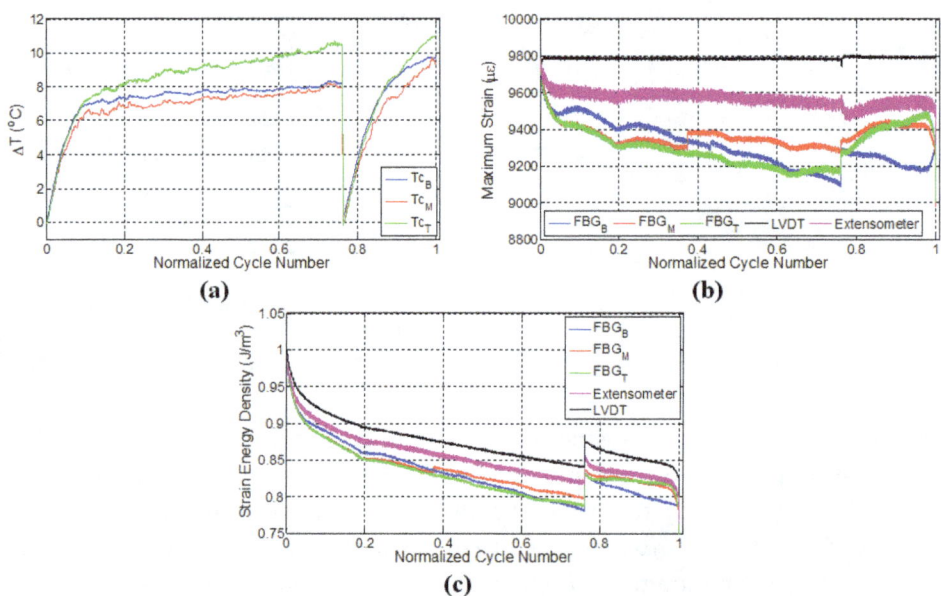

Figure 5. Evolution of (**a**) temperature; (**b**) strain and (**c**) strain energy density for specimen L3.

Temperature variations of the specimens showed similar behaviour as in the case of specimen L2 possessing distinct thermal regimes corresponding to the fatigue stages, namely initial rapid increase followed by a gradual increase at a smaller rate, implying that similar damage mechanisms are occurring inside L3, as well. In comparison to the bottom and middle thermocouples, the top thermocouple measures a significant temperature rise in the second thermal stage until the termination of the load, which can be attributed to the higher degree of deformations and related heat generation due to the friction between the surfaces of micro cracks or damage acting as an additional heat source within the specimen. Consistent with the results of specimen L2, the rate of temperature rise during the second loading was higher compared to the first fatigue loading, as the friction between the crack surfaces causes heating within the stress concentration regions already formed during the first loading, generating additional heat during the second loading.

Following the analysis of the strain variations for respective sensors, one can see that the application of constant displacement onto the specimen also causes a decline in strain measured by the extensometer, which is around 300 $\mu\varepsilon$. Such a reduction possibly emanates from the non-uniform elongation in the gage length of the specimen, causing less strain transfer to the region falling into the gage length of the extensometer in response to the same imposed global displacement. Another important observation is that the decline in maximum FBG strains is higher than what is read from the extensometer strains. This is likely due to the damage formed in the surrounding of the FBG sensors, which can reduce the effective strain transfer between the local matrix housing the FBG and fibres, thereby causing FBG sensors to read less strain. The reason behind larger drops in the strain of FBG sensors with respect to the extensometer strain can also be related to the difference in the gage lengths of the extensometer (50 mm) and FBG sensors (1 mm), as FBG sensors measure local strains. Again, consistent with the results of preceding experiments, strains measured by different FBG sensors showed different strain quantities due to the non-uniform elongations and damage formation along the specimen. Similar to the previous experiment, the sudden jump in the strain upon the application of the second fatigue loading is caused by thermal strain. At the second fatigue loading, FBG strains behaved rather differently compared to the initial loading, and especially close to failure, there are significant variations in the measured strains pointing to the subsequent incoming catastrophic failure. The maximum strains of bottom FBG sensor significantly increased at the last 1000 cycles (5%) of the fatigue loading, signifying the occurrence of serious damage in the vicinity of the sensor. The specimen failed at a location about 2 cm above the lower grip and about 1.5 cm away from the bottom FBG sensor (Figure 2c). It is important to note that the failure occurred towards the locations of two FBG sensors with the highest strains throughout the experiment. Figure 5c illustrates the variation of strain energy density again calculated via taking the product of the strain read from a given sensor and stress applied by the test machine. There is a good synchronization among the regimes of the temperature, FBG strain and strain energy density variations, keeping in mind that forces applied to the specimen at each cycle vary considerably as the experiments are carried out at constant global strain.

3.4. Specimens E1 and E2

Two additional experiments were conducted under constant strain, but this time using an extensometer mounted on the specimens until the failure, where the pertinent specimens are denoted with E1 and E2. Similar to the results obtained from experiments with LVDT control, the decrease in the maximum strain values is also noted for both E1 and E2, wherein FBG strains experience distinctive phases during the fatigue loading process.

Experimental results for the specimen E1 are provided in Figure 6. For this specimen, the imposed strain ratio for the fatigue loading was 0.55. Specific to this specimen, the interval between the subsequent FBG sensors is 6 cm. The specimen failed from the grip location close to the bottom FBG sensor, as can be inferred from Figure 6b, such that the failure is expected to occur in the region near the FBG sensor pair with the largest strain values. Figure 6a shows the local temperature variations along the specimen. Here, the temperatures of all three locations increase sharply due to the autogenous heating, then gradually and slowly drop and then level off since the rate of heat generation is balanced by the rate of heat transfer to the environment, and heat generation related to deformation/damage is not high enough to increase the temperature. The presence of internal damage will alter the effective thermal conductivity of the composite specimen since the discontinuities at cracks act as insulating media. Expectedly, in specimens with larger damage density, the heat generation rate due to damage accumulation or crack formation would be higher than the heat removal rate by the ambient environment in accordance with the $k\partial T/\partial n = h\Delta T$, where $\Delta T = (T_b - T_\infty)$ and T_b and T_∞ are temperatures of specimen's surface and the ambient environment; k is the thermal conductivity; h is the heat transfer coefficient; and $\partial/\partial n$ is the spatial derivative along the normal direction. As one can immediately conclude, the smaller the crack density, the smaller the heat generation and the larger the thermal conductivity and the conductive heat flux to the boundary that can be removed from the surface through convection. Therefore, in this specimen, the crack density is envisaged to be smaller than those in the three previous test specimens. This conclusion can be further substantiated referring to smaller deviation of FBG read strains from the strain of the control sensor with respect to former specimens, noting that the higher the damage density, the larger the relaxation of the local strain. Moreover, the maximum temperature attained at the end of the first thermal phase is smaller than those in the former experiments. Finally, after around the 16,000th cycle (83% of the total number of cycles), temperature starts to increase again due to the rate of increase of surfaces via crack growth and related heat generation. Temperature variations for specimen E1 have a rather good correlation with the three fatigue phases, especially for the top thermocouple.

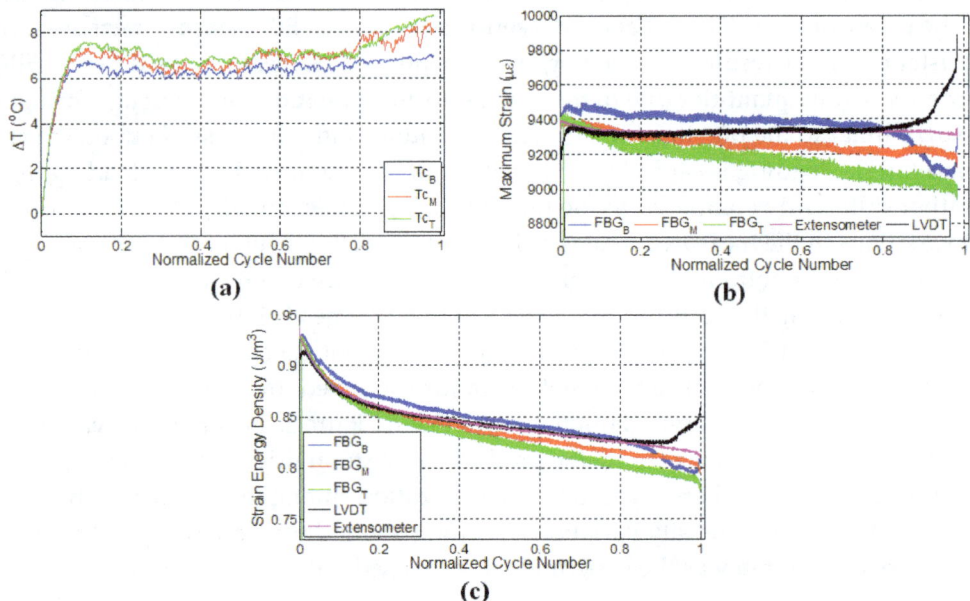

Figure 6. Evolution of (a) temperature; (b) strain and (c) strain energy density for specimen E1.

Strain variations for the extensometer, LVDT and FBG sensors are provided in Figure 6b. Recall that using the previously-determined calibration coefficient, the LVDT-recorded displacement was converted into strain. As seen from Figure 6b, until nearly 80% of the fatigue life, the strain values of both extensometer and the LVDT are nearly the same, while FBG sensors have rather different strain values, which are attributed to strain relaxation due to the damage in the vicinity of FBG sensors. After this, the strain of LVDT starts to increase, implying significant deformation or elongation in the specimen to impose the desired strain onto the gage length region of the extensometer (50 mm). In parallel, the strains of the bottom and middle FBG sensors also indicate some notable variations at this stage. Particularly, the strain of the bottom FBG sensor begins to experience a sharper decline with respect to the other FBG sensors, which coincides with the onset of the sudden increase in the LVDT strain. We think that this is related to the formation of major damages and elongations in the vicinity of the bottom FBG sensor signalling the initiation of the third fatigue phase followed by catastrophic failure, which is again consistent with the failure location that occurred at the lower grip from the tab section. A similar behaviour was noted for the bottom FBG sensor of the specimen L3 where failure took place in the similar section of the specimen. Being consistent with all previous cases, the location of the failure is in the vicinity of a region of a pair of FBGs with higher strains throughout the fatigue experiment. Such an outcome is entirely consistent with the scenario that the strain relaxed regions have already failed locally and lost their ability to contribute to the "load-bearing" action and that the rest of the volume with increased strain carries the load, and specimen fracture will occur at these regions. Note that no such conclusion can be drawn from the data of the external extensometer data. For this experiment, the recorded strain trends of the FBG sensors in Figure 6b follow the trend of the strain energy density variation calculated based on the constant strain recorded by the extensometer shown in Figure 6c. Unlike LVDT-controlled experiments, the deviation in FBG strains with respect to the extensometer and LVDT strains is smaller, which is due to the fact that the gage length of the extensometer is much smaller than the gage length of the specimen that LVDT takes into account. Hence, the specimen exhibits a much more uniform strain field, leading to smaller deviation in FBG strains with respect to the imposed/intended global strain. Despite this, the common feature of this specimen with the others before failure is that one of FBG strain signals starts to deviate from the others prominently, in addition to an observable rise in local temperatures.

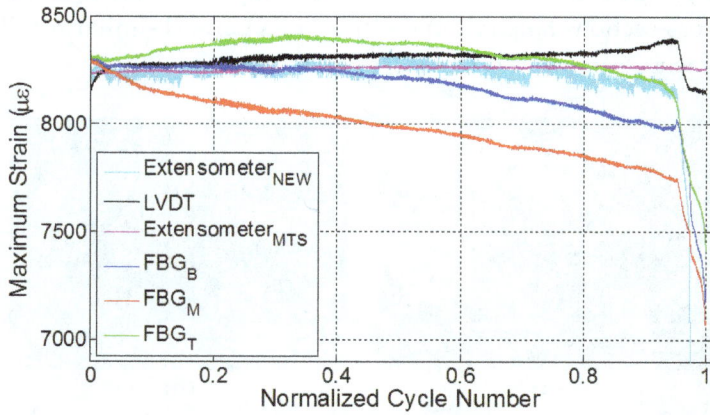

Figure 7. Evolution of strain for specimen E2.

One should keep in mind that the specimen E2 was manufactured by using the VI method and tested under extensometer control with the strain ratio of 0.5. Unlike the RTM manufactured specimens, this specimen failed at a lower cycle number, which might be attributed to the one-sided rough surface due to presence of peel ply, known to be the source of crack initiation points, or unnoticed possible defects, which are likely formed during the manufacturing. However, the failure cycle of the specimen is irrelevant within the focus of this study. In this test, in order to have further information on the strain variations along the specimen, a second extensometer with a smaller gage length (10 mm) was mounted onto the middle section of the specimen between the two ends of the first extensometer, which inputs the strain data into fatigue testing machine, as shown in Figure 2b. The results of this experiment are presented in Figure 7. The specimen failed at a location between the bottom and top FBG sensors, as shown in Figure 2c. The strain of the second extensometer follows the strain of the first extensometer until ten thousand cycles and, thereafter, deviates from the constant strain value of the first extensometer, highlighting the importance of the gage length and the local dependence of the strain measurement in the specimen experiencing non-uniform deformation. In similitude with previous results, FBG-recorded strains showed a descent over the length of the fatigue experiment. For this experiment, the third phase is quite distinctive in comparison to previous experiments, such that there are abrupt changes and a considerable decline in all FBG-measured strains, as well as in the strain recorded by the second extensometer. It is actually interesting to note that L1, L2, L3 and E1 exhibited a behaviour where at least one FBG would indicate a local increase in strain, concurrent with a local stress increase, whereas no such behaviour took place in this sample. While it is a possibility that the fracture might have started from a location far from any FBG, the "abrupt local strain variations", namely the third stage before fracture is common, although different in character, with other experiments.

3.5. Microscopic Analysis

During the analysis of the strains acquired by FBG sensors, LVDT and extensometers, we observed that, under constant amplitude, high strain fatigue loadings, the maximum local strains sensed by 1 mm-long FBG sensors can significantly be different from global strain values measured over a larger gage length, and nearly in all experiments, FBG-recorded strains decline. In addition, FBG sensors at different locations have different strains due to non-uniform strain fields or deformation states in discrete regions of specimens. At first sight, one may argue that this decline might be associated with debonding of the sensing part of FBGs and the deterioration of the FBG sensors. The integrity of FBG sensors and their bonding with the host material is very crucial: to check whether debonding of the sensors might have been the case for sudden strain relaxation in some of the FBGs in the experiments, we used SEM and optical microscopy to examine the cross-sections of composite specimens (Figure 8a) and optical microscopy (Figure 8b). No noticeable debonding was detected, as can be seen in Figure 8, which presents the cross-sections taken from the specimen L2, and similar results were obtained for the other failed specimens.

(a)   **(b)**

Figure 8. (**a**) Perpendicular and (**b**) longitudinal cross-sections of optical fibres around FBG regions.

The above observation points out the fact that local strain variations are due to matrix cracking, fibre-matrix debonding and fibre breakages in the vicinity of the FBG sensors, not related to the debonding of FBGs from the matrix. Such cracks, as they accumulate, probably contribute to the observed strain relaxations around the sensor regions, thereby affecting the strain sensed by the sensor. This is not a negative outcome; in fact, the progressive damage modes in the fatigue experiments are well captured by the FBG sensors, such that the variation of strains recorded by FBG sensors can follow the three fatigue phases of the composite corresponding to different internal damage densities. It is again inferred here that it is not possible to observe this effect via the data of the externally-mounted strain gages. In light of all of our experimental findings, it is inferred that the strain relaxation in the vicinity of FBG sensors is one of the contributing reasons for the observed reduction in the maximum strains measured by FBG sensors and is a novel way of predicting an approaching failure. In one of our previous studies [5], FBG-embedded composite specimens with the same constituent materials and stacking configurations as in the case of this study were subjected to a constant strain high cycle fatigue loading (i.e., with the strain ratio of 0.27). It was observed that the maximum strains of the FBG sensors did not experience a severe decline compared to the results of the current work, which suggests that the damage was more limited.

In such experiments, another crucial point that requires special attention is the gage length of the dynamic extensometer used for the fatigue testing can influence the evaluation of the results and, therefore, the damage mechanism dramatically. An extensometer measures the strains within its gage length, i.e., the region between its pins and fatigue testing system impose constant strains only along this region. As a result, the remaining parts of the 150-cm gage length towards the grip sections can show very distinct strain behaviour compared to the middle part of the specimen. It should also be noted that the FBG sensors located at the centre of the gage length experience a decrease in the strain even though they are located within the extensometer gage length range for which the global strain is set to be constant. Thus, local and global strain values can differ significantly from each other, and FBG sensors directly probe the local strain that is reduced or enhanced by the formation of defects. Keeping in mind that strain gradients will reach a maximum near a crack, local readings obtained by FBGs are much more precise and reliable in terms of monitoring the health of the composite and predicting remaining useful lifetime.

At the initial fatigue stage, a sharp rise in the temperature of the composite (first thermal stage) is observed followed by a gradual transition to a linear curve (second thermal stage). The nature of the temperature variation in the second thermal stage differed among the specimens presented in this work. In some specimens (i.e., E1 and E2), the temperature showed a very subtle and gradual decrease followed by levelling off, while in some others (i.e., L1, L2, L3), the temperature was rising at a smaller rate than the one for the first thermal stage. This difference in results can be attributed to the fact that in the latter case, the deformation level was higher and, correspondingly, the rate of heat generation was higher than the rate of heat given off to the surrounding. This effect is also reflected in the strain behaviour of the specimens, such that specimens showing a significant temperature rise in the second thermal stage tend to have higher reductions in the FBG-measured strains due to the stress relaxations caused by the higher degree of deformations. One may thus infer that those specimens showing a temperature rise are likely to have more cracks/deformations than the others. The third thermal stage was generally associated with the sharp rise in temperature, closer to the failure of the specimen despite a decrease in applied load to sustain constant strain.

4. Conclusions

Composite specimens containing three sequential FBG sensors embedded along their gage length were exposed to constant, high strain fatigue loads. Considerable differences among the strains recorded by FBG, LVDT and extensometer sensors were observed as the fatigue loading continued, demonstrating that the local and global behaviour of composite materials can be dramatically different. It was found that composite specimens exposed to the same global strain did not necessarily give rise to the same magnitude of strains at the local level. It was demonstrated that such a response from FBG sensors comes from the heterogeneous nature of composites, which leads to a non-uniform strain distribution. While this might not come as a surprise, the interesting strain relaxation behaviour in the vicinity of FBG sensors close to failure via the formation of various damage modes such as matrix cracking and fibre-matrix debonding appears to be related to local heating of the specimen, which was also monitored. Contrary to what one might expect, the highest temperatures were measured at the locations where the lowest strains (or strain relaxation) were recorded, implying that frictional rubbing of surfaces around cracks led to such an outcome. It has also been shown unambiguously that the temperature variations in response to fatigue loading show three different stages, which contribute to damage-induced heat generation. Upon application of the second fatigue loading as in the case of specimens L2 and L3, the temperature rises to the same temperature levels in a much shorter time during the second loading compared to the first fatigue loading, directly revealing the contribution of the friction between the crack surfaces. One important conclusion that is drawn from the study is that the significant deviations in local strains and a noticeable increase in temperature of the composite regardless of location can be used as a signal of an oncoming catastrophic failure. Sensor gage length is also shown to be a crucial factor to consider in such experiments, as constant strains imposed using an extensometer do not necessarily induce similar strains over other strain sensors.

Acknowledgments: The authors gratefully acknowledge the funding provided by The Scientific and Technological Research Council of Turkey (TUBITAK) for the project 112M357.

Author Contributions: All authors conceived of and designed the experiments together. Esat Selim Kocaman, Erdem Akay and Cagatay Yilmaz performed the experiments and analysed the data. Esat Selim Kocaman, Ibrahim Burc Misirlioglu and Mehmet Yildiz wrote the paper.

Conflicts of Interest: The authors declare no conflict of interest.

References

1. Keulen, C.J.; Yildiz, M.; Suleman, A. Damage Detection of Composite Plates by Lamb Wave Ultrasonic Tomography with a Sparse Hexagonal Network Using Damage Progression Trends. *Shock Vib.* **2014**, *2014*, 949671.

2. Cusano, A.; Cutolo, A.; Albert, J. *Fiber Bragg Grating Sensors: Recent Advancements, Industrial Applications and Market Exploitation*; Bentham Science Publishers: Sharjah, UAE, 2011.

3. Othonos, A.; Kalli, K. *Fiber Bragg Gratings: Fundamentals and Applications in Telecommunications and Sensing*; Artech House: Norwood, MA, USA, 1999.

4. Luyckx, G.; Voet, E.; Lammens, N.; Degrieck, J. Strain measurements of composite laminates with embedded fibre Bragg gratings: Criticism and opportunities for research. *Sensors* **2011**, *11*, 384–408.

5. Keulen, C.J.; Akay, E.; Melemez, F.F.; Kocaman, E.S.; Deniz, A.; Yilmaz, C.; Boz, T.; Yildiz, M.; Turkmen, H.S.; Suleman, A. Prediction of fatigue response of composite structures by monitoring the strain energy release rate with embedded fibre Bragg gratings. *J. Intell. Mater. Syst. Struct.* **2016**, *27*, 17–27.

6. Keulen, C.J.; Yildiz, M.; Suleman, A. Multiplexed FBG and etched fibre sensors for process and health monitoring of 2-&3-D RTM components. *J. Reinf. Plast. Compos.* **2011**, *30*, 1055–1064.

7. Yildiz, M.; Ozdemir, N.G.; Bektas, G.; Keulen, C.J.; Boz, T.; Sengun, E.F.; Ozturk, C.; Menceloglu, Y.Z.; Suleman, A. An experimental study on the process monitoring of resin transfer moulded composite structures using fibre optic sensors. *J. Manuf. Sci. Eng. Trans. ASME* **2012**, *134*, 044502.

8. Murukeshan, V.M.; Chan, P.Y.; Ong, L.S.; Seah, L.K. Cure monitoring of smart composites using fibre Bragg grating based embedded sensors. *Sens. Actuator A Phys.* **2000**, *79*, 153–161.

9. Yilmaz, C.; Akalin, C.; Kocaman, E.S.; Suleman, A.; Yildiz, M. Monitoring Poisson's ratio of glass fibre-reinforced composites as damage index using biaxial fibre Bragg grating sensors. *Polym. Test.* **2016**, *53*, 98–107.

10. Okabe, Y.; Yashiro, S.; Tsuji, R.; Mizutani, T.; Takeda, N. Effect of thermal residual stress on the reflection spectrum from FBG sensors embedded in CFRP composites. In *Nondestructive Evaluation and Health Monitoring of Aerospace Materials and Civil Infrastructures, Proceedings of the Society of Photo-Optical Instrumentation Engineers (SPIE), Conference on Nondestructive Evaluation and Health Monitoring of Aerospace Materials and Civil Infrastructures, Newport Beach, CA, USA, 18–19 March 2002*; Gyekenyesi, A.L., Shepard, S.M., Huston, D.R., Aktan, A.E., Shull, P.J., Eds.; International Society for Optics and Photonics: Bellingham, WA, USA, 2002; Volume 4704, pp. 59–68.

11. Sorensen, L.; Gmur, T.; Botsis, J. Long FBG sensor characterization of residual strains in AS4/PPS thermoplastic laminates. In Proceedings of the Smart Structures and Materials 2004: Smart Sensor Technology and Measurement Systems, San Diego, CA, USA, 15–18 March 2004; Udd, E., Inaudi, D., Eds.; Volume 5384, pp. 267–278.

12. Gebremichael, Y.M.; Li, W.; Boyle, W.J.O.; Meggitt, B.T.; Grattan, K.T.V.; McKinley, B.; Fernando, G.F.; Kister, G.; Winter, D.; Canning, L.; et al. Integration and assessment of fibre Bragg grating sensors in an all-fibre-reinforced polymer composite road bridge. *Sens. Actuator A Phys.* **2005**, *118*, 78–85.

13. Zhou, Z.; Liu, W.; Huang, Y.; Wang, H.; He, J.; Huang, M.; Ou, J. Optical fibre Bragg grating sensor assembly for 3D strain monitoring and its case study in highway pavement. *Mech. Syst. Signal Process.* **2012**, *28*, 36–49.

14. Kocaman, E.S.; Keulen, C.J.; Akay, E.; Yildiz, M.; Turkmen, H.S.; Suleman, A. An experimental study on the effect of length and orientation of embedded FBG sensors on the signal properties under fatigue loading. *Sci. Eng. Compos. Mater.* **2016**, *23*, 711–719.

15. Kocaman, E.S.; Yilmaz, C.; Deniz, A.; Yildiz, M. The performance of embedded fibre Bragg grating sensors for monitoring failure modes of foam cored sandwich structures under flexural loads. *J. Sandw. Struct. Mater.* **2016**, 1–25, doi:10.1177/1099636216664777.

16. Maher, M.H.; Tabrizi, K.; Prohaska, J.D.; Snitzer, E. Fiber Bragg gratings for civil engineering applications. In *Laser Diodes and Applications II, Proceedings of the Society of Photo-Optical Instrumentation Engineers (SPIE), Laser Diodes and Applications II Conference, San Jose, CA, USA, 29–31 January 1996*; Linden, K.J., Akkapeddi, P.R., Eds.; International Society for Optics and Photonics: Bellingham, WA, USA, 1996; Volume 2682, pp. 298–302.

17. Bullock, D.; Dunphy, J.; Hufstetler, G. Embedded Bragg grating fibre optic sensor for composite flexbeams. In *Fiber Optic Smart Structures and Skins V, Proceedings of the Society of Photo-Optical Instrumentation Engineers (SPIE), 5th Annual SPIE Smart Structures and Skins Conference, Boston, MA, USA, 8–9 September 1992*; Claus, R.O., Rogowski, R.S., Eds.; International Society for Optics and Photonics: Bellingham, WA, USA, 1992; Volume 1798, pp. 253–261.

18. Chen, B.X.; Maher, M.H.; Nawy, E.G. Fiberoptic Bragg grating sensor for nondestructive evaluation of composite beams. *J. Struct. Eng. ASCE* **1994**, *120*, 3456–3470.

19. De Waele, W.; Degrieck, J.; Moerman, W.; Taerwe, L.; De Baets, P. Feasibility of integrated optical fibre sensors for condition monitoring of composite structures—Part I: Comparison of Bragg-sensors and strain gauges. *Insight* **2003**, *45*, 266–271.

20. Degrieck, J.; De Waele, W.; Verleysen, P. Monitoring of fibre-reinforced composites with embedded optical fibre Bragg sensors, with application to filament wound pressure vessels. *NDT E Int.* **2001**, *34*, 289–296.

21. Friebele, E.J.; Askins, C.G.; Putnam, M.A.; Fosha, A.A.; Florio, J.; Donti, R.P.; Blosser, R.G. Distributed strain sensing with fibre Bragg grating arrays embedded in CRTM^{TM} composites. *Electron. Lett.* **1994**, *30*, 1783–1784.

22. Friebele, E.J.; Askins, C.G.; Putnam, M.A.; Heider, P.E.; Blosser, R.G.; Florio, J.; Donti, R.P.; Garcia, J. Demonstration of distributed strain sensing in production scale instrumented structures. In *Industrial and Commercial Applications of Smart Structures Technologies—Smart Structures and Materials 1996, Proceedings of the Society of Photo-Optical Instrumentation Engineers (SPIE), Smart Structures and Materials 1996 Conference—Industrial and Commercial Applications of Smart Structures Technologies, San Diego, CA, USA, 27–29 February 1996*; Crowe, C.R., Ed.; International Society for Optics and Photonics: Bellingham, WA, USA, 1996; Volume 2721, pp. 118–124.

23. Kahandawa, G.C.; Epaarachchi, J.; Wang, H.; Lau, K.T. Use of FBG sensors for SHM in aerospace structures. *Photonic Sens.* **2012**, *2*, 203–214.

24. Takeda, N.; Yashiro, S.; Okabe, T. Estimation of the damage patterns in notched laminates with embedded FBG sensors. *Compos. Sci. Technol.* **2006**, *66*, 684–693.

25. Doyle, C.; Martin, A.; Liu, T.; Wu, M.; Hayes, S.; Crosby, P.A.; Powell, G.R.; Brooks, D.; Fernando, G.F. In-situ process and condition monitoring of advanced fibre-reinforced composite materials using optical fibre sensors. *Smart Mater. Struct.* **1998**, *7*, 145–158.

26. De Baere, I.; Luyckx, G.; Voet, E.; Van Paepegem, W.; Degrieck, J. On the feasibility of optical fibre sensors for strain monitoring in thermoplastic composites under fatigue loading conditions. *Opt. Lasers Eng.* **2009**, *47*, 403–411.

27. Shin, C.S.; Chiang, C.C. Fatigue damage monitoring in polymeric composites using multiple fibre Bragg gratings. *Int. J. Fatigue* **2006**, *28*, 1315–1321.

28. Takeda, N. Characterization of microscopic damage in composite laminates and real-time monitoring by embedded optical fibre sensors. *Int. J. Fatigue* **2002**, *24*, 281–289.

29. Takeda, S.; Aoki, Y.; Ishikawa, T.; Takeda, N.; Kikukawa, H. Structural health monitoring of composite wing structure during durability test. *Compos. Struct.* **2007**, *79*, 133–139.

30. Harris, B. *Fatigue in Composites: Science and Technology of the Fatigue Response of Fibre-Reinforced Plastics*; Woodhead Publishing: Cambridge, UK, 2003.

31. Natarajan, V.; GangaRao, H.V.S.; Shekar, V. Fatigue response of fabric-reinforced polymeric composites. *J. Compos. Mater.* **2005**, *39*, 1541–1559.

32. Akay, E.; Yilmaz, C.; Kocaman, E.S.; Turkmen, H.S.; Yildiz, M. Monitoring Poisson's ratio degradation of FRP composites under fatigue loading using biaxially embedded FBG sensors. *Materials* **2016**, *9*, 781.

33. Jiang, L.; Wang, H.; Liaw, P.K.; Brooks, C.; Klarstrom, D.L. Characterization of the temperature evolution during high-cycle fatigue of the ULTIMET superalloy: Experiment and theoretical modeling. *Metall. Mater. Trans. A Phys. Metall. Mater. Sci.* **2001**, *32*, 2279–2296.

34. Holmes, J.W.; Shuler, S.F. Temperature rise during fatigue of fibre-reinforced ceramics. *J. Mater. Sci. Lett.* **1990**, *9*, 1290–1291.

35. Zettl, B.; Mayer, H.; Ede, C.; Stanzl-Tschegg, S. Very high cycle fatigue of normalized carbon steels. *Int. J. Fatigue* **2006**, *28*, 1583–1589.

36. Wong, A.K.; Kirby, G.C. A hybrid numerical experimental-technique for determining the heat dissipated during low-cycle fatigue. *Eng. Fract. Mech.* **1990**, *37*, 493–504.

37. Jacobsen, T.K.; Sorensen, B.F.; Brondsted, P. Measurement of uniform and localized heat dissipation induced by cyclic loading. *Exp. Mech.* **1998**, *38*, 289–294.

38. Pandey, K.N.; Chand, S. Analysis of temperature distribution near the crack tip under constant amplitude loading. *Fatigue Fract. Eng. Mater. Struct.* **2008**, *31*, 316–326.

39. Naderi, M.; Khonsari, M.M. Thermodynamic analysis of fatigue failure in a composite laminate. *Mech. Mater.* **2012**, *46*, 113–122.

40. Naderi, M.; Khonsari, M.M. On the role of damage energy in the fatigue degradation characterization of a composite laminate. *Compos. Part B Eng.* **2013**, *45*, 528–537.

41. Amiri, M.; Khonsari, M.M. Life prediction of metals undergoing fatigue load based on temperature evolution. *Mater. Sci. Eng. A Struct. Mater. Prop. Microstruct. Process.* **2010**, *527*, 1555–1559.

42. Zhang, L.; Liu, X.S.; Wu, S.H.; Ma, Z.Q.; Fang, H.Y. Rapid determination of fatigue life based on temperature evolution. *Int. J. Fatigue* **2013**, *54*, 1–6.

43. Renshaw, J.; Chen, J.C.; Holland, S.D.; Thompson, R.B. The sources of heat generation in vibrothermography. *NDT E Int.* **2011**, *44*, 736–739.

44. Stinchcomb, W.W.; Duke, J.C.; Henneke, E.G.; Reifsnider, K.L. In *Mechanics of Nondestructive Testing*; Springer: New York, NY, USA, 1980.

45. Lang, R.W.; Manson, J.A.; Hertzberg, R.W. Mechanisms of fatigue fracture in short glass fibre-reinforced polymers. *J. Mater. Sci.* **1987**, *22*, 4015–4030.

Permissions

All chapters in this book were first published in MATERIALS, by MDPI; hereby published with permission under the Creative Commons Attribution License or equivalent. Every chapter published in this book has been scrutinized by our experts. Their significance has been extensively debated. The topics covered herein carry significant findings which will fuel the growth of the discipline. They may even be implemented as practical applications or may be referred to as a beginning point for another development.

The contributors of this book come from diverse backgrounds, making this book a truly international effort. This book will bring forth new frontiers with its revolutionizing research information and detailed analysis of the nascent developments around the world.

We would like to thank all the contributing authors for lending their expertise to make the book truly unique. They have played a crucial role in the development of this book. Without their invaluable contributions this book wouldn't have been possible. They have made vital efforts to compile up to date information on the varied aspects of this subject to make this book a valuable addition to the collection of many professionals and students.

This book was conceptualized with the vision of imparting up-to-date information and advanced data in this field. To ensure the same, a matchless editorial board was set up. Every individual on the board went through rigorous rounds of assessment to prove their worth. After which they invested a large part of their time researching and compiling the most relevant data for our readers.

The editorial board has been involved in producing this book since its inception. They have spent rigorous hours researching and exploring the diverse topics which have resulted in the successful publishing of this book. They have passed on their knowledge of decades through this book. To expedite this challenging task, the publisher supported the team at every step. A small team of assistant editors was also appointed to further simplify the editing procedure and attain best results for the readers.

Apart from the editorial board, the designing team has also invested a significant amount of their time in understanding the subject and creating the most relevant covers. They scrutinized every image to scout for the most suitable representation of the subject and create an appropriate cover for the book.

The publishing team has been an ardent support to the editorial, designing and production team. Their endless efforts to recruit the best for this project, has resulted in the accomplishment of this book. They are a veteran in the field of academics and their pool of knowledge is as vast as their experience in printing. Their expertise and guidance has proved useful at every step. Their uncompromising quality standards have made this book an exceptional effort. Their encouragement from time to time has been an inspiration for everyone.

The publisher and the editorial board hope that this book will prove to be a valuable piece of knowledge for researchers, students, practitioners and scholars across the globe.

List of Contributors

Ryo Nakanishi, Mudasir Ahmad Yatoo, Keiichi Katoh and Brian K. Breedlove
Department of Chemistry, Graduate School of Science, Tohoku University, 6-3 Aza-Aoba, Aoba-ku, Sendai, Miyagi 980-8578, Japan; muda.amu@gmail.com (M.A.Y.); kkatoh@m.tohoku.ac.jp (K.K.)

Masahiro Yamashita
WPI Research Center, Advanced Institute for Materials Research, Tohoku University, 2-1-1 Katahira, Aobaku, Sendai 980-8577, Japan

Teresa Casimiro and Ana Aguiar-Ricardo
LAQV-REQUIMTE, Departamento de Química, Faculdade de Ciências e Tecnologia, Universidade NOVA de Lisboa, Campus de Caparica, Caparica 2829-516, Portugal

Marta C. Silva
LAQV-REQUIMTE, Departamento de Química, Faculdade de Ciências e Tecnologia, Universidade NOVA de Lisboa, Campus de Caparica, Caparica 2829-516, Portugal
BIOSCOPE Research Group, UCIBIO@REQUIMTE, Chemistry Department, Faculty of Science and Technology, University NOVA of Lisbon, Caparica Campus, Caparica 2829-516, Portugal

Ana Sofia Silva
LAQV-REQUIMTE, Departamento de Química, Faculdade de Ciências e Tecnologia, Universidade NOVA de Lisboa, Campus de Caparica, Caparica 2829-516, Portugal
CICS-UBI, Health Sciences Research Center, Faculdade de Ciências da Saúde, Universidade da Beira Interior, Av. Infante D. Henrique, Covilhã 6200-506, Portugal

Javier Fernandez-Lodeiro and Carlos Lodeiro
BIOSCOPE Research Group, UCIBIO@REQUIMTE, Chemistry Department, Faculty of Science and Technology, University NOVA of Lisbon, Caparica Campus, Caparica 2829-516, Portugal
PROTEOMASS Scientific Society, Rua dos Inventores, Madam Parque, Caparica Campus, Caparica 2829-516, Portugal

Hyun-Sang Yoo, Ji-Hyeon Bae, Eun-Bin Bae, Chang-Mo Jeong and Jung-Bo Huh
Department of Prosthodontics, Dental Research Institute, Institute of Translational Dental Sciences, BK21 PLUS Project, School of Dentistry, Pusan National University, Yangsan 50612, Korea

Se-Eun Kim
Department of Veterinary Surgery, College of Veterinary Medicine, Chonnam National University, Gwangju 61186, Korea

So-Yeun Kim
Department of Prosthodontics, Pusan National University Hospital, Pusan 49241, Korea

Kyung-Hee Choi and Keum-Ok Moon
Tissue Biotech Institute, Cowellmedi Co., Ltd., Busan 46986, Korea

Jinhua Huang, Guang Ran, Jianxin Lin, Qiang Shen, Penghui Lei, Xina Wang and Ning Li
College of Energy, Xiamen University, Xiamen 361102, China

Liyang Lin
Chongqing Key Laboratory of Heterogeneous Material Mechanics, College of Aerospace Engineering, Chongqing University, Chongqing 400044, China; jack_linliyang@cqu.edu.cn

Qibin Li
Chongqing Key Laboratory of Heterogeneous Material Mechanics, College of Aerospace Engineering, Chongqing University, Chongqing 400044, China
Key Laboratory of Low-grade Energy Utilization Technology & System, Ministry of Education, College of Power Engineering, Chongqing University, Chongqing 400044, China

Yinsheng Yu and Chao Liu
Key Laboratory of Low-grade Energy Utilization Technology & System, Ministry of Education, College of Power Engineering, Chongqing University, Chongqing 400044, China

Yilun Liu
State Key Laboratory for Strength and Vibration

Linas Jonušauskas, Darius Gailevičius and Mangirdas Malinauskas
Department of Quantum Electronics, Faculty of Physics, Vilnius University, Saul˙etekio Ave. 10, Vilnius LT-10223, Lithuania

Lina Mikoliūnaitė, Danas Sakalauskas and Simas Šakirzanovas
Department of Applied Chemistry, Vilnius University, Naugarduko Str. 24, Vilnius LT-03225, Lithuania

Saulius Juodkazis
Center for Micro-Photonics, Faculty of Engineering and Industrial Sciences, Swinburne University of Technology, Hawthorn 3122, Australia
Melbourne Center for Nanofabrication, Australian National Fabrication Facility, Clayton 3168, Australia

Lin Liu, Ting Sun and Huizhu Ren
Henan Province of Key Laboratory of New Optoelectronic Functional Materials, College of Chemistry and Chemical Engineering, Anyang Normal University, Anyang 455000, China

Hailong Dong, Ana Kuzmanoski and Claus Feldmann
Karlsruhe Institute of Technology (KIT), Institut für Anorganische Chemie, Engesserstrasse 15, 76131 Karlsruhe, Germany

Tobias Wehner and Klaus Müller-Buschbaum
Institute of Inorganic Chemistry, University of Würzburg, Am Hubland, D-97074 Würzburg, Germany

Antonio S. Gliozzi and Marco Scalerandi
Department of Applied Science and Technology, Condensed Matter and Complex Systems Physics Institute, Politecnico di Torino, 10129 Torino, Italy; aitouarabi@gmail.com (M.A.O.)

Mohand Ait Ouarabi
Department of Applied Science and Technology, Condensed Matter and Complex Systems Physics Institute, Politecnico di Torino, 10129 Torino, Italy; aitouarabi@gmail.com (M.A.O.)
Laboratoire de Physique des Matériaux, Université des Sciences et de la Technologie Houari Boumediene, BP 32 El Alia, Bab Ezzouar 16111, Algeria

Fouad Boubenider
Laboratoire de Physique des Matériaux, Université des Sciences et de la Technologie Houari Boumediene, BP 32 El Alia, Bab Ezzouar 16111, Algeria

Paola Antonaci
Department of Structural, Geotechnical and Building Engineering, Politecnico di Torino, 10129 Torino, Italy

Xianfeng Wang, Peipei Sun, Ningxu Han and Feng Xing
Guangdong Provincial Key Laboratory of Durability for Marine Civil Engineering, College of Civil Engineering, Shenzhen University, Shenzhen 518060, Guangdong, China

Eric A. Jägle, Philipp Kürnsteiner and Dierk Raabe
Department Microstructure Physics and Alloy Design, Max-Planck-Institut für Eisenforschung GmbH, Max-Planck-Strasse 1, 40237 Düsseldorf, Germany

Zhendong Sheng
Department Microstructure Physics and Alloy Design, Max-Planck-Institut für Eisenforschung GmbH, Max-Planck-Strasse 1, 40237 Düsseldorf, Germany
Institut für Eisenhüttenkunde, Rheinisch-Westfälische Technische Hochschule Aachen, Intzestrasse 1, 52072 Aachen, Germany

Sörn Ocylok and AndreasWeisheit
Competence Area Additive Manufacturing and Functional Layers, Fraunhofer Institut für Lasertechnik, Steinbachstrasse 15, 52074 Aachen, Germany

Davide Ricci, Michele M. Nava and Manuela T. Raimondi
Department of Chemistry, Materials and Chemical Engineering "Giulio Natta", Politecnico di Milano, 20133 Milano, Italy

Tommaso Zandrini, Giulio Cerullo and Roberto Osellame
Istituto di Fotonica e Nanotecnologie (IFN)-CNR and Department of Physics, Politecnico di Milano, 20133 Milano, Italy

Esat Selim Kocaman
Faculty of Engineering and Natural Sciences, Sabanci University, Tuzla, 34956 Istanbul, Turkey

Erdem Akay and Halit Suleyman Turkmen
Faculty of Aeronautics and Astronautics, Ayazaga Campus, Istanbul Technical University, Maslak, 34469 Istanbul, Turkey

Cagatay Yilmaz, Ibrahim Burc Misirlioglu and Mehmet Yildiz
Faculty of Engineering and Natural Sciences, Sabanci University, Tuzla, 34956 Istanbul, Turkey
Integrated Manufacturing Technologies Research and Application Center, Sabanci University, Tuzla, 34956 Istanbul, Turkey
Sabanci University-Kordsa Global, Composite Technologies Center of Excellence, Istanbul Technology Development Zone, Sanayi Mah. Teknopark Blvd. No: 1/1B, Pendik, 34906 Istanbul, Turkey

Afzal Suleman
Mechanical Engineering Department, University of Victoria, Victoria, BC V8W 2Y2, Canada

Shengwen Qi
Key Laboratory of Shale Gas and Geoengineering, Institute of Geology and Geophysics, Chinese Academy of Sciences, Beijing 100029, China

Index

www.ingramcontent.com/pod-product-compliance
Lightning Source LLC
Chambersburg PA
CBHW050456200326
41458CB00014B/5198